Veterinary Entomology

Veterinary Entomology

Arthropod Ectoparasites of Veterinary Importance

Richard Wall
School of Biological Sciences, The University of Bristol, UK

and

David Shearer
School of Veterinary Science, The University of Bristol, UK

CHAPMAN & HALL
London · Weinheim · New York · Tokyo · Melbourne · Madras

Published by Chapman & Hall, 2–6 Boundary Row, London SE1 8HN, UK

Chapman & Hall, 2–6 Boundary Row, London SE1 8HN, UK

Chapman & Hall GmbH, Pappelallee 3, 69469 Weinheim, Germany

Chapman & Hall USA, 115 Fifth Avenue, New York, NY 10003, USA

Chapman & Hall Japan, ITP-Japan, Kyowa Building, 3F, 2-2-1 Hirakawacho, Chiyoda-ku, Tokyo 102, Japan

Chapman & Hall Australia, 102 Dodds Street, South Melbourne, Victoria 3205, Australia

Chapman & Hall India, R. Seshadri, 32 Second Main Road, CIT East, Madras 600 035, India

First edition 1997

Printed in Great Britain by T.J. International Ltd, Padstow, Cornwall

ISBN 0 412 61510 X

A Catalogue record for this book is available from the British Library

Library of Congress Catalog Number: 96–71695

♾ Printed on permanent acid-free text paper, manufactured in accordance with ANSI/NISO Z39.48-1992 (Permanence of Paper).

CONTENTS

PREFACE

Although usually treated as unified subject, in many respects the two components of what is broadly described as 'medical and veterinary entomology' are clearly distinct. As is usual, the term entomology is used loosely here to refer to both insects and arachnids. In medical entomology blood-feeding Diptera are of paramount importance, primarily as vectors of pathogenic disease. Most existing textbooks reflect this bias. However, in veterinary entomology ectoparasites such as the mites, fleas or dipteran agents of myiasis assume far greater prominence and the most important effects of their parasitic activity may be mechanical damage, pruritus, blood loss, myiasis, hypersensitivity and dermatitis, in addition to vector-borne pathogenic disease. Ectoparasite infestation of domestic and companion animals, therefore, has clinical consequences necessitating a distinct approach to diagnosis and control.

The aim of this book is to introduce the behaviour, ecology, pathology and control of arthropod ectoparasites of domestic animals to students and practitioners of veterinary medicine, animal husbandry and applied biology. Since the book is directed primarily at the non-entomologist, some simplification of a number of the more involved entomological issues has been deemed necessary to improve the book's logical structure and comprehensibility, and keep its length within limits. A reading list is presented at the end of each chapter to act as a stepping-stone into the specialist literature. In particular, the recognition guides in each chapter need to be used with a degree of caution since they are not comprehensive and, where more detailed identification is required, specialist keys should be consulted. Similarly, in the discussion of the control of ectoparasites, the principles of control and ectoparasiticide use have been emphasized; any attempt to provide a detailed recipe for control might be inappropriate in different parts of the world and would go out of date rapidly. It should be stressed that all insecticides should be used in strict accordance with the manufacturer's local instructions.

This book focuses primarily on the arthropod ectoparasites of the temperate northern hemisphere. To some extent this distinction has been blurred slightly, since so many ectoparasite species have been transported worldwide with humans and domestic animals. But, for the most part, important ectoparasites from tropical and sub-

tropical habitats have been described only briefly and would require their own volume to do them justice.

ACKNOWLEDGEMENTS

I first would like to thank Shelagh Adam for her patient support and encouragement throughout the writing of this book. Long dinner-table monologues on the vagaries of mite classification and nomenclature would have driven lesser constitutions to despair – or domestic violence. I also would like to express my immense gratitude to my parents, Shirley and Douglas, for the considerable help and guidance over the years, without which the writing of this book would not have possible – literally.

Richard Wall *August 1996*

I would like to thank my wife Alison for her support and encouragement during the writing of this book. I would also like to thank my colleagues who have encouraged me to pursue an interest in veterinary dermatology and made me realize the need for such a book on veterinary ectoparasites.

David Shearer *August 1996*

We would like to thank Les Strong, Shelagh Adam, Alison Shearer and Roger Avery who read all, or parts of, the manuscript of this book and helped to eliminate many factual and textual errors. Neverthless, any remaining mistakes are all ours.

R.W. and D.S.

1

The importance and diversity of arthropod ectoparasites

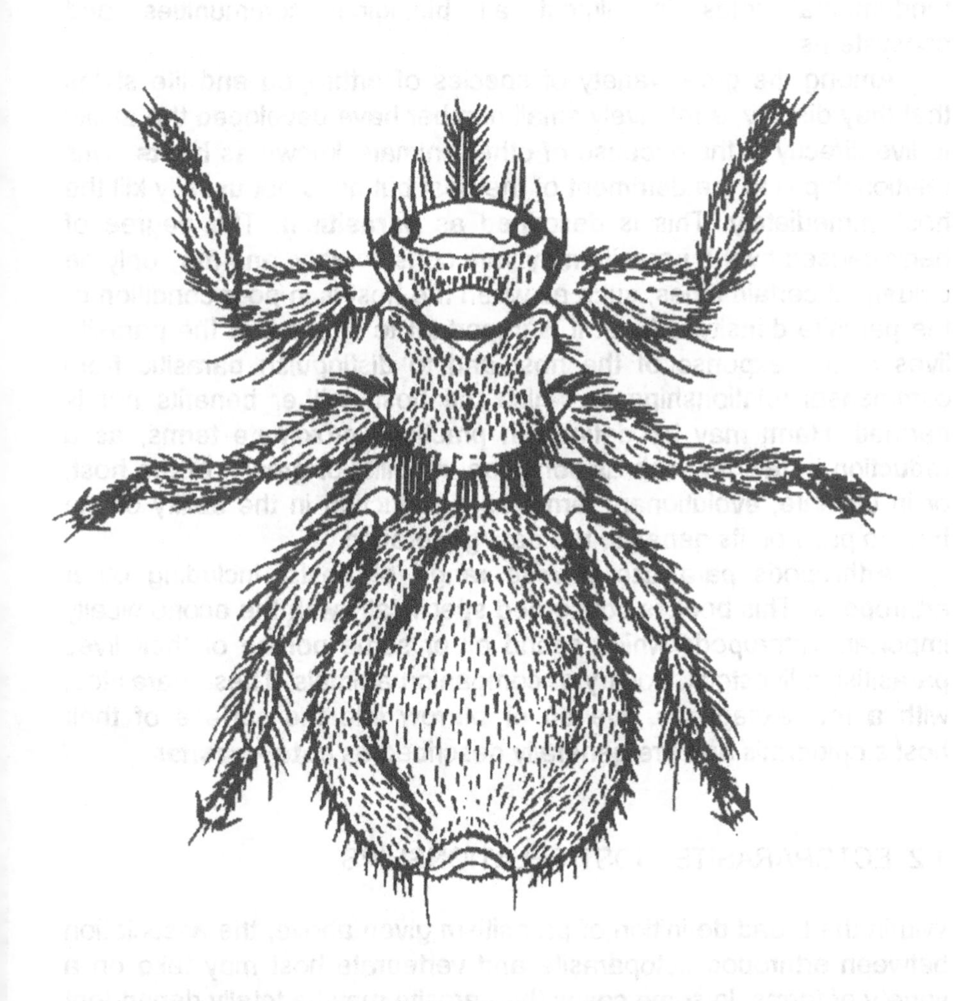

Sheep ked, *Melophagus ovinus* (from Lane and Crosskey, 1993).

1.1 INTRODUCTION

The arthropods are a bewilderingly diverse assemblage of invertebrates, containing over 80% of all known animal species and occupying almost every known habitat. They include such familiar animals as flies, crabs, centipedes and spiders as well as a plethora of small and little-known groups.

There are more species of arthropod than all other animals on earth combined; over a million species have been described and millions more may be awaiting description or discovery. Dazzlingly beautiful, behaviourally complex and ecologically essential, they play fundamental roles in almost all biological communities and ecosystems.

Among the great variety of species of arthropod and life styles that they display, a relatively small number have developed the ability to live directly at the expense of other animals, known as **hosts**. This relationship is to the detriment of the host but does not usually kill the host immediately. This is described as **parasitism**. The degree of harm caused by the parasite may vary considerably, and may only be evident at certain times, such as when the host is in poor condition or the parasite density is high. It is important to stress that the parasite lives at the expense of the host and to distinguish parasitic from commensal relationships, in which the host neither benefits nor is harmed. Harm may be defined in practical, proximate terms, as a reduction in factors such as condition, mobility or growth of the host, or in ultimate, evolutionary terms as a reduction in the ability of the host to pass on its genes to the next generation.

Arthropods parasitize a wide range of hosts, including other arthropods. This book is concerned specifically with the economically important arthropods which spend all or some portion of their lives parasitising livestock, poultry or companion animals. These parasites, with a few exceptions, live on or burrow into the surface of their host's epidermis and are generally described as **ectoparasites**.

1.2 ECTOPARASITE–HOST RELATIONSHIPS

Within the broad definition of parasitism given above, the association between arthropod ectoparasite and vertebrate host may take on a variety of forms. In some cases the parasite may be totally dependent on the host, in which case the parasitism is described as **obligatory**.

Alternatively, the parasite may feed or live only occasionally on the host, without being dependent on it, in which case the parasitism is described as **facultative**.

The host provides a number of important resources for the ectoparasite. Most vitally, the host supplies a source of food, which may be blood, lymph, tears or sweat or the debris of skin, hair or feathers. The host's body also provides the environment in which many ectoparasites live, generating warmth, moisture and, within the skin or hair, a degree of protection from the external environment. The host, may also provide transportation from place to place for the parasite, a site at which to mate and, in many cases, the means of transmission from host to host.

Despite the benefits of a close association with the host, there is considerable variation in the amount of time spent on the host by various species of ectoparasite. Some ectoparasites, such as many of the species of lice for example, live in **continuous** association with their host throughout their life cycle and are therefore highly dependent on the host. The majority of ectoparasites, however, have only **intermittent** contact with their host, and are free-living for the major portion of their life cycles. In some cases, ectoparasites, such as many of the species of mite, are highly **host specific**; only one host species is exploited and, in some instances, the parasite can exist only on one defined area of the host's body. Other species are able to exploit a wider range of hosts.

Whether a pest is an obligatory or facultative ectoparasite, lives in continuous or intermittent association with its host, or is host specific or a generalist, is of interest from a biological perspective and has major implications for both the control of ectoparasites and the treatment of ectoparasite-associated disease.

1.3 ECTOPARASITE DAMAGE

As a result of their activity, arthropod ectoparasites may have a variety of direct and indirect effects on their hosts. Direct harm caused may be due to:

- **Blood loss**: although each individual ectoparasite only removes a small volume of blood from a host, in large numbers the blood removed by feeding may be directly debilitating and anaemia is common in heavily infested hosts. In one study in the USA over 90 kg of blood was estimated to have been removed by ticks from a cow over a single season. Similarly, the feeding of horse flies has been estimated to be responsible for the loss of up to 0.5 litres of blood per day from cattle. Two hundred fleas feeding on a kitten may be capable of removing up to 10% of the animal's blood over a period of several days.

- **Myiasis**: the infestation of the living tissues with fly larvae causes direct damage to carcasses or skin.
- **Skin inflammation and pruritus**: various skin infestations caused by arthropod activity cause **pruritus** (itching), often accompanied by hair and wool loss (**alopecia**) and occasionally by skin thickening (**lichenification**). The presence of ectoparasites on or burrowing into the skin can stimulate keratinocytes to release cytokines (e.g. IL-1) which leads to epidermal hyperplasia and cutaneous inflammation. The antigens produced by ectoparasites (e.g. salivary and faecal) can in some individuals stimulate an immune response leading to hypersensitivity. *Sarcoptes scabiei* infestation in the dog, for example, leads to an IgE-mediated type I hypersensitivity which is manifested in severe cutaneous inflammation and pruritus.
- **Toxic and allergic responses**: caused by antigens and anticoagulants in the saliva of blood-feeding arthropods.

The behaviour of ectoparasites also may cause harm indirectly, again particularly when they are present at high density, causing:

- **Disturbance**: the irritation caused, particularly by flies as they attempt to feed or oviposit, commonly results in a variety of behaviours such as head shaking, stamping, skin twitching, tail switching or scratching. Cattle under persistent attack from flies may congregate in a group with their heads facing the centre. Sheep under attack from nasal bot flies may be seen pressing their nostrils to the ground before running short distances and repeating the action. These activities may result in reduced growth and loss of condition because the time spent in avoidance behaviour is lost from grazing or resting. Of interest is the observation that different individual hosts may vary considerably in their behavioural response to blood-feeding flies and that some, possibly the more passive individuals, tend to be most heavily attacked.

- **Self-wounding**: the activity of particular ectoparasites, such as warble flies, may cause dramatic avoidance responses in the intended host, known as **gadding**. The madly panicking animal may cause serious self-injury following collision with fences and other objects.
- **Social nuisance**: large populations of flies may breed in animal dung, particularly in and around intensive husbandry units. The activity of flies may result in considerable social and legal problems, especially where suburban developments have encroached on previously rural areas. Adult flies and their faeces may also decrease the aesthetic appearance and value of farm facilities and produce, such as hens' eggs, and cause irritation and annoyance to employees.

In addition to direct effects, one of the most important roles of ectoparasites is in their action as **vectors** of pathogens. These pathogens include protozoa, bacteria, viruses, cestodes (tapeworms) and nematodes (round worms). Pathogens such as bacteria and viruses may be transmitted directly to new hosts, the ectoparasite acting as a **mechanical vector**. These pathogens may be picked up on the body, feet or mouthparts when the ectoparasites feeds. Mechanical transmission usually has to occur within a few hours of the original contact with the infected host, because the survival of most pathogens is relatively limited when exposed outside their host.

Alternatively, for many protozoa, tapeworms and nematodes, the pathogens need to go through specific stages of their life cycle in the body of the arthropod ectoparasite. In these cases the arthropod serves as an **intermediate host** and is known as a **biological vector**. Again the vector acquires the pathogen from an infected animal when it feeds. After development of the pathogen in the vector, the vector becomes infective and can transmit the pathogen when it next feeds. In contrast to mechanical transmission, biological transmission requires a period of time between acquisition of the pathogen and the maturation of infection. The vector may then remain infective for the remainder of its life. The effects of pathogens on the vector are largely unknown and this may be a fruitful area for future research; those studies which have been possible suggest that there may be measurable costs to the vector for carrying a heavy infection.

A pathogen may reside and multiply in alternative vertebrate hosts which are immune or only mildly infected by it. For example, the bacterium *Yersinia pestis*, which causes bubonic plague known as Black Death, is endemic in wild rodent populations. However, in domestic rats and humans, to which it is transmitted by fleas, it is highly pathogenic. Such alternate hosts are known as **reservoirs** of disease.

The direct damage caused by most ectoparasites is directly proportional to their abundance. This is not the case, however, for disease vectors, where even very low numbers of infected vectors may cause considerable economic and welfare problems.

Although relatively few in number, through their direct and indirect effects on their hosts, the various species of arthropod ectoparasite have had, and continue to exert, a major impact on the history of humans and their domesticated animals.

> Swords and lances, arrows, machine guns and even high explosives have had far less power over the fates of nations than the typhus louse, the plague flea and the yellow fever mosquito. Civilisations have retreated from the plasmodium of malaria and armies have crumbled into rabbles under the onslaught of cholera spirilla or of dysentery and typhoid baccili. Huge areas have been devastated by the trypanosome that travels on the wings of the tsetse fly ... war ... conquest ... civilisation have merely set the stage for these more powerful agents of human tragedy. (Zinser, 1934)

1.4 THE EVOLUTION OF ECTOPARASITE–HOST RELATIONSHIPS

Insects and related arthropods probably arose at least 500 million years ago, 300 million years before warm-blooded vertebrates. Unfortunately, the poor geological record for insects gives us little direct evidence of how parasitism evolved. Nevertheless, it would appear likely that, over time, as the terrestrial vertebrates appeared on earth several species of arthropod were able to exploit the new resource and opportunities created.

Parasitism probably evolved at least twice, and possibly several times, independently in different arthropod groups, depending on the relationship between the ectoparasite and its host. One route may have involved arthropods which were pre-adapted to living with vertebrates. These arthropods initially may have fed on general vertebrates. These arthropods initially may have fed on general organic matter, and then moved to scavenging detritus, such as skin or hair, present in a vertebrate lair or nest. From here, coupled with the generalized feeding habits, it is only a short evolutionary step for the ectoparasite to move on to the host to feed on skin and hair and, in some cases, to facultative and obligate blood-feeding.

The second route to ectoparasitism may have involved arthropods which had existing adaptations which allowed them to feed on vertebrates. These arthropods may have had mouthparts already adapted for biting, rasping or sucking. They were perhaps liquid feeders, which occasionally opportunistically fed on blood in wounds, or they may have been active predators, in the adult or pre-adult stages, perhaps of other arthropods. Again, from taking the occasional meal from a vertebrate some subsequently may have switched to depending on blood as a food source.

These two evolutionary pathways involve similar adaptations, but they have led to very different relationships with the host. The generally accepted viewpoint throughout most of the twentieth century has been that commensalism or very mild parasitism is the inevitable eventual evolutionary end product of host–parasite co-evolution. It was thought that parasites would be selected to minimize the damage that they did to the host and that virulent (damaging) parasites were more recently evolved and were poorly adapted. This was because more virulent parasites might quickly weaken and damage the host and, if the host were to die, either as a direct result of the parasitism or perhaps because the weakened host were to succumb to disease or predation, the ectoparasite would lose the benefits of a predictable food supply, protection from the external environment and its means of dispersal. However, more recent work has shown that there may be good evolutionary reasons to expect quite variable levels in the damage caused by parasites, which is the outcome of the behaviour and ecology of both the parasite and host, and the way in which these two animals interact.

Ectoparasites, such as many of the lice or mites, which live in relatively permanent association with their hosts, are usually small, with relatively low mobility. The risks and uncertainties associated with living without their host and having to find another meal are sufficiently high that for these animals excessive virulence, leading to the death or debilitation of the host, might result in their own death and failure to reproduce. Hence, these ectoparasites have become obligate, host-specific specialists, in normal circumstances, doing minimal damage and, in some cases, existing almost as commensals, e.g. species of *Demodex* mites. In many cases these arthropods may well have followed the first evolutionary route to ectoparasitism; even before becoming ectoparasites they were pre-adapted to living in close association with vertebrates and their survival and dispersal depended on the continued existence and health of the vertebrate with which they were associated.

In contrast, any ectoparasite that (1) was relatively mobile, so that it could find a new host quickly and efficiently, (2) was relatively resistant to the adverse effects of climate, so that it could survive without its host and (3) had a broad range of relatively abundant hosts on which it could feed, might be expected to evolve higher levels of virulence to maximize the amount that could be 'extracted' from a host as quickly as possible. For these ectoparasites, the death of the host would be of little importance since they could survive well independently and find a new host quickly when needed. Arthropods with these characteristics, which inflict relatively high levels of damage, such as many of the blood-feeding flies and ticks, may have followed the second evolutionary route to parasitism, starting off as free-living scavengers or predators which subsequently became opportunistic feeders on vertebrates.

Of course, varying degrees of virulence between the extremes described will have evolved, depending on the precise interactions between the ectoparasite and host. The associations seen today between ectoparasites and hosts are also the outcome of the response of the host to the ectoparasite. From the perspective of the host, the attention of any parasites is, by definition, unwelcome. As arthropods evolved new ways of exploiting their hosts more effectively, the hosts also evolved strategies to combat the activities of the ectoparasites. These range from immune responses to behaviours such as grooming, periodic changing of nest sites or bedding, or even seasonal mass migrations to avoid areas of high parasite density. Hosts that were better able to tolerate their ectoparasites, minimize the damage caused or developed ways of ridding themselves of ectoparasites, survived longer and produced more offspring than other hosts. However, over time ectoparasites have also been selected to try to get round or exploit these host responses. Hence, over the millions of years, a constant evolutionary battle has been waged, in which arthropods have been evolving to exploit vertebrate hosts and in which hosts have been co-evolving to mitigate the effects of parasites on their fitness.

However, in relatively recent history, the associations between ectoparasites and their hosts have been subjected to a number of dramatic changes which have resulted in a substantial shift in the nature of their relationships.

1.5 A MODERN AND GROWING PROBLEM

During the late mesolithic and early neolithic periods, 10,000–20,000 years ago, livestock and companion animals were first domesticated and farmed by humans (Fig. 1.1). This development has continued to the present day and has been combined with rapidly growing human populations, expansion and settlement in new areas, increased rates of human movement worldwide and increasing urbanization.

- A massive increase in human populations has necessitated a growing intensification of animal husbandry, not only to meet the immediate demands for food and animal products such as wool and leather, but also to provide draught animals and fulfil the multitude of roles that animals play in human society. More and more animals have been reared at ever greater stocking densities. This presents ectoparasites with a super-abundance of hosts. The higher host density increases the potential for ectoparasite transmission and allows ectoparasites, adapted to experience huge mortalities associated with finding a new host, to build up massive population densities in very short periods of time.

- The artificial selection of livestock, poultry and companion animals for domestication and high productivity has been associated in many cases with a reduction in resistance to ectoparasite damage and the exaggeration of features which confer greater susceptibility to ectoparasite infestation. For example, the outer coat of primitive sheep is stiff and hairy and covers a woolly undercoat which only grows in winter. The outer hairs are known as kemps. In highly domesticated sheep these kemps are absent and the fleece consists entirely of the woolly undercoat which grows all year round. Selection for a longer, thick fleece has increased the susceptibility of sheep to various types of disease and ectoparasite, particularly blowfly myiasis.

Fig. 1.1 Detail from Tomb 3, Beni Hasan, Egypt (from Newberry, 1893).

- **With the increasing global movement of human populations, domestic animals have been transported into new areas of the world where they are attacked by endemic ectoparasites to which they have little or no resistance. This has been the case particularly with the introduction of domestic cattle, *Bos taurus*, into areas where they are attacked by a wide variety of ectoparasites and ectoparasite-borne diseases, which previously existed only on indigenous Bovidae or other Artiodactyla. The movement of humans and domestic animals has also allowed the introduction of ectoparasites into areas in which they were previously absent, such as sucking lice (Anoplura) introduced into Australia with sheep. By the year 2000 the majority of the world population will live in urban areas. This growing urbanization of humans and their associated companion animals provides arthropods and arthropod-borne diseases with a large, concentrated pool of potential hosts and enables ectoparasites to transfer between individual hosts more readily. In addition, in houses, which they frequently share with their companion animals, humans have created conditions in which many species of arthropod pest survive and flourish, often in areas of the world where they would otherwise perish. This is especially the case for fleas and mites, where modern houses can provide them with carefully controlled temperature, humidity and lighting, a protective microhabitat and a regular supply of hosts.**

1.6 ARTHROPOD STRUCTURE AND FUNCTION

To someone unfamiliar with invertebrate morphology and physiology, arthropods can seem like creatures from outer space. They have many anatomical features which are often analogous to those of vertebrates but which are totally dissimilar in structure and function. In the following sections, a brief overview of a range of key arthropod features is presented, with specific reference to the ectoparasites of veterinary interest. This is by no means a comprehensive examination of the subject. A huge range of variation exists in the morphology and physiology of this diverse phylum and for a more detailed treatment the reader is referred to more specialist texts at the end of the chapter.

1.6.1 Arthropod segmentation

Arthropods are **metameric**, that is they are divided into segments. However, within a number of arthropod classes, particularly the arachnids and the crustaceans, there has been a tendency for segmentation to become reduced and, in the mites for example, it has almost disappeared. Reduction in segmentation, has occurred through loss, fusion or the tendency for segments to become dramatically changed in structure to fulfil specific functions, often associated with feeding, oviposition or mating. However, even in those arthropods that have almost lost their segmentation it can still be seen in the embryo.

A characteristic feature of many arthropod groups is the division of the body into clusters of segments, such as the head, thorax and abdomen (Fig. 1.2). This is known as **tagmatization**. Each **tagma** contains a specific set of segments and is specialized for functions different from those of the other tagmata.

1.6.2 The arthropod exoskeleton

The **exoskeleton** is the outer covering which provides support and protection to the living tissues of arthropods. In many respects it is one of the keys to the success of the phylum, but it also imposes many limitations.

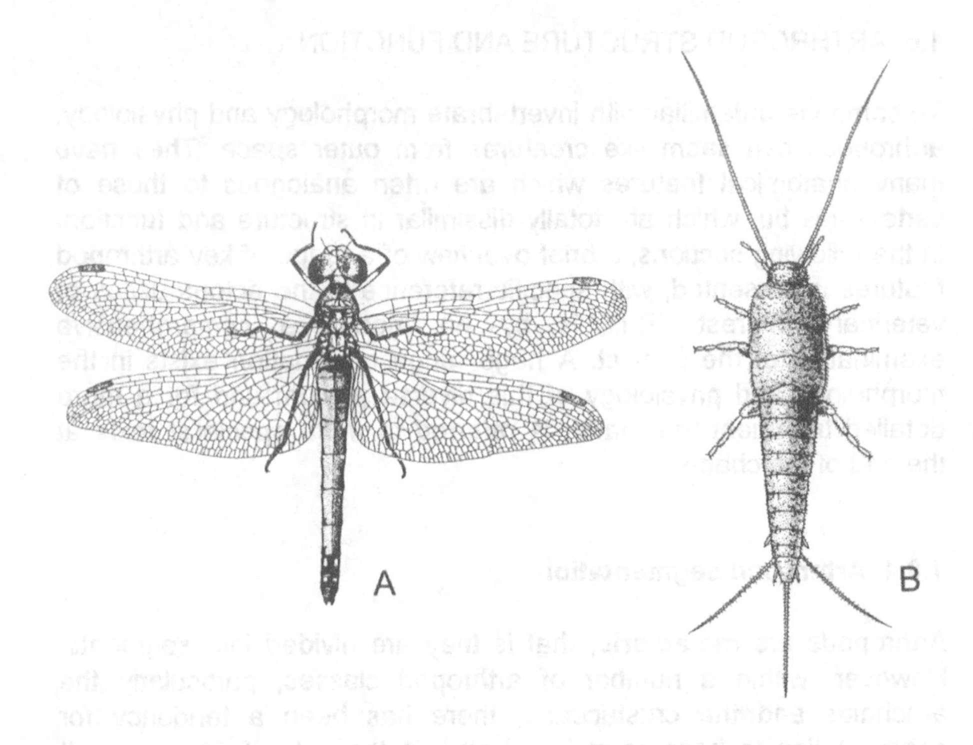

Fig. 1.2 A winged damselfly (A) and primitively flightless silverfish (B), showing division of the body into head, thoracic and abdominal segments (reproduced from Gullan and Cranston, 1994).

The exoskeleton is non-cellular. Instead'it is composed of a number of layers of **cuticle** which are secreted by a single outer cell layer of the body known as the **epidermis** (Fig. 1.3). The outer layer of cuticle, the **epicuticle** is composed largely of proteins and, in many arthropods, is covered by a waxy layer. The next two layers are the outer **exocuticle** and the inner **endocuticle**. Both are composed of a protein and a polysaccharide called **chitin**, which has long, fibrous molecules containing nitrogen. These are bound together into a stable, complex glycoprotein. In addition, the exocuticle may be tanned, or **sclerotized**, with quinones giving it extra strength from additional cross-linkages formed with the cuticular proteins.

The cuticle is often penetrated by fine pore canals which allow the passage of secretions from the epidermis to the surface. The cuticle has many outgrowths in the form of scales, spines, hairs and bristles. These outgrowths fall into two categories: those which are

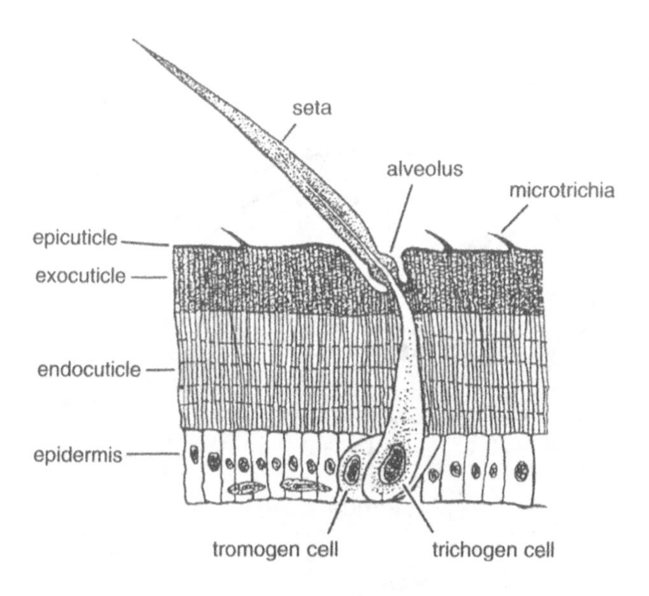

Fig. 1.3 Diagrammatic section through the arthropod integument (after Davies, 1988).

simply fine foldings of the outer layer of the cuticle (**microtrichiae**) and those which are articulated, such as setae (**macrotrichiae**). Microtrichiae can be very fine and give the arthropod distinctive patterns of shading. The articulated setae are attached to the cuticle by a thin membrane in a pit knows as the **alveolus** (Fig. 1.3). Setae are hollow outgrowths of the epicuticle and exocuticle, secreted by a **trichogen cell**. The socket is secreted by a **tormogen** cell.

Movement is made possible by the division of the cuticle into separate plates, called **sclerites.** Primitively these plates are confined to segments and the cuticle of each segment is divided into four primary plates: a **dorsal tergum**, two **lateral pleura** and a **ventral sternum** (Fig. 1.4). However, this pattern has frequently disappeared because of either fusion or subdivision of the segments.

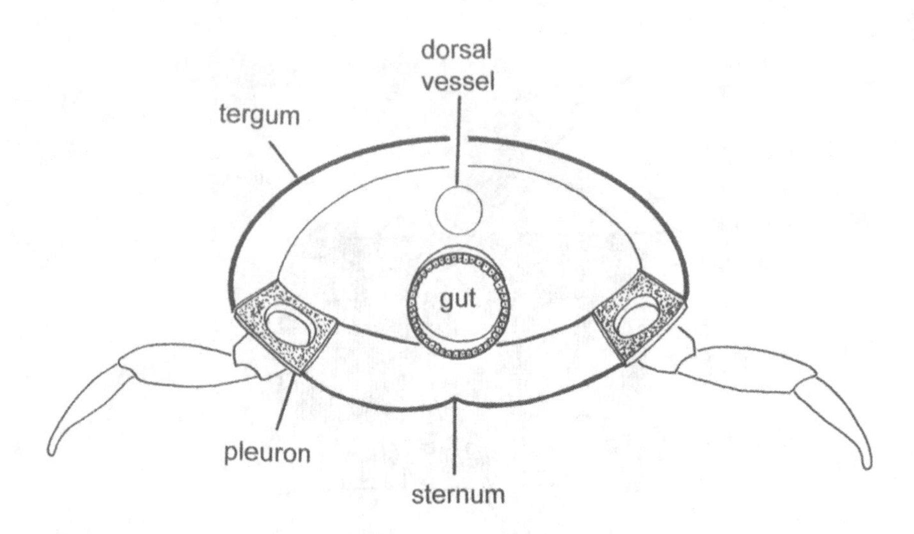

Fig. 1.4 Cross-section through the exoskeleton of a generalized arthropod, showing the tergum, pleuron and sternum of a single segment (after Barnes, 1974).

Plates are connected by flexible **articular membranes**, where the cuticle is not sclerotized and is flexible. These joints allow the body to move. In most arthropods the articular membrane is folded beneath the segment in front (Fig. 1.5). The muscles attach on the inside of the exoskeleton, the opposite of the vertebrate body plan. Muscles are often attached to rod-like invaginations of the cuticle called **apodemes** (Fig. 1.5). The soft, flexible cuticle present at the joints of the adult arthropod exoskeleton also occurs in the integument of larval arthropods.

The colours of most arthropods are produced by the deposition of brown, yellow, orange and red melanin pigments within the cuticle. However, iridescent greens and purples result from structural features of the cuticle itself, such as microtrichia, which selectively scatter or reflect light of specific wavelengths.

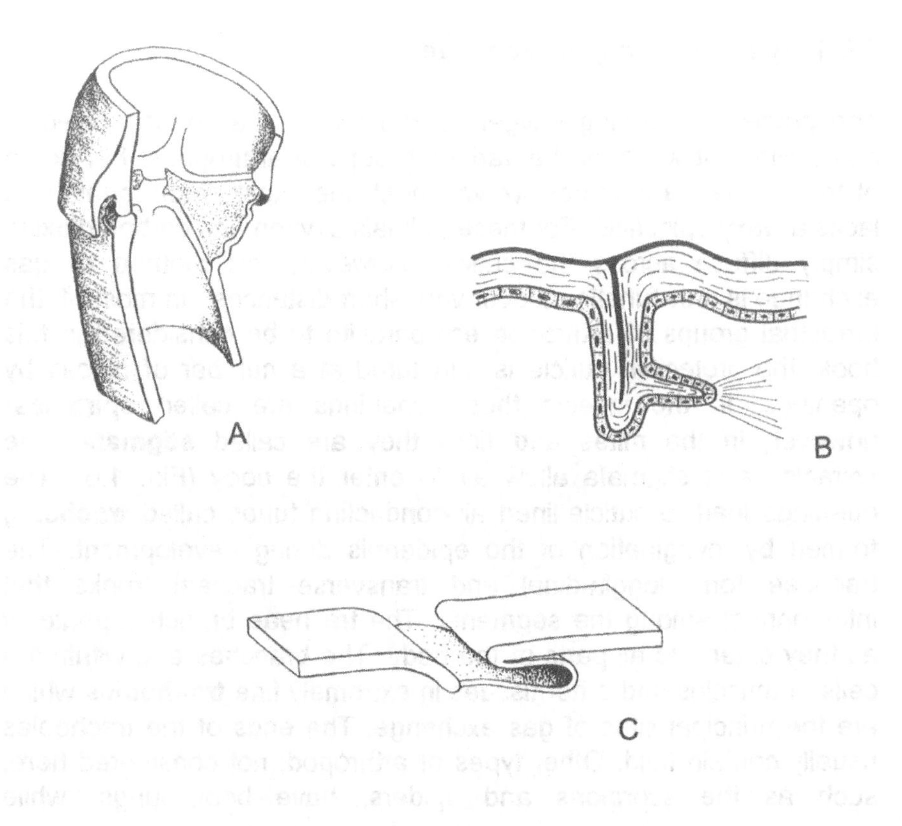

Fig. 1.5 (A) Articulation of a generalized arthropod leg joint. (B) An apodeme. (C) Intersegmental articulation, showing articular membrane folded beneath the segmental exoskeleton. (After Barnes, 1974, from Vandel, 1949.)

1.6.3 Jointed legs

The name arthropod is derived from the ancient greek *arthron,* meaning joint, and *pous,* meaning foot. Primitively each arthropod segment bears a pair of leg-like appendages. However, the number of appendages has frequently been modified through loss or structural differentiation. The cuticular skeleton of the legs is divided into tube-like segments connected to one another by articular membranes, creating joints at each junction (Fig. 1.5). The legs are usually six-segmented.

1.6.4 Spiracles and gas exchange

The process of getting oxygen to the tissues has been solved in many different ways by the various groups of arthropods. For some of the smallest arthropods we will meet, the exoskeleton is thin and lacks a waxy epicuticle. For these animals oxygen and carbon dioxide simply diffuse across the cuticle. However, this method of gas exchange is only functional over very short distances. In most of the terrestrial groups of arthropod ectoparasite to be considered in this book, the protective cuticle is punctured at a number of places by openings. In the insects these openings are called **spiracles;** however, in the mites and ticks they are called **stigmata**. The spiracles and stigmata allow air to enter the body (Fig. 1.6). The openings lead to cuticle-lined air-conducting tubes called **tracheae,** formed by invagination of the epidermis during development. The tracheae form longitudinal and transverse tracheal trunks that interconnect among the segments. The tracheae branch repeatedly as they extend to all parts of the body. The branches end within the cells of muscles and other tissues in extremely fine **tracheoles** which are the principal sites of gas exchange. The ends of the tracheoles usually contain fluid. Other types of arthropod, not considered here, such as the scorpions and spiders, have book lungs, while

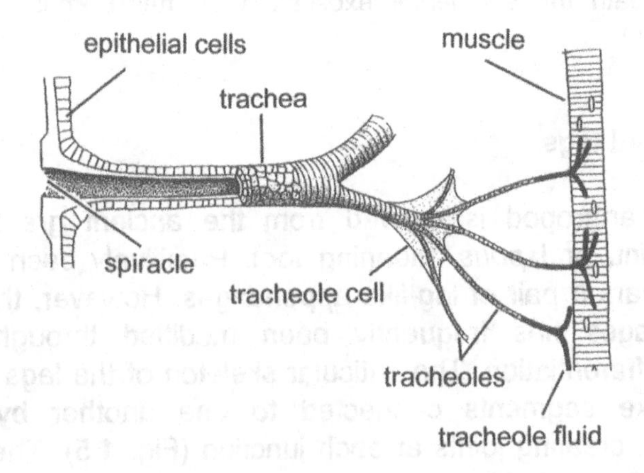

Fig. 1.6 A spiracle, trachea and tracheoles (reproduced from Barnes, 1974).

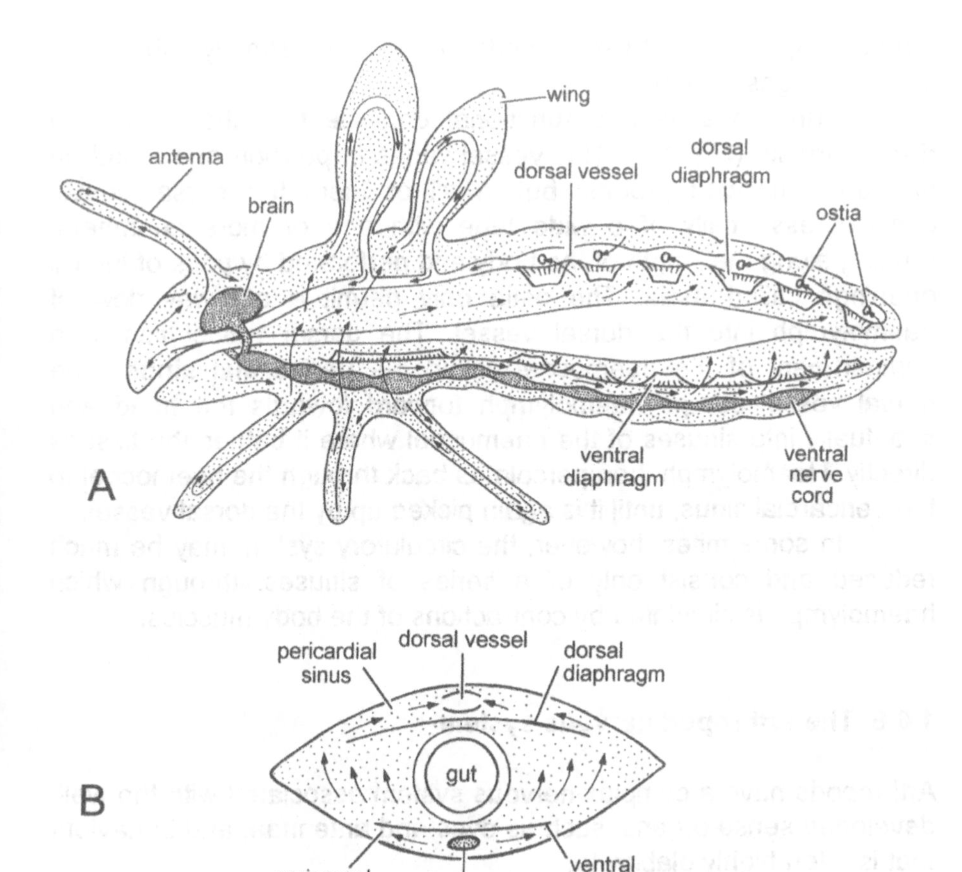

Fig. 1.7 Generalized arthropod circulatory system. (A) Longitudinal section through the body. (B) Transverse section through the abdomen. (Reproduced from Gullan and Cranston, 1994, after Wigglesworth, 1972.)

the aquatic crustaceans have gills.

1.6.5 The arthropod circulatory system

The arthropod circulatory system is relatively simple, consisting of a series of central cavities or sinuses, called a **haemocoel**, separated by muscular septa (Fig. 1.7). The haemocoel contains blood, called

haemolymph. In contrast to vertebrates, the haemolymph is not involved in gas exchange.

Arthropods have a functional equivalent of the heart, the **dorsal vessel** (Fig. 1.7). This vessel varies in position and length in different arthropod groups, but in all of them the dorsal vessel consists essentially of a wide tube with one or more chambers, running along the length of the body and perforated by pairs of lateral openings called **ostia**. The ostia only permit a one-way flow of haemolymph into the dorsal vessel. The dorsal vessel lies in a compartment of the haemocoel called the **pericardial sinus.** The dorsal vessel pumps haemolymph forward towards the head and eventually into sinuses of the haemocoel where it bathes the tissues directly. Haemolymph then percolates back through the haemocoel to the pericardial sinus, until it is again picked up by the dorsal vessel.

In some mites, however, the circulatory system may be much reduced and consist only of a series of sinuses, through which haemolymph is circulated by contractions of the body muscles.

1.6.6 The arthropod nervous system

Arthropods have a complex nervous system associated with the well-developed sense organs, such as eyes and antennae, and behaviour that is often highly elaborate.

The central nervous system consists of a dorsal brain in the head which is connected by a pair of nerves around the foregut to a series of ventral nerve cord ganglia (Fig. 1.7). In the embryo each segment gives rise to a pair of ganglia which then fuse to form a single ganglion and this pattern can still be seen in primitive arthropods. The ganglia are connected between segments by pairs of connective nerves. In more advanced arthropods ganglia may be fused. In blowflies, for example, there is only a single thoracic ganglion and no abdominal ganglia, and in the mites and ticks only a single cephalothoracic ganglion.

1.6.7 Digestion and absorption

The gut of an arthropod is essentially a tube that runs from mouth to anus (Fig 1.8). Nutrients are absorbed across the gut wall directly into the haemolymph. The precise shape of the gut varies between arthropods, various outpockets or large digestive glands being present, depending on the precise nature of their diet.

In general, the gut is divided into three sections: the foregut, midgut and hindgut (Fig. 1.8). The foregut and hindgut consist of invaginations of the exoskeleton at the mouth and anus, respectively; therefore they are lined with cuticle. The foregut is concerned primarily with the ingestion and storage of food. The latter usually taking place in the **crop**. Between the foregut and the midgut is a valve called the **proventriculus**. In some arthropods, the proventriculus is armed with teeth and functions in the crushing and grinding of food. The midgut is the principal site of digestion and absorption. It has a cellular lining which secretes digestive enzymes. Absorption takes place largely in the anterior of the midgut, in large outpockets called **gastric caeca**. The hindgut terminates in an expanded region, the **rectum**, which functions in the absorption of water and the formation of faeces. Nitrogenous wastes are eliminated from the haemocoel by long, thin projections called the **Malpighian tubules**, which open into the gut at the junction of the mid- and the hindgut. Mites have only one pair of Malpighian tubules whereas there may be large numbers present in insects.

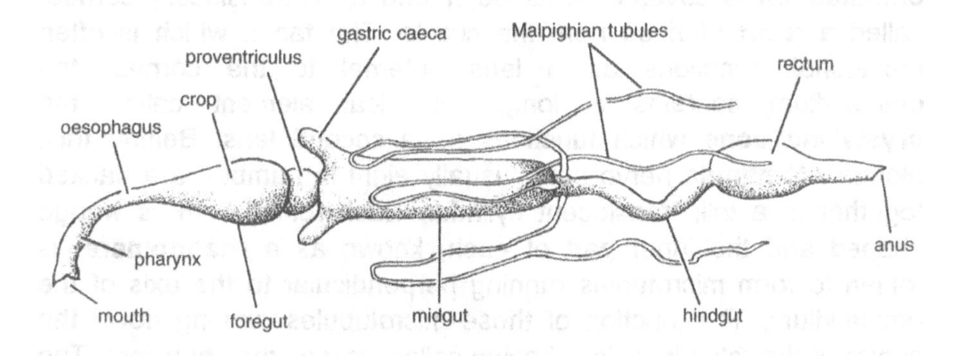

Fig. 1.8 Generalized digestive tract of an arthropod, showing the fore-, mid- and hindgut (reproduced from Barnes, 1974).

1.6.8 Arthropod sense organs

The sensory receptors of arthropods are usually associated with modifications of the chitinous exoskeleton. One common type of receptor is connected with hairs, bristles and setae. The bristle may be designed so that it acts as a mechanoreceptor, movement triggering the receptor at its base. Alternatively, the bristle may carry a chemoreceptor at its tip. Other common modifications for receptors are slits or pits in the exoskeleton. These may house chemoreceptors or the opening may be covered by a membrane with a nerve ending attached to its underside, to detect vibrations. Such receptors may be scattered over the body or concentrated on appendages such as the legs or antennae.

Most arthropods have eyes, but these can vary greatly in complexity. Some contain only a few photoreceptors. For example, in the **stemmata** of larval holometabolous insects and the **ocelli** of larval and adult hemimetabolous insects, a corneal lens overlies from 1 to 1000 sensory cells (Fig. 1.9). These simple eyes do not form images but are very sensitive at low light intensities and to changes in light intensity. Other types of arthropod eye, known as compound eyes, are large and complex with thousands of retinal cells (Fig. 1.9).

The **compound eyes** of insects and many crustaceans are composed of many long, cylindrical units. Each unit, called an **ommatidium** is covered at its outer end by a translucent cornea, called a **facet**, derived from the cuticle. The facet, which is often hexagonal, functions as a lens. Internal to the cornea, the ommatidium contains a long, cylindrical element called the **crystalline cone** which functions as a second lens. Behind this, elongated **retinula** nerve cells, usually eight in number, are packed together in a tall, translucent cylinder. Each retinula cell is wedge shaped and the inner part of each, known as a **rhabdomere**, is folded to form microtubles running perpendicular to the axis of the ommatidium. The junction of these microtubules running down the centre of the retinula cells is known collectively as the **rhabdom**. The retinula cells contain black or brown photosensitive molecules of a protein–retinene complex called **rhodopsin**.

The retinula nerve cells are also surrounded by a ring of light-absorbing pigment cells which screen the light entering each ommatidium from its neighbour.

The rhabdomeres of an ommatidium function as a single

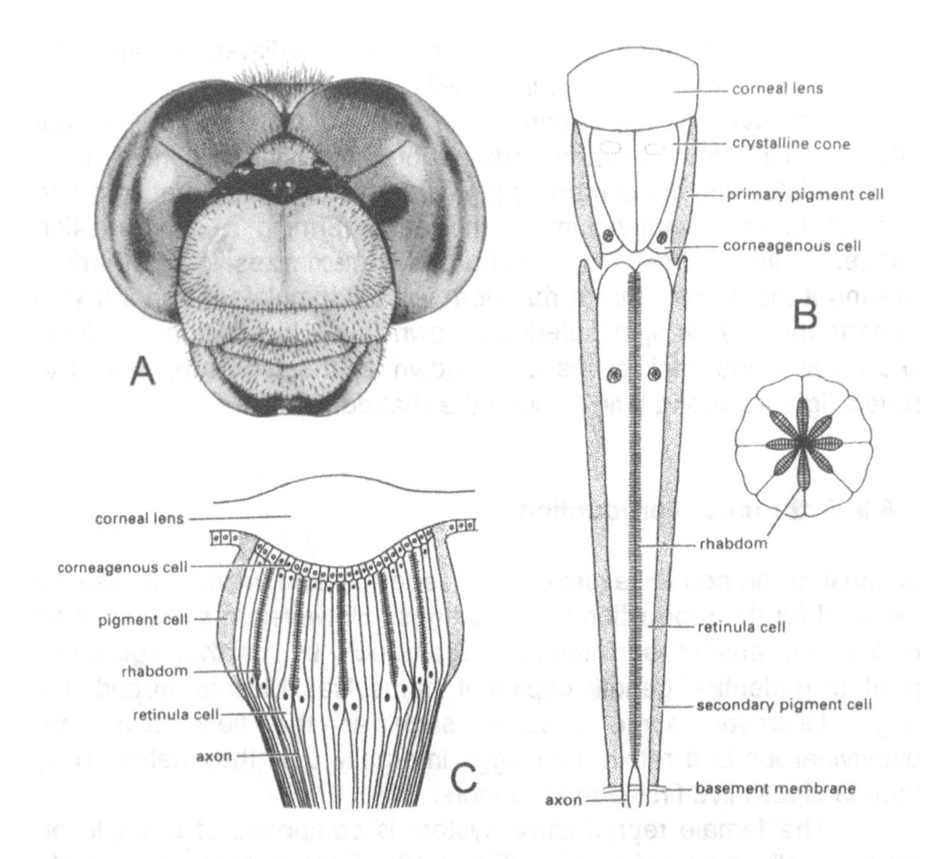

Fig. 1.9 (A) Head of a dragon fly, showing two large compound eyes and, between them, three ocelli. (B) Longitudinal section through an ommatidium with an enlargement of a transverse section. (C) Longitudinal section through an ocellus. (Reproduced from Gullan and Cranston, 1994.)

photoreceptor unit and transmit a signal that represents a single point of light. Individual ommatidia cannot form a detailed image and only overall light intensity is registered. Each ommatidium points in a slightly different direction. The image formed by a compound eye, therefore, represents a series of apposed points of light of different intensities, termed an apposition image. The detail available depends on the number of ommatidia present. There is no mechanism for accommodation and the principal function of a compound eye is in detecting movement as an image passes from one ommatidium to the next. This is assisted in many arthropods by the fact that the total corneal surface is highly convex, resulting in a wide visual field. In addition, many arthropods have colour vision, mediated by variations in the visual pigment in the retinula cells.

However, an apposition image does not work well at low light intensity. Therefore, in arthropods adapted for living in conditions of low light intensity, the screening pigment is retracted so that light can pass from one ommatidium to the next, forming a superposition image. While this image is less sharp, this maximizes light gathering, making it more likely that a rhabdom will be stimulated than if it was dependent only on light entering its own facet. In addition, a mirror-like layer at the back of the eye, known as the **tapetum**, serves to reflect light a second time through the rhabdom.

1.6.9 Arthropod reproduction

In most arthropods the sexes are separate and mating is usually required for the production of fertile eggs. However, in some species males are absent and females reproduce by **parthenogenesis**, producing identical genetic copies of themselves. Most arthropods lay eggs. However, some species, such as the flesh flies, are ovoviviparous and retain their eggs internally until they hatch. They then larviposit live first-stage maggots.

The female reproductive system is composed of a single or, more usually, a pair of **ovaries** (Fig. 1.10). Each ovary is divided into egg tubes, or **ovarioles**. The ovarioles join the lateral **oviduct** which in turn meets a median oviduct. This often ends in an **ovipositor**. A portion of the median oviduct may be expanded to receive the **aedeagus** during copulation.

The male reproductive system is usually composed of a pair or **testes**, each subdivided into a set of sperm tubes or follicles in which the formation of sperm takes place (Fig. 1.10). The follicles join the **vas deferens**, which is often expanded into the **seminal vesicle** which stores the sperm. The vas deferens join a common ejaculatory duct which ends in the external genitalia, with an intromittent organ, the penis, or **aedeagus**. Accessory glands produce secretions which may form a packet, called a **spermatophore**, that encloses the sperm and protects it during insemination.

Sperm may be delivered directly to the female during copulation or, as in some species of mite, the spermatophore is deposited on the ground and the female is induced to walk over and

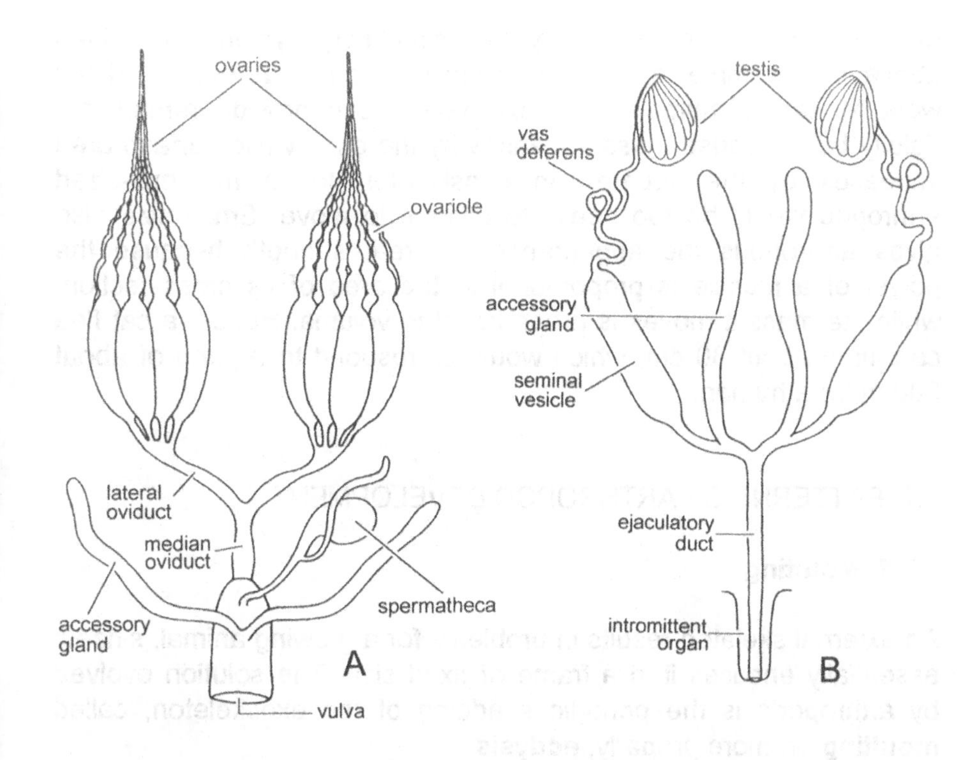

Fig. 1.10 Generalized (A) female and (B) male reproductive systems (from Gullan and Cranston, 1994, after Snodgrass, 1935).

pick up the spermatophore with her genital opening. Sperm are usually stored by the female in simple seminal receptacles or more complex organs called **spermathecae**. As an ovulated egg passes down the median oviduct it is fertilized by sperm released from the spermathecae. Accessory glands join the median or lateral oviducts and in many species produce secretions that coat and protect the eggs.

1.6.10 Arthropod size

The patterns of arthropod anatomy and physiology are intimately related to their size. The largest terrestrial insects and spiders weigh no more than about 100 g and the smallest are less than 0.25 mm in length. Only marine forms have managed to attain relatively large sizes; the Japanese spider crab, *Macrocheira,* may have a leg-span of over 3.5 m. The respiratory and circulatory systems described above, for example, are both efficient for arthropods but would not work for larger animals. The exoskeleon can provide remarkable rigidity but, because mass increases by the cube while surface area increases by the square, an exoskeleton for a mammal-sized arthropod would be too heavy to allow it to move. Small size also gives arthropods the appearance of great strength because the power of a muscle is proportional to the area of its cross-section, while the mass it moves is proportional to volume. Hence, a cat flea can jump about 30 cm, which would correspond to a jump of about 300 m for a human.

1.7 PATTERNS OF ARTHROPOD DEVELOPMENT

1.7.1 Moulting

An external skeleton results in problems for a growing animal, since it essentially encases it in a frame of fixed size. The solution evolved by arthropods is the periodic shedding of the exoskeleton, called **moulting** or, more properly, **ecdysis**.

Before the old skeleton is shed the epidermis detaches itself from the old cuticle (**apolysis**) and secretes a new epicuticle. The new epicuticle is soft and wrinkled at this stage. Enzymes (chitinases and proteinases), are then produced which pass through the new epicuticle and begin to erode the old endocuticle; the exocuticle is not affected. Muscle attachments and nerve connections are unaffected and the animal can continue to behave normally. Following digestion of the endocuticle new undifferentiated tissue, known as **procuticle**, is also produced. Exocuticle is absent along specific paths known as moulting lines. The old skeleton splits along these predetermined lines and the animal pulls out of the old encasement.

The soft, whitish exoskeleton of the newly moulted animal is stretched, often by the ingestion of air or water. Once expanded, quinones cross-link cuticular proteins of the new procuticle, particularly in its outer layers, forming exocuticle. This cross-linking results in hardening and darkening of the cuticle. Endocuticle continues to be deposited, on a daily cycle, for some time after moulting, producing daily growth lines that can be used to estimate age in some species of insect. The internal tissues of the animal may then be expanded to fill the new frame.

The stages between moults are known as **stages**, or **stadia**, and the form of the stadium as the **instar.** The duration of each stadium becomes longer as the animal becomes progressively older.

1.7.2 Simple and complex life cycles

In the arthropods of veterinary interest, growth and maturation from egg to adult may be accomplished via a number of different developmental paths. In most, the juvenile stadia broadly resemble the adult, except that the genitalia and, where appropriate, wings are not developed. The juveniles, usually called **nymphs**, are similar to the adults in appearance, feeding habits and habitat. The animal makes new cuticle and sheds the old one at intervals throughout development, typically four or five times, increasing in size before the emergence of the adult. This is often described as a simple life cycle with incomplete or partial metamorphosis, known as **hemimetabolous metamorphosis** (Fig. 1.11). In general, the same cells or tissues that make larval structures go on to make the same structures in the adults after the final moult.

In other arthropods, however, particularly some higher insects, there has been a trend towards increasing functional and

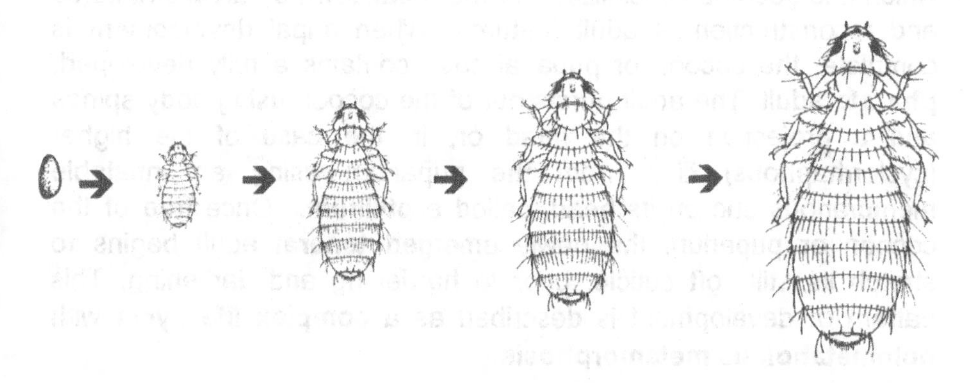

Fig. 1.11 Life cycle of the louse, *Menopon gallinae*, displaying hemimetabolous metamorphosis and passing through three nymphal stages prior to emergence as a reproductive adult (modified from Herms and James, 1961).

structural divergence in juvenile and adult stages (Fig. 1.12). The juvenile instar, which may be referred to as a **larva**, **maggot**, **grub** or **caterpillar**, has become concerned primarily with feeding and growth and may bear no physical resemblance to the adult. In contrast, the adult, or **imago**, has become the specialized reproductive and dispersal instar. In the juvenile stages the cuticle is usually soft and pliable and is not differentiated into hardened plates. These stages depend on a hydrostatic skeleton, provided by fluid pressure in the haemocoel, for support and movement. To reach the adult form, the larva must undergo complete metamorphosis, during which the entire body is reorganized and reconstructed.

The transformation between the juvenile and the adult is made possible by the incorporation of a pupal instar, which acts as a bridge between juvenile and adult instars. The juvenile feeds, moults and grows until it has reached its final juvenile stadium. In many species of fly (Chapter 4), the cuticle of the final larval stage contracts and tans to form a protective shell, the **puparium**. In other insects, such as the fleas (Chapter 6), the larva may spin a protective cocoon of silk produced by the salivary glands, prior to a final moult within the cocoon. The **pupa** lies within the puparium or cocoon. The pupa does not feed and is generally immobile. However, it is metabolically very active as old larval tissues and organs are lost or remoulded and replaced by adult organs. During the process of pupation, tissues undergo histolysis and are reassembled in the adult form. The pupa is probably a highly modified final juvenile stage, which has become specialized for the breakdown of larval structures and reconstruction of adult features. When pupal development is complete, the cocoon or puparial case contains a fully developed, **pharate** adult. The adult bursts out of the cocoon using body spines and a projection on the head or, in the case of the higher (cyclorrhaphous) flies, from the puparium using an inflatable membranous sac on its head, called a **ptilinum**. Once free of the cocoon or puparium the newly emerged **teneral** adult begins to stretch its still soft cuticle, prior to hardening and darkening. This pattern of development is described as a **complex life cycle** with **holometabolous metamorphosis**.

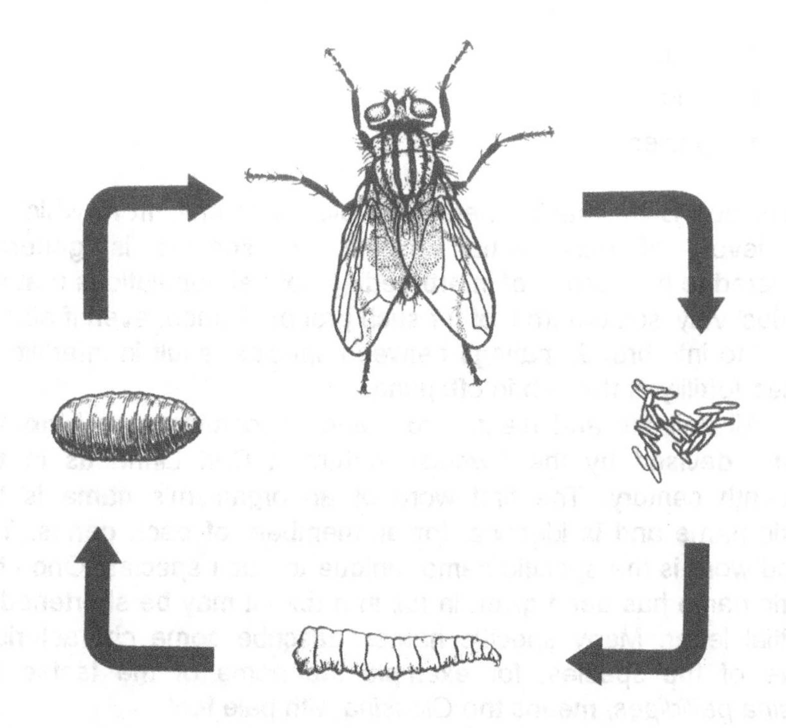

Fig. 1.12 Life cycle of a flesh fly, *Sarcophaga*, displaying holometabolous metamorphosis, with the egg giving rise to maggot-like larva, pupa and finally reproductive adult (modified from Evans, 1984).

1.8 THE CLASSIFICATION OF DIVERSITY

To make sense of the diversity of animal species, they are classified into biological units. These units are usually based on similarities in morphological characters, but may increasingly be based on isoenzyme electrophoresis and DNA analysis. The structure of the classification aims to describe biologically meaningful groups, usually attempting to represent evolutionary pathways. There are six basic categories into which organisms are classified:

- **phylum**
- **class**
- **order**

- family
- genus
- species

The species is the basic operational biological unit, from which all other levels of classification ascend. A species is generally considered to be a group of interbreeding natural populations that are reproductively isolated from other such groups. Hence, even if able or induced to interbreed, matings between species result in infertility or reduced fertility of the hybrid offspring.

All animals and plants are named according to a **binomial system**, devised by the Swedish naturalist Carl Linnaeus in the eighteenth century. The first word of an organism's name is the generic name and is identical for all members of each genus. The second word is the specific name, unique to each species. Once the generic name has been given in full in a text, it may be shortened to its initial letter. Many specific names describe some characteristic feature of the species, for example the name of the tsetse fly, *Glossina pallidipes,* means the *Glossina* with pale feet.

Even within a species not all populations are exactly the same and often there may be considerable variation associated with geographical, environmental, seasonal and genetic factors. Within a species, geographically isolated populations may be classified as sub-species, with each sub-species showing slight morphological differences but still being capable of interbreeding normally where populations overlap.

At the other end of the spectrum, the complexities of animal taxonomy have brought about the need to introduce numerous intermediate ranks in the classification, such as the sub-genus, sub-family, sub-order and also, for particularly complicated groups, a range of other terms such as species complex, tribe and super-family. There is no fixed limit to the number of categories that can be used. It is important to remember, however, that these groups are artificial creations of the taxonomist who is attempting to give order to the confusing diversity of arthropod forms. There is, therefore, no single 'correct' classification and, indeed, arthropod systematics is often the subject of heated debate!

It is helpful to note that the names of super-families usually end in '-oidea', families in '-idae' and sub-families in '-inae'.

1.9 THE ORIGINS OF ARTHROPODS

As described previously in this chapter, the phylum Arthropoda can be well defined by the presence of seven features:

- segmented bodies
- exoskeleton
- jointed limbs
- tagmatization
- dorsal blood vessel
- haemocoel
- ventral nerve cord

However, within the phylum there is considerable variation in morphology and the evolutionary origins of the different groups are far from clear.

There is general agreement that the arthropods probably evolved from some primitive **polychaete** stock or an ancestor common to both. Polychaetes are a class of metameric annelid worms, with a pair of paddle-like appendages on each segment. However, there has been considerable debate about whether the arthropods arose from a single common ancestor (monophyletic origin) or arose from a number of more or less related ancestors (polyphyletic origin).

The initial monophyletic view was based on analysis of morphological similarities and suggested that all arthropods arose from a basic **trilobite** or pre-trilobite stem (Fig 1.13). The trilobites were believed to be the most primitive of all known arthropods. Once abundant in the oceans 500 million years ago, they are now extinct. Trilobites had oval, flattened bodies, usually about 3–10 cm in length, with a thickened dorsal cuticle and ventral appendages. The majority of trilobites appear to have been bottom dwellers and crawled over mud and sand using their walking legs. It was proposed that, from the primitive trilobites, one line may have led to a sub-phylum known as the **Mandibulata**, containing the **crustaceans** (e.g. crabs, shrimps, barnacles), **myriapods** (millipedes and centipedes) and **insects** (e.g. flies, locusts and ants). A second line may have led to a sub-phylum known as the **Chelicerata**, containing an aquatic group, the **merestomes** (e.g. horseshoe crabs) and a terrestrial group, the **arachnids** (e.g. spiders, scorpions, mites and ticks).

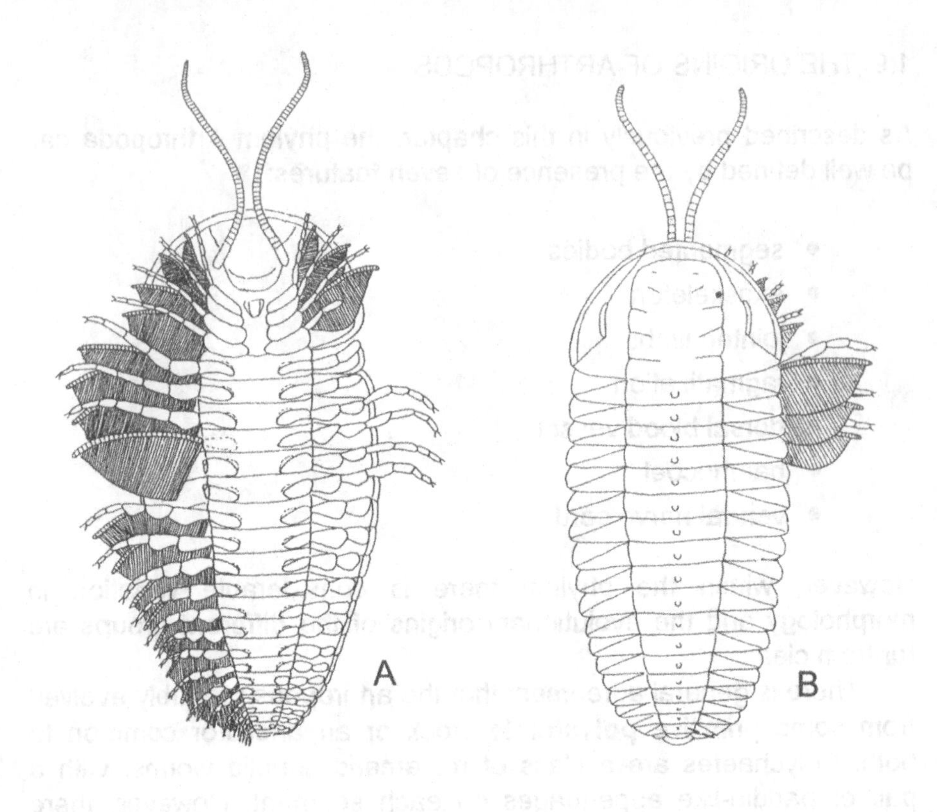

Fig. 1.13 The trilobite *Triarthrus eatoni*, (A) dorsal view (B) ventral view (reproduced from Barnes, 1974).

However, while there was general agreement that the chelicerates constitute a natural group, as do the insects and the myriapods, the crustaceans did not appear to fit easily within the insect–myriapod assemblage. There was growing acceptance that insect–myriapod mandibles are structurally different to crustacean mandibles and that the superficial similarities reflect convergent evolution, i.e. these features have arisen independently in each group, rather than reflecting any close phylogenetic ancestry.

The monophylectic view, therefore, was largely replaced by a polyphylectic scheme, due largely to the work of Manton. This suggested that, given the diversity of the arthropod groups, they must have arisen as a number of independent branches from the basic polychaete stock. From marine polychaete ancestors, several groups may have invaded land independently, giving rise to ancestral forms

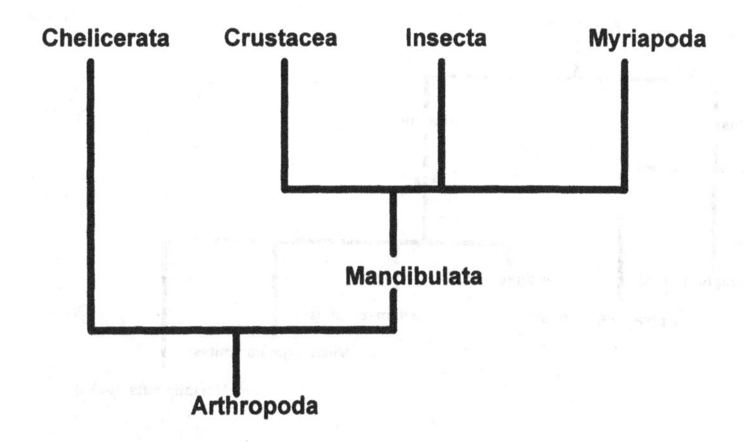

Fig. 1.14 Classification of the major classes of arthropod.

from which the myriapod–insect assemblage arose. Independently marine, perhaps bottom-dwelling, groups may have given rise to the crustaceans, the trilobites and chelicerates.

However, more recently the use of DNA analysis and cladistic techniques has given renewed support for a monophyletic origin of arthropods from annelid worms. Nevertheless, there is still considerable debate about the precise relationships of the various arthropod groups within the phylum. For example, analysis of ribosomal DNA places the Myriapoda closer to the Chelicerata than the Insects (Hexapoda) and Crustacea. In contrast, studies of mitochondrial DNA support the the view, outlined above, that the Crustacea, Myriapoda and Insecta may be considered together in one sub-phylum Mandibulata, with the Chelicerata as a second sub-phylum; which is the relationship between the various groups presented here (Fig. 1.14).

1.10 LIVING ARTHROPOD GROUPS

Only two classes, the Arachnida and the Insecta, contain species of major veterinary importance (Fig. 1.15).

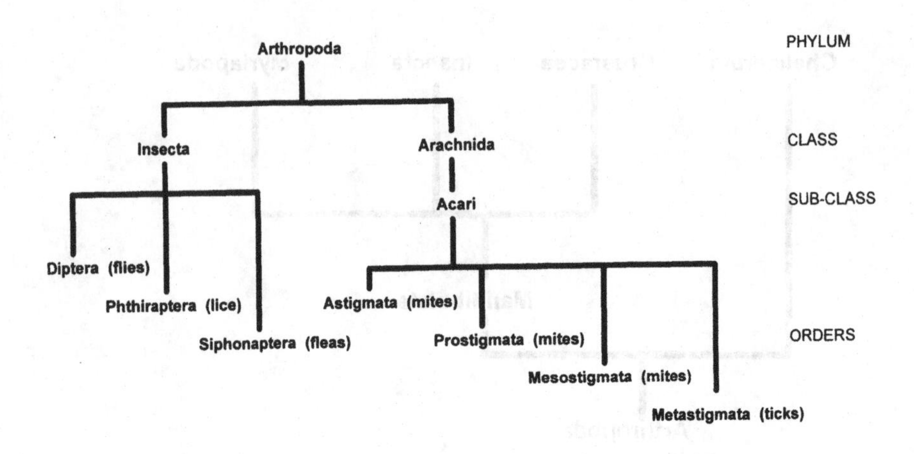

Fig. 1.15 Classification of the arthropod orders of veterinary importance.

1.10.1 Arachnids

Members of the class Arachnida are a highly diverse group of largely carnivorous, terrestrial, chelicerate arthropods. They are characterized by having the body divided into two parts, the **cephalothorax** and the **abdomen**. The unsegmented cephalothorax is usually covered dorsally by a solid carapace. In primitive forms the abdomen is divided into two; however, in most forms this segmentation has been lost.

On the cephalothorax the first pair of appendages, which are positioned in front of the mouth and which are used in feeding, are called **chelicerae**. The name Chelicerata comes from the ancient Greek *chele,* meaning claw, and *keras,* meaning horn. The mouthparts do not have true jaws. The second pair of appendages appear behind the mouth and are called **pedipalps**. Their precise structure and function varies from order to order. The arachnids do not possess antennae or wings and they only have simple eyes.

In the class Arachnida there is only one group of major veterinary importance, the sub-class **Acari** (sometimes also called Acarina), containing the mites and ticks (Fig. 1.16). Other major sub-classes or orders of arachnid include the scorpions (Scorpiones), spiders (Araneae) and pseudoscorpions (Pseudoscorpiones).

The sub-class Acari is an extremely diverse assembly, grouped together more from taxonomic convenience than true phylogenetic homogeneity. They are the most abundant of the arachnids; over 25,000 species have been described to date. They are usually small, averaging about 1 mm in length. However, some ticks may be over 3 cm in length. The cephalothorax and abdomen are broadly fused and abdominal segmentation is inconspicuous or absent, so that the body appears sack-like. The pedipalps are short, sensory structures associated with the chelicerae in a discrete structure called a **gnathosoma**. The body posterior to the gnatho-

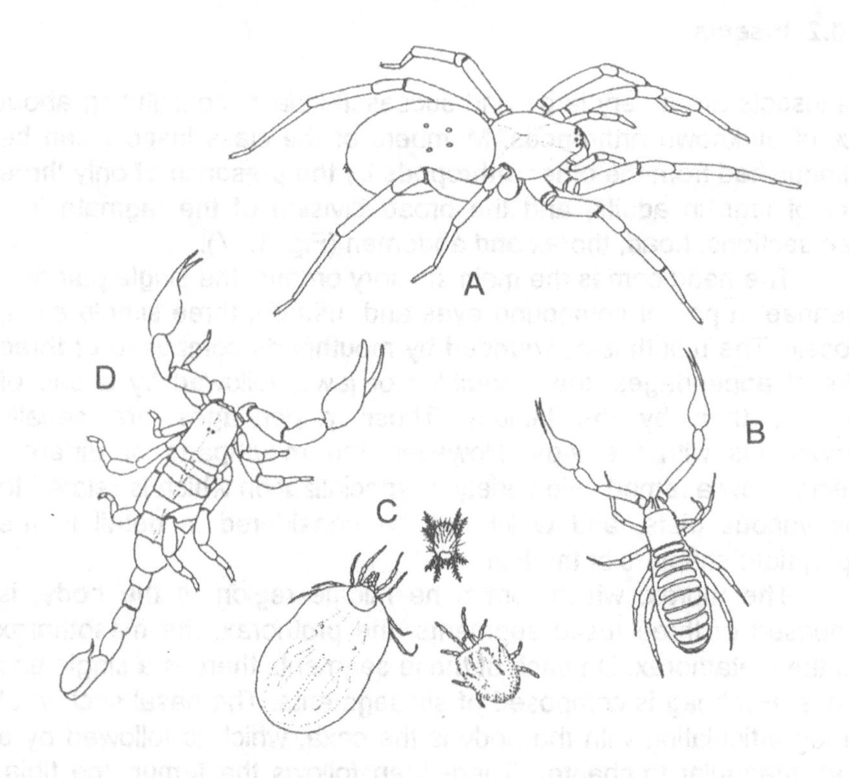

Fig. 1.16 Common orders of the class Arachnida: (A) a spider (Araneae), (B) a pseudoscorpion (Pseudoscorpiones), (C) mites and ticks (Acari) and (D) a scorpion (Scorpiones) (reproduced from Roberts, 1995).

soma is known as the **idiosoma**. There are four life-cycle stages: six-legged larva, eight-legged nymph and eight-legged adult. Within the larval and nymphal instars there may be more than one stadium.

In the adult, the idiosoma is subdivided into the region that carries the legs, the podosoma, and the area behind the last pair of legs, the opisthosoma. The legs are six-segmented and are attached to the podosoma at the coxa, also known as the epimere. This is then followed by the trochanter, femur, genu, tibia and tarsus.

The Acari are divided into seven orders, of which only four include parasitic forms: the Metastigmata, the ticks, also known as the Ixodida, and the orders Mesostigmata, Prostigmata and Astigmata, collectively forming the group known as mites (Fig. 1.15). The mites and ticks and the veterinary problems they cause will be considered in detail in chapters 2 and 3, respectively.

1.10.2 Insects

The insects are a very large and successful class, constituting about 90% of all known arthropods. Members of the class Insecta can be distinguished from the other arthropods by the presence of only three pairs of legs in adults, and the broad division of the tagmata into three sections: head, thorax and abdomen (Fig. 1.17).

The head carries the main sensory organs: the single pair of antennae, a pair of compound eyes and, usually, three simple eyes, or ocelli. The mouth is surrounded by mouthparts composed of three pairs of appendages: the mandibles or jaws, followed by a pair of maxillae, then by the labium. These appendages are serially homologous with the legs. However, the mouthparts of different insects show a remarkable variety of specialization which is related to their various diets, and which will be considered in detail in the appropriate chapters of the text.

The thorax, which forms the middle region of the body, is composed of three fused segments: the prothorax, the mesothorax and the metathorax. On each of these segments there is a single pair of legs. Each leg is composed of six segments. The basal section of the leg articulating with the body is the coxa, which is followed by a short, triangular trochanter. There then follows the femur, the tibia, one to five segments of the tarsus and, finally, the pretarsus composed of a pair of claws. The legs of insects are generally

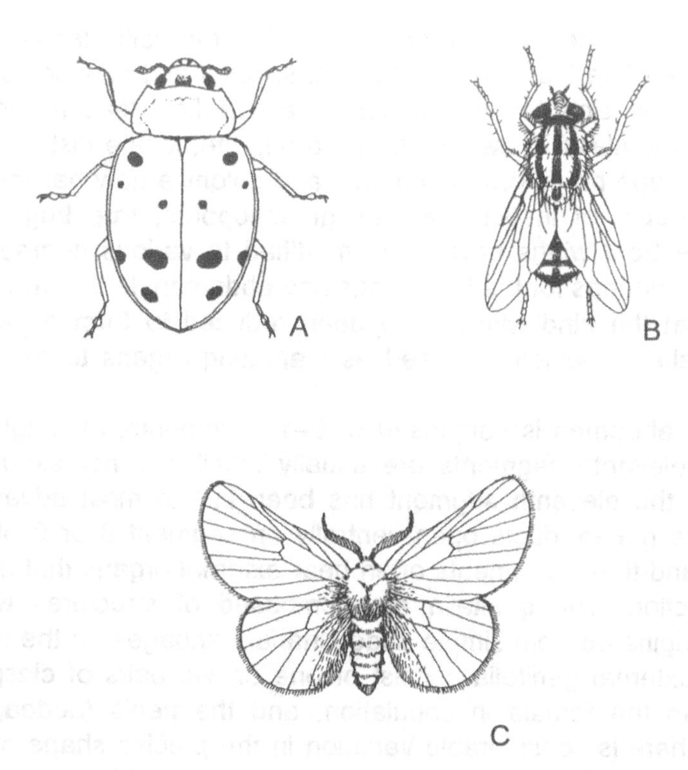

Fig. 1.17 Common orders of the class Insecta: (A) a beetle (Coleoptera), (B) a fly (Diptera) and (C) a moth (Lepidoptera) (reproduced from Evans, 1984).

adapted for walking or running but, as we will see, some are modified for specialized functions such as jumping (fleas) or clinging to the hairs of their host's body (lice).

Many groups of insect have two pairs of wings articulating with the mesothorax and metathorax. Some groups of primitive insects have never developed wings while others, such as the fleas and lice, which once had wings, have now lost them completely. Others, such as some of the hippoboscids, which we will meet in Chapter 4, have wings for only a short time as adults, after which they are shed. The wing consists of a network of sclerotized veins which enclose regions of thin, transparent cuticle called cells. The veins act as a framework to brace and stabilize the wing and carry haemolymph and nerves. The arrangement of the veins tends to be characteristic of various groups of insect species and so is important in identification and taxonomy. Wings are a key reason for the success of the class; allowing insects to migrate, locate distant food sources, escape predators, find mates and colonize new habitats. In several groups of insect, such as grasshoppers, true bugs and beetles, the front wings have been modified to various degrees as protective coverings for the hind wings and abdomen. In the true flies (the Diptera) the hind wings have been reduced to form a pair of club-like halteres, which are used as stabilizing organs to assist in flight.

The abdomen is composed of 9–11 segments, although the tenth and eleventh segments are usually small and not externally visible and the eleventh segment has been lost in most advanced groups. The genital ducts open ventrally on segment 8 or 9 of the abdomen and these segments often bear external organs that assist in reproduction. The genitalia are composed of structures which probably originated from simple abdominal appendages. In the male, the basic external genitalia consist of one or two pairs of claspers, which grasp the female in copulation, and the penis (aedeagus). However, there is considerable variation in the precise shape of the male genitalia in various groups of insect and these differences may be important in the identification of species. In the female the tip of the abdomen is usually elongated to form an ovipositor.

Within the class Insecta there are generally considered to be 26 orders, of which only three, the flies (Diptera), fleas (Siphonaptera) and lice (Phthiraptera) are of veterinary importance. Adult flies and their veterinary importance will be discussed in Chapter 4 and the problems caused by fly larvae in Chapter 5. Fleas and lice will be considered in chapters 6 and 7, respectively.

1.10.3 Other living arthropod classes

Crustacea

There are probably more than 26,000 species of Crustacea. Almost all are aquatic and the majority are marine. The class includes such familiar animals as the crabs, lobsters, shrimps, crayfish and woodlice, as well as many thousands of tiny planktonic species that play an essential role in marine food webs (Fig. 1.18).

Their bodies are organized into a head and segmented thorax, often combined in a cephalothorax, and posterior abdomen. Anteriorly the head bears five pairs of appendages, the first two of which are the antennae. The third pair of appendages are the mandibles which flank the mouth and behind these are two pairs of feeding appendages. Behind the head, the thorax is often covered by a carapace which arises as a fold of the head and which may overhang the sides of the body.

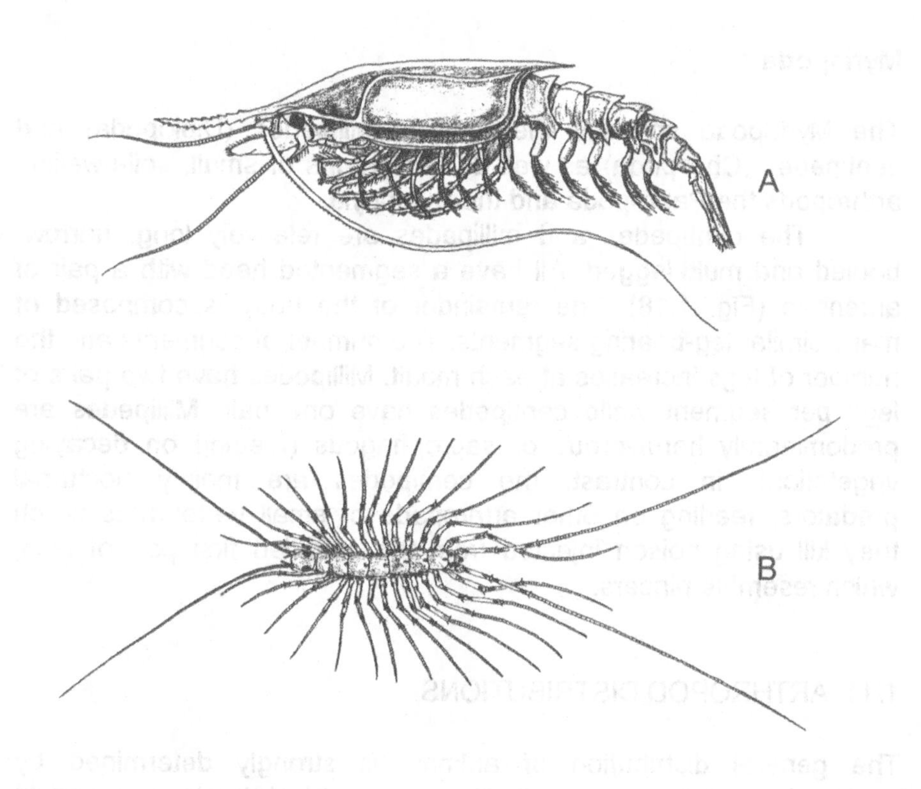

Fig. 1.18 The shrimp-like crustacean, *Gnathophausia* (A) and the myriapod, centipede *Scutigera coleoptera* (B) (from Barnes, 1974, after Snodgrass, 1935).

Primitively, each of the many segments of the throrax and abdomen carries a pair of modified appendages. Crustacean appendages are described as **biramous**, since typically they are composed of an inner and an outer branch. However, the structure is very variable since the segments have undergone various degrees of fusion and reduction and the thoracic and abdominal appendages are often modified to perform specific specialized functions, such as swimming, crawling, sperm transmission or egg brooding.

The Crustacea are of little or no veterinary importance, with the exception perhaps of copepod crustaceans, known as 'fish lice'. Copepod crustaceans are specialized ectoparasites which attach to the gill filaments or fins and feed by piercing and sucking.

Myriapoda

The Myriapoda contains the familiar millipedes (Diplopoda) and centipedes (Chilopoda) as well as two groups of small, soil-dwelling arthropods the Pauropoda and the Symphyla.

The centipedes and millipedes are relatively long, narrow-bodied and multi-legged. All have a segmented head with a pair of antennae (Fig. 1.18). The remainder of the body is composed of many similar leg-bearing segments. The number of segments and the number of legs increases at each moult. Millipedes have two pairs of legs per segment while centipedes have one pair. Millipedes are predominantly herbivorous or saprophagous (feeding on decaying vegetation). In contrast, the centipedes are mainly nocturnal predators, feeding on other arthropods or small vertebrates which they kill using poison injected from the modified first pair of legs, which resemble pincers.

1.11 ARTHROPOD DISTRIBUTIONS

The general distribution of animals is strongly determined by geography and climate. In reflecting geographical divisions, the world is often divided into six zoogeographical regions, each region containing its characteristic animal and plant species (Fig 1.19). Each region is usually isolated by physical boundaries such as deserts, mountains and oceans. Each region contains many species

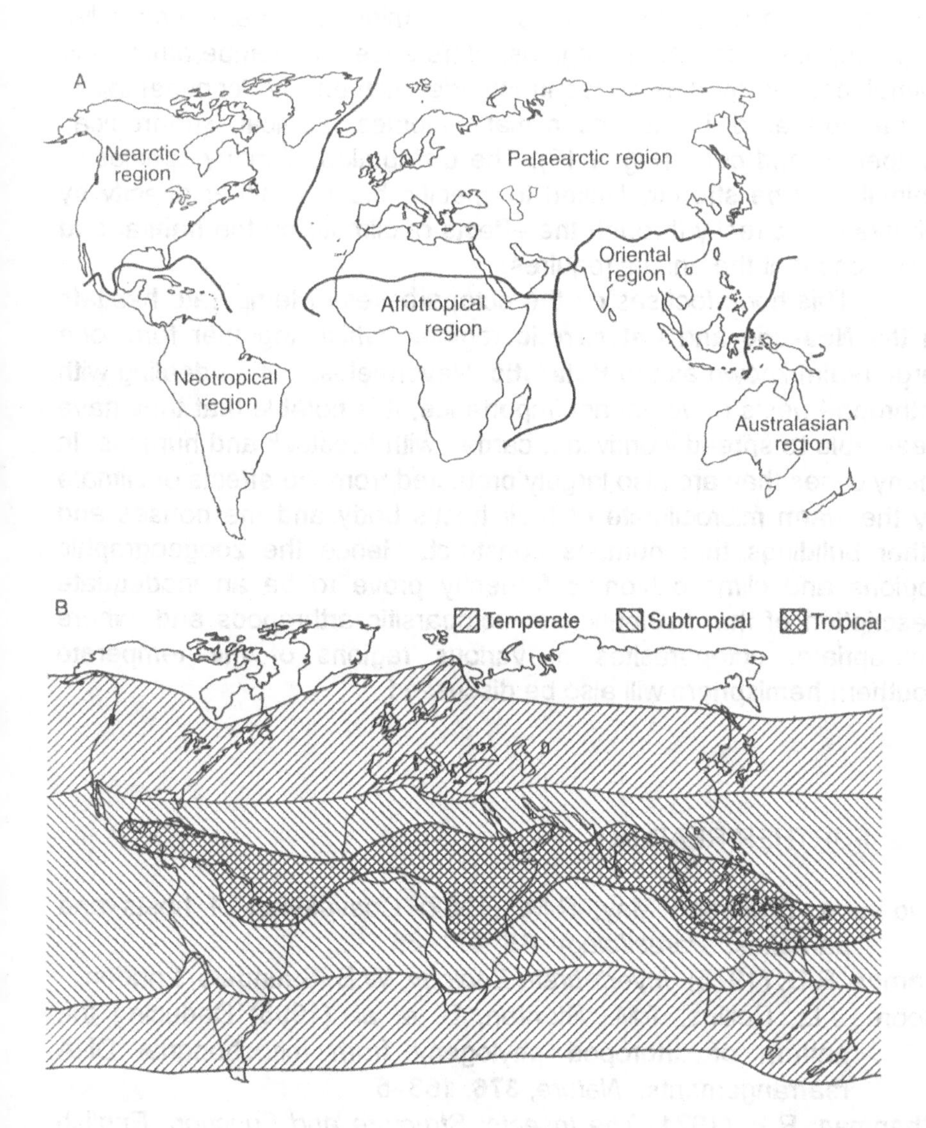

Fig. 1.19 (A) World zoogeographic zones; (B) world climatic biomes.

of animal and plant which are **endemic** (native to the region) or which are **indigenous** (have dispersed or migrated there naturally). Superimposed on these regions, differences in temperature and rainfall caused by differences in latitude, altitude and 'continentality', create further divisions into climatic biomes: tropical, sub-tropical, temperate and polar (Fig. 1.19). The distribution of many species of animal may be strongly limited to specific biomes, either directly by climate or indirectly through the effects of climate on the habitat and resources that the animal requires.

This book focuses on the ectoparasites of temperate habitats in the **Nearctic** and **Palaearctic** regions, which together form one large realm known as the **Holarctic**. Nevertheless, when dealing with arthropod pests of veterinary importance, it is notable that thay have been able to spread worldwide, carried with livestock and humans. In many cases thay are also largely protected from the effects of climate by the warm microclimate of their host's body and the houses and other buildings that humans construct. Hence the zoogeographic regions and climatic biomes freqently prove to be an inadequate description of the distribution of ectoparsitic arthropods and, where appropriate, ectoparasites of various regions of the temperate southern hemisphere will also be discussed.

1.12 FURTHER READING

Anderson, R.M. and May, R.M. (1982) Coevolution of hosts and parasites. *Parasitology*, **85**, 411–26.

Barnes, R.D. (1974) *Invertebrate Zoology*, W.B. Saunders, London.

Boore, J.L., Collins, T.M., Stanton, D. *et al.* (1995) Deducing the pattern of arthropod phylogeny from mitochondrial DNA rearrangements. *Nature*, **376**, 163–5.

Chapman, R.F. (1971) *The Insects: Structure and Function*, English Universities Press, London.

Clutton-Brock, J. (1987) *A Natural History of Domesticated Mammals*, Cambridge University Press, Cambridge.

Davies, R.G. (1988) *Outlines of Entomology*, Chapman & Hall, London.

Evans, H.E. (1984) *Insect Biology. A Textbook of Entomology*, Adison Wesley, Massachusetts.

Ewald, P.W. (1983) Host–parasite relations, vectors and the evolution of disease severity. *Annual Review of Ecology and Systematics*, **14**, 465–85.
Ewald, P.W. (1993) The evolution of virulence. *Scientific American*, **268**, 56–62.
Ewald, P.W. (1995) The evolution of virulence: a unifying link between parasitology and ecology. *Journal of Parasitology*, **81**, 659–69.
Fallis, A.M. (1980) Arthropods as pests and vectors of disease. *Veterinary Parasitology*, **6**, 47–73.
Freidrich, M. and Tautz, D. (1995) Ribosomal DNA phylogeny of the major extant arthropod classes and the evolution of myriapods. *Nature*, **376**, 165–7.
Gullan, P.J. and Cranston, P.S. (1994) *The Insects. An Outline of Entomology*, Chapman & Hall, London.
Harwood, R.F. and James, M.T. (1979) *Entomology in Human and Animal Health,* Macmillan, New York.
Herms, W.B. and James, M.T. (1961) *Medical Entomology*, MacMillan, New York.
Kettle, D.S. (1984) *Medical and Veterinary Entomology,* Croom Helm, London.
Kim, K.C. (1985) *Coevolution of Parasitic Arthropods and Mammals*, John Wiley and Sons, Chichester.
Lane, R.P. and Crosskey, R.W. (1993) *Medical Insects and Arachnids*, Chapman & Hall, London.
Lenski, R. and May, R.M. (1994) The evolution of virulence in parasites and pathogens: a reconciliation between two conflicting hypotheses. *Journal of Theoretical Biology*, **169**, 253–65.
Manton, S.M. (1973) Arthropod phylogeny – a modern synthesis. *Journal of Zoology*, **171**, 111–30.
Marshall, A.G. (1985) *The Ecology of Ectoparasitic Insects*, Academic Press, London.
May, R.M. and Anderson, R.M. (1990) Parasite–host coevolution. *Parasitology*, **100**, 89–101.
Poulin, R. (1996) The evolution of life history strategies in parasitic animals. *Advances in Parasitology*, **37**, 107–34.
Roberts, M.J. (1995) *Spiders of Britain and Northern Europe*, Harper Collins, London.
Snodgrass, R.E. (1935) *Principles of Insect Morphology*, McGraw-Hill, New York.

Waage, J.E. (1979) The evolution of insect/vertebrate associations. *Biological Journal of the Linnean Society*, **12**, 187–224.

Walker, A. (1994) *Arthropods of Humans and Domestic Animals. A Guide to Preliminary Identification*, Chapman & Hall, London.

Wigglesworth, V.B. (1972) *The Principles of Insect Physiology*, Chapman & Hall, London.

Zinser, H. (1934) *Rats, Lice and History*, Little, Brown, Boston.

2

Mites (Acari)

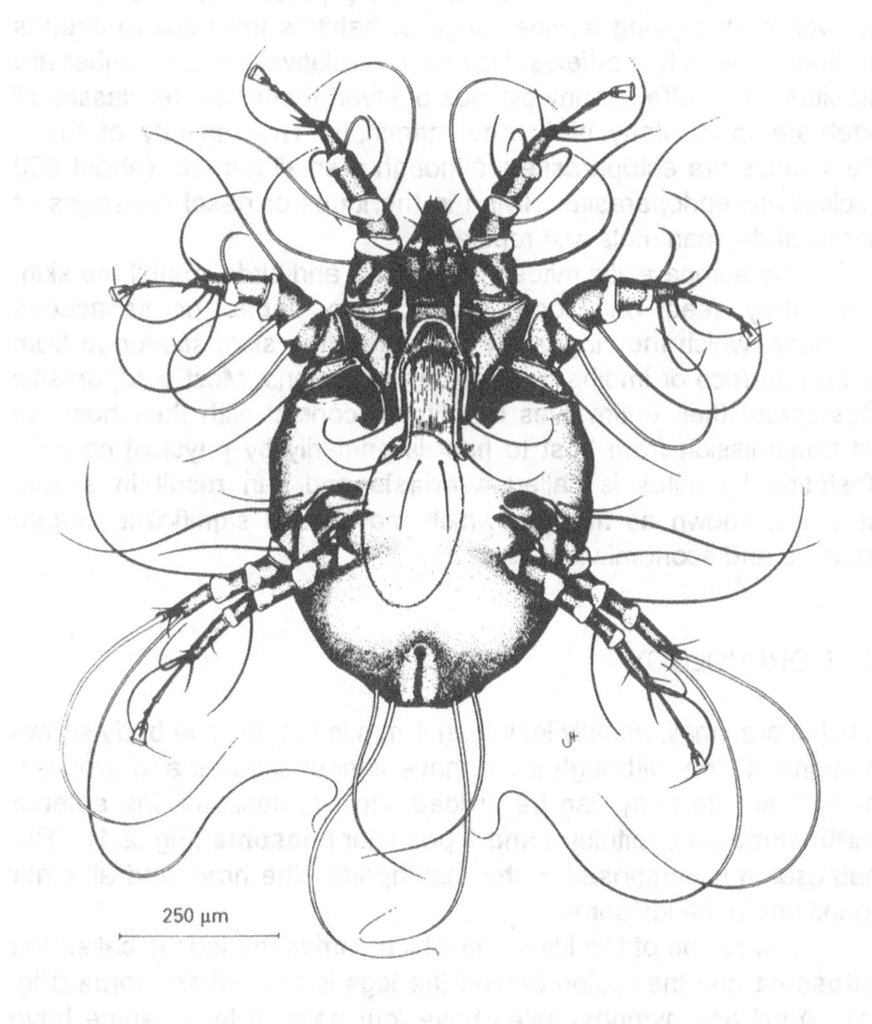

Adult female sheep scab mite, *Psoroptes ovis*, ventral view (from Kettle, 1984).

2.1 INTRODUCTION

The mites are a huge and disparate group of almost 30,000 species, with possibly another 450,000 species waiting to be described. Superficially they may appear simple and somewhat uninteresting but in reality they are extremely variable both in structure and habit and, as ectoparasites, are responsible for significant problems in domestic animals.

The majority of mites are free-living predators, herbivores or detritivores, occupying a wide range of habitats from soil to oceans and from deserts to ice-fields. However, a relatively small number are parasites. They affect many classes of invertebrate and all classes of vertebrate, particularly birds and mammals. The majority of these mite species are ectoparasites, although a small number (about 500 species) are endoparasites, living in the lungs or nasal passages of various birds, mammals and reptiles.

The ectoparasitic mites of mammals and birds inhabit the skin, where they feed on blood, lymph, skin debris or sebaceous secretions, which they ingest by puncturing the skin, scavenge from the skin surface or imbibe from epidermal lesions. Most ectoparasitic mites spend their entire lives in intimate contact with their host, so that transmission from host to host is primarily by physical contact. Infestation by mites is called **acariasis** and can result in severe dermatitis, known as **mange**, which may cause significant welfare problems and economic losses.

2.2 MORPHOLOGY

All mites are small, usually less than 1 mm in length. The body shows no segmentation, although it can have various sutures and grooves. The typical mite body can be divided into two sections, the anterior **gnathosoma** (or capitulum) and a posterior **idiosoma** (Fig. 2.1). The gnathosoma is composed of the mouthparts. The brain and all other organs are in the idiosoma.

The region of the idiosoma which carries the legs is called the **podosoma** and the region behind the legs is the **opisthosoma** (Fig. 2.1). Adult and nymphal mites have four pairs of legs, larvae have only three pairs. In adults and nymphs the legs are arranged in two sets, two pairs of anterior legs and two pairs of posterior legs. The

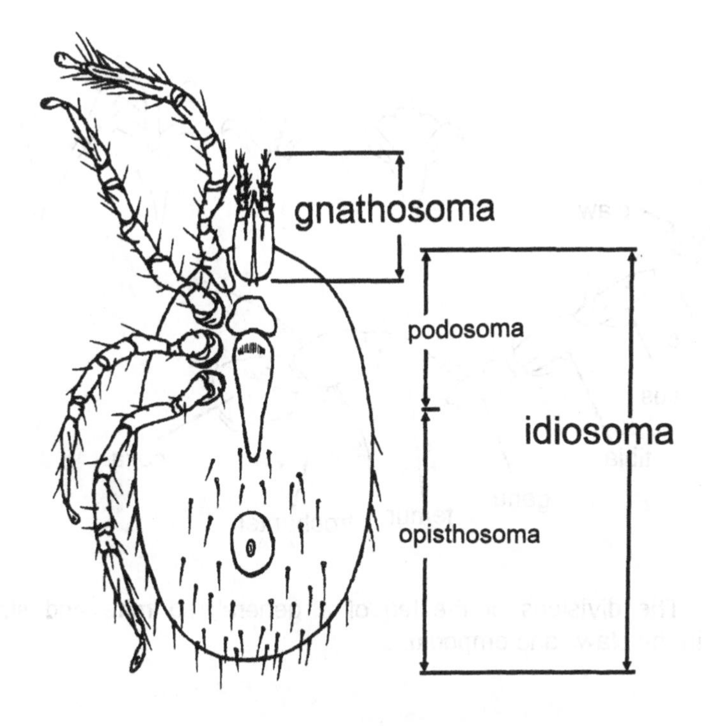

Fig. 2.1 Divisions of the body of a generalized mite.

first pair of legs are often modified to form sensory structures or to assist with capturing prey, and frequently are longer and more slender than the others. The legs are usually six-segmented (Fig. 2.2) and are attached to the body at the **coxa**, also known as the **epimere**. This is then followed by the **trochanter**, **femur**, **genu**, **tibia** and **tarsus**. At the end of the tarsus, may be a **pretarsus** and this may bear the **ambulacrum**, which is usually composed of paired claws and, between them, a structure known as the **empodium**. The empodium can be highly variable in form, being membranous or resembling a filamentous hair, pad, sucker or claw (Fig. 2.2). In some of the parasitic astigmatid mites, the claws are absent and on some of the legs the ambulacral organs consist of stalked **pretarsi** which may be expanded terminally into membranous bell- or sucker-like discs, known as **pulvilli.**

The idiosoma may be soft, wrinkled and unsclerotized.

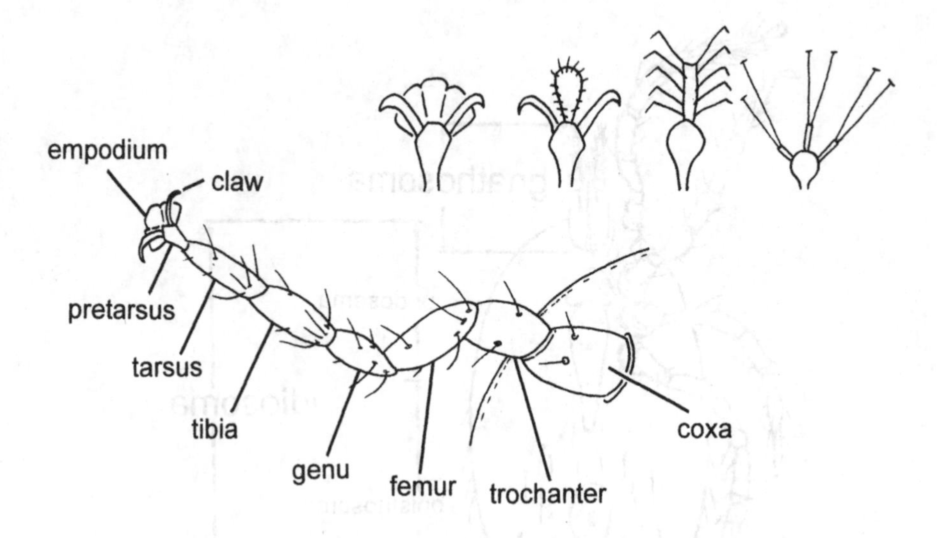

Fig. 2.2 The divisions of the leg of a generalized mite and structural variation in the claws and empodia.

However, many mites may have two or more sclerotized dorsal shields and two or three ventral shields: the **sternal**, **genitoventral** and **anal** shields (Fig. 2.3). These may be important features for mite identification. The genitoventral shield, located between the last two pairs of legs, bears the genital orifice.

In many small mites, particularly the astigmatid mites, respiration takes place directly through the integument. In others, between one and four pairs of respiratory openings, called **stigmata**, are the found on the idiosoma. As will be discussed, the presence or absence of stigmata and their position on the body is extremely important in the higher classification of mites. The stigmata lead to a complex branching system of **tracheae**. The circulatory system is reduced and in most groups consists of a network of sinuses through which blood is circulated by contraction of body muscles. The stigmata, particularly in the mesostigmatid mites, are usually associated with elongated sclerotized processes of unknown function, known as the **peritremes**.

Sexual differentiation is usually not obvious in the larva and nymph. Adult males possess a pair of testes and vasa deferentia which extend from each testis to open through the median gonopore or through a chitinous penis which can project through the genital orifice of the female. In the adult female there is usually a single ovary which is connected to the genital orifice by an oviduct. Sperm is transmitted directly from male to female in some mites. However, indirect sperm transfer is known to occur in many species. A spermatophore is produced and transferred to the genital orifice of the female by the chelicerae or the legs. In other species the male produces a stalked capsule containing a droplet of semen. Females may be attracted by semiochemicals produced by the spermatophore or by a trail laid down by the male. After encountering a droplet the female removes the sperm from its stalked cup with her genital orifice.

Eyes are usually absent and, hence, most mites are blind. Where they are present, however, in groups such as the trombidiformes, the eyes are simple. Hairs, or setae, many of which are sensory in function, cover the idiosoma of many species of mite.

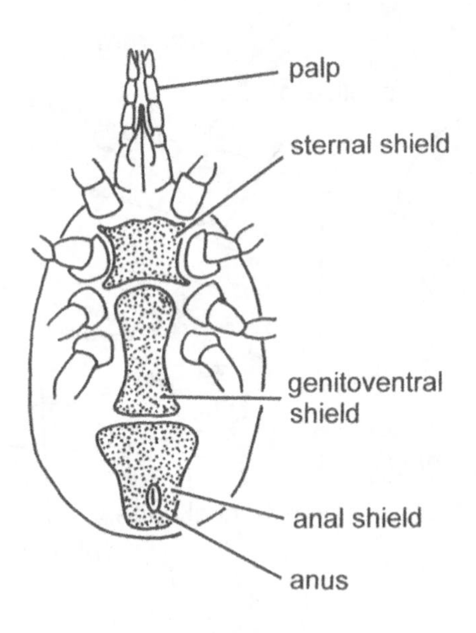

Fig. 2.3 Ventral shields of a generalized mesostigmatid mite.

The number, position and size of the setae are extremely important in identification of mite species.

The gnathosoma (capitulum) is the highly specialized feeding apparatus (Fig. 2.4). It is essentially a food-carrying tube connected to the oesophagus. The gnathosoma carries a pair of **palps**, which are simple sensory organs that help the mite to locate its food. The palps are one- or two-segmented in most Astigmata and Prostigmata and five- or six-segmented in the Mesostigmata. The last segment of the palps usually carries a claw-like structure, the **palpal claw** or **apotele**. Between the palps lies a pair of three-segmented **chelicerae**. These are used for tearing, grasping or piercing, and their precise structure may be highly modified, depending on the feeding habits of the various species of mite: stylet-like when used for piercing, or claw-like when used for tearing tissues. At their tip the chelicerae may also carry structures called **chelae**, which may be claw-like or long and stylet-like. Between the chelicerae is the **buccal**

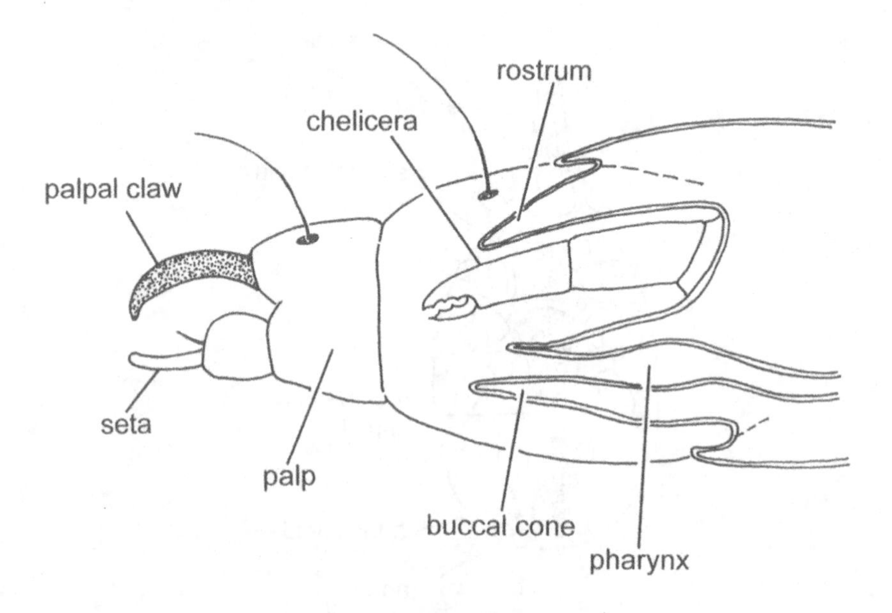

Fig. 2.4 Longitudinal section through the gnathosoma (capitulum) of a generalized mite.

cone. The buccal cone and chelicerae fit within a socket-like anterior chamber in the opisthosoma. This is formed by a dorsal projection of the body wall, called the rostrum, and ventrally and laterally by the enlarged coxae of the palps. The palps are attached to both sides of the chamber and the chelicerae to the back wall of the chamber. The buccal cone can be extended and retracted in some species. The muscular pharynx is the primary pumping organ used for ingestion. When mites feed they usually do not show dramatic increases in body size seen in other blood-feeding parasites, such as the ticks.

In the mesostigmatid mites the fused expanded coxal segments of the palps at the base of the gnathosoma are known as the **basis capituli.** The chelicerae are enclosed within the basis capituli. A structure called the **hypostome** projects forward from the ventral surface of the basis capituli. However, unlike ticks (Chapter 3), mesostigmatid mites have no teeth projecting from the hypostome.

2.3 LIFE HISTORY

Female mites produce relatively large eggs, from which a small, six-legged **larva** hatches (Fig. 2.5). A few species are ovoviviparous, producing live offspring. The larva moults to become an eight-legged **nymph**. There may be between one and three nymphal stages, known respectively as the **protonymph**, **deutonymph** and **tritonymph**. At least one of these nymphal stages is usually inactive and development proceeds without feeding. The nymph then moults to become an eight-legged **adult**.

The mites are small and highly adaptable animals capable of living in a wide range of habitats. The number of eggs produced per female is highly variable but lifetime reproductive outputs may be as low as 16 eggs per female. Nevertheless, the life cycle of many parasitic species may be completed in less than 4 weeks and in some species may be as short as 8 days. Hence, these mites have the potential for explosive increases in their population size.

2.4 PATHOLOGY

In many cases, the activity of mites may have no obvious effect on

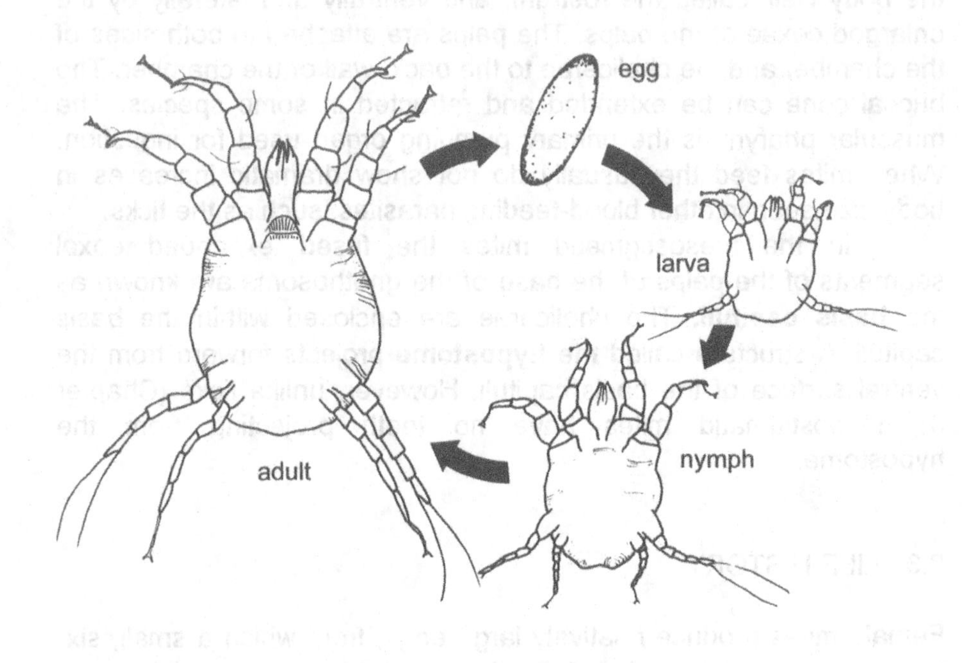

Fig. 2.5 Generalized life cycle of a psoroptid mite, with only a single nymphal stage shown (redrawn from Walker, 1994).

the host and, indeed, some species of mite such as *Demodex* may be considered to be a normal part of the skin fauna. However, on domestic, farmed and captive wild animals explosive increases in mite populations can occur with great rapidity.

Problems with mite infestation and dramatic increases in mite populations occur more commonly in animals in poor condition and are more often seen at the end of winter or in early spring. Some forms of mange, such as demodectic mange, are the result of underlying disease or immunosuppression. The clinical signs of erythema, pruritus and scale or crust formation are due to the inflammatory response of the skin and resulting excoriation. This response is stimulated by feeding, burrowing or the production of antigenic material by the mite. The keratinocytes release cytokines (especially IL-1) in response to non-specific damage, which diffuses into the dermis leading to cutaneous inflammation. In addition, mite antigens *de novo* or processed by antigen-presenting cells can be carried in the dermal lymphatics to local lymph nodes where an immunological response occurs. This response can be humoral or cell mediated, resulting in either a protective immune response or hypersensitivity. In particular, mite faecal antigens are thought to be important in the production of hypersensitivity which amplifies the innate inflammatory response. An example of such a response is canine *Sarcoptes scabiei* infestation (sarcoptic mange) in which an IgE-mediated hypersensitivity develops.

Mite infestation can result in:

- direct epidermal damage leading to inflammation; this results in skin erythema, pruritus, scale formation, lichenification (thickening) and crust (inflammatory exudate) formation;
- the production of cutaneous hypersensitivity (especially type I hypersensititivity);
- loss of blood or other tissue fluids;
- mechanical or biological transmission of pathogens.

2.5 CLASSIFICATION

The classification of the mites is extremely complex and uncertain and a variety of names for the various groupings are used. In general, mites and ticks are considered to belong to the sub-class **Acari**, in the class **Arachnida**. The Acari may be divided into two orders, the **Parasitiformes** and the **Acariformes**. The Acariformes do not have visible stigmata posterior to the coxae of the second pair of legs and the coxae are often fused to the ventral body wall. The Parasitiformes possess one to four pairs of lateral stigmata posterior to the coxae of the second pairs of legs and the coxae are usually free.

In general, seven sub-orders of Acari are recognized, of which four include parasitic forms: the **Metastigmata** or Ixodida (ticks) and three sub-orders, known as mites, the **Mesostigmata**, **Prostigmata** and **Astigmata**. The Metastigmata and Mesostigmata are usually grouped into the order of **Parasitiformes** and the Prostigmata and Astigmata into the **Acariformes**. The ticks will be dealt with in Chapter 3. As the names Mesostigmata, Prostigmata, and Astigmata suggest, the higher classification of mites is also based primarily on the presence or absence of stigmata and their position on the body.

The classification and identification of mites may be complicated by the fact that individuals within a species may be highly variable morphologically and behaviourally. As a result, the precise status of a number of specific and sub-specific groupings is unclear and the subject of ongoing debate. This situation may be resolved in the future as more sophisticated genetic techniques become available for characterizing species.

2.5.1 Astigmata

The sub-order **Astigmata** (= Acaridida, = Sarcoptiformes) is a large group of relatively similar mites. They are all weakly sclerotized; stigmata and tracheae are absent and respiration occurs directly through the cuticle. The order includes the families **Sarcoptidae**, **Psoroptoidae** and **Knemidocoptidae** which are of major veterinary importance because they contain the most common mite species which cause mange and scab. Also of interest are species of the **Listerophoridae**, which are ectoparasites of rodents, and species of the **Cytodidae** and **Laminosioptidae,** which live in the respiratory tracts of birds and mammals.

2.5.2 Prostigmata

The **Prostigmata** (= Actinedida, = Trombidiformes) is a large and heterogeneous sub-order which exists in a diversity of forms and occupies a wide range of habitats. Many species of prostigmatid mite, such as the spider mites, are important pests of plants. Others are predatory, feeding on a range of invertebrates. Prostigmatid mites usually have stigmata which open on the gnathosoma or the anterior part of the idiosoma, which is known as propodosoma, hence giving the sub-order its name. There are over 50 families of which four contain important ectoparasitic species: the **Trombiculidae**, **Demodicidae**, **Cheyletiellidae** and **Psorergatidae**. A number of other families, such as the **Pyemotidea**, contain species which are not ectoparasites but which are of minor veterinary interest because of the allergic responses they induce in animals which come into contact with them.

2.5.3 Mesostigmata

The **Mesostigmata** (= Gamesida) is a large and successful group, the majority of which are predatory, but a relatively small number of species are important ectoparasites of birds and some mammals. Mesostigmatid mites have stigmata which are located above the coxae of the second, third or fourth pair of legs. They are generally large, ranging from 200 μm to 2 mm in length. There is usually one large, sclerotized shield on the dorsal surface and a series of smaller shields in the midline on the ventral surface. They have legs which are longer and positioned more anteriorly than in other mites, like those of ticks. In appearance they look somewhat spider-like. Some mesostigmatid mites may be host specific, but most species can parasitize a range of hosts opportunistically. They can usually survive for several months between feeds. These two features make their control difficult. There are two main families of veterinary importance, the **Macronyssidae** and **Dermanyssidae** and a number of families, such as **Halarachnidae**, **Rhinonyssidae** and **Laelapidae**, which are of minor interest because they contain species which live in the respiratory tracts of birds and mammals.

2.6 RECOGNITION OF MITES OF VETERINARY IMPORTANCE

The identification of mites is problematic. However, since mites in general tend to be relatively host specific, a good first practical indication of the likely identity of any species in question can be the species of host and the location of the mite on that host. The following is a general guide to the adults of the most common species and genera of ectoparasitic mites likely to be encountered (modified from from Baker *et al.*, 1956; Varma, 1993).

2.6.1 Guide to the sub-orders of Acari

1 Hypostome of the gnathosoma without backwardly directed barbs. Stigmata present or absent, when present not opening on stigmatal plates; if stigmata lateral to coxae 2 and 3 then with peritremes. Tarsi of first pair of legs without sensory pit

2

Hypostome of the gnathosoma with backwardly directed barbs. Stigmatal shields present behind coxae of the fourth pair of legs or laterally above the coxae of legs 2 or 3: stigmata without peritremes. Tarsi of the first pair of legs with a sensory pit **METASTIGMATA (IXODIDA)**

2 Idiosoma without conspicuous shields. Legs with coxae fused to body wall. Palps without an apotele **3**

Idiosoma with sclerotized areas forming distinct shields (darkened brown colour). Legs with free coxae articulated to the idiosoma. Palps with an apotele **MESOSTIGMATA**

3 Stigmata absent. Palps small, inconspicuous and pressed against the sides of the hypostome. Legs usually with three claws and with a complex pulvillus (varying from pad-like to trumpet like). Body never worm-like **ASTIGMATA**

Palps usually well developed. Chelicerae usually adapted for piercing, sometimes pincer-like. Legs with one or two claws, without a complex pulvillus. Body sometimes worm-like. Stigmata present or absent; when present positioned between the bases of the chelicerae or on the upper surface of the propodosoma **PROSTIGMATA**

2.6.2 Guide to species and families of veterinary importance

1 Stigmata absent posterior to the second pair of legs **2**

Stigmata present as one lateral pair between the bases of legs II and IV **3**

2 Legs without claws; palps with two segments; stigmata absent **14**

Some or all legs with claws; palps with more than two segments; stigmata present or absent, when present they open on the gnathosoma or the anterior part of the idiosoma **20**

3 Genital plate rudimentary or absent **4**
Genital plate well defined **5**

4 Genital plate present although rudimentary; in lungs of canary
***Sternostoma tracheacolum* (Rhinonyssidae)**
Genital plate absent; palps elongated with five segments; in nasal passage of dogs
***Pneumonyssus caninum* (Halarachnidae)**

5 Chelicerae long and whip-like; chelae at tips absent or very small **6**
Chelicerae not long and whip-like, shorter and stronger; chelae blade-like at tips **7**

6 Dorsal surface of body with one shield; anal shield not egg shaped and with anal opening at posterior end (Fig. 2.21); parasite of birds
***Dermanyssus gallinae* (Dermanyssidae)**

7 Dorsal shield not nearly covering dorsal body surface; genitoventral shield narrowed posteriorly; chelicerae with toothless chelae **8**
Dorsal shield virtually covering dorsal body surface; genitoventral shield not narrowed posteriorly; chelicerae usually with toothed chelae **10**

8 Dorsal shield broad, its setae short **9**
Dorsal shield narrow and tapering posteriorly, its setae long (Fig. 2.20B); parasite of rats, mice, hamsters
***Ornithonyssus bacoti* (Macronyssidae)**

9 Sternal shield with two pairs of setae (Fig. 2.20A); parasite of birds ***Ornithonyssus sylviarum* (Macronyssidae)**
Sternal shield with three pairs of setae; parasite of birds
***Ornithonyssus bursa* (Macronyssidae)**

10 Genitoventral shield widened posteriorly, with more than one pair of setae **11**

Genitoventral shield not widened posteriorly, one pair of setae; on small rodents, weasels and moles

Hirstionyssus isabellinus **(Laelapidae)**

11 Body densely covered in setae **12**

Body with few setae (these arranged in transverse rows) **13**

12 Genitoventral shield with pear-shaped outline; on rodents

Haemogamasus pontiger **(Laelapidae)**

Genitoventral shield with large subcircular outline; on rodents

Eulaelaps stabularis **(Laelapidae)**

13 Genitoventral shield with concave posterior margin, surrounding anterior part of anal shield; on rodents

Laelaps echidninus **(Laelapidae)**

14 Legs short and stubby; genital opening of female a transverse slit paralleling body striations; dorsal striations broken by strong pointed scales; dorsal setae strong and spine-like; anus terminal (Fig. 2.6); on mammals

Sarcoptes scabiei **(Sarcoptidae)**

Dorsal setae not spine-like **15**

Legs not short and stubby **17**

15 Anus terminal; tarsi claw-like, with terminal setae **16**

Anus dorsal; dorsal striations broken by many pointed scales; dorsal setae simple, not spine-like (Fig. 2.7B); on rats and guinea-pigs ***Trixicarus caviae*** **(Sarcoptidae)**

Anus dorsal; dorsal striations not broken by pointed scales; dorsal setae simple, not spine-like; tarsi with long pretarsi on legs I and II (Fig. 2.7A); on cats

Notoedres cati **(Sarcoptidae)**

16 Dorsal striations simple, unbroken (Fig. 2.11); on poultry

Knemidocoptes laevis gallinae **(Knemidocoptidae)**

Dorsal striations broken, forming scale-like pattern; on poultry
Knemidocoptes mutans (Knemidocoptidae)

Dorsal striations broken, forming scale-like pattern; on caged birds
Knemidocoptes pilae (Knemidocoptidae)

17 Pretarsi with short stalks **18**

In the adult female, pretarsi of I, II and IV with three-jointed long stalks; tarsi III with two long terminal whip-like setae; legs of equal sizes; genital opening an inverted U. In the adult male, pretarsi on legs I, II and III with three-jointed long stalks; long setae on legs IV which are smaller than others (Fig. 2.8); on domestic mammals **_Psoroptes_ spp. (Psoroptidae)**

18 In the adult female, tarsi I, II and IV with short-stalked pretarsi; tarsi III with a pair of long terminal whip-like setae; legs I and II stronger than the others; legs III shortest; legs IV with long slender tarsi; genital opening almost a transverse slit. In the adult male all legs with short-stalked pretarsi; fourth pair of legs short (Fig. 2.9); on domestic animals
Chorioptes bovis (Psoroptidae)

Legs I and II with short-stalked pretarsi; legs III and IV with a pair of terminal whip-like setae; legs IV much reduced; genital opening transverse (Fig. 2.10); found in the ears of cats and dogs **_Otodectes cynotis_ (Psoroptidae)**

19 Mouthparts not well developed, reduced; small oval, nude mites; all tarsi with pretarsi (Fig. 2.14A); in the tissues of birds
Cytodites nudus (Cytoditidae)

Mouthparts well developed; elongated mites; body setae long; tarsi I and II claw-like distally; tarsi III and IV with long, spatulate pretarsi (Fig. 2.14B); in the tissues of birds
Laminosioptes cysticola (Laminosioptidae)

20 Body not unusually elongated, with setae **21**

Body unusually elongated and crocodile-like with annulations, without setae (Fig. 2.15); in skin pores of mammals
Demodex species (Demodicidae)

21 Gnathosoma and palps conspicuous; body with feathery setae; three pairs of legs when attached to host (larval forms) (Fig. 2.17) species of **Trombiculidae**

Gnathosoma and palps conspicuous; body not with feathery setae; stigma opening at base of chelicerae **22**

Gnathsoma and palps inconspicuous; body with simple non-feathery setae; not ectoparasitic

***Pyemotes tritici* (Pyemotidae)**

22 Palps with thumb-claw complex **23**

Palps without thumb-claw complex **24**

23 Chelicerae fused with rostrum to form cone; palps opposable, with large distal claws; peritreme obvious, M-shaped on gnathosoma (Fig. 2.16)

On rabbits ***Cheyletiella parasitivorax* (Cheyletiellidae)**

On cats ***Cheyletiella blakei* (Cheyletiellidae)**

On dogs ***Cheyletiella yasguri* (Cheyletiellidae)**

24 Legs normal, for walking **25**

First pair of legs highly modified for clasping hairs of host; body elongate, with transverse striations; on mice and rats **Myobidae**

Legs I and II and tarsi IV adapted for clasping hairs (Fig. 2.13); on guinea-pigs

***Chirodiscoides caviae* (Listrophoridae)**

Legs III and IV of female modified for clasping hairs (Fig. 2.12); on mice ***Myocoptes musculinus* (Listrophoridae)**

25 Small, round mites with short stubby, radiating legs, each with a strong hook; female with two pairs of posterior setae, male with a single pair of posterior setae; (Fig. 2.18)

On sheep ***Psorergates ovis* (Psorergatidae)**

On mice ***Psorergates simplex* (Psorergatidae)**

2.7 ASTIGMATA

2.7.1 Sarcoptidae

The members of the family Sarcoptidae are burrowing astigmatid mites, which are parasitic throughout their lives. Their morphology and ecology are highly adapted to a life of intimate contact with their host. The empodium is claw-like and the pulvillus is borne on a stalk-like pretarsus. Paired claws on the tarsus are absent. They have circular bodies with the ventral surface somewhat flattened. The coxae are sunk into the body, creating a characteristic 'short-legged' appearance; from a dorsal view, the legs only just project beyond the edge of the body and the posterior two pairs may not be visible at all. The chelicerae are usually adapted for cutting or sucking. The cuticle may be covered by fine striations. There are three genera of veterinary importance: *Sarcoptes*, *Notoedres* and *Trixicarus*.

(a) *Sarcoptes*

There is believed to be only a single species of *Sarcoptes,* the itch mite, *Sarcoptes scabiei.* Nevertheless, there are a number of host-adapted varieties distiguished by the presence or absence of patches of dorsal and/or ventral spines. Each population may be highly adapted to its particular host, and strains from one host may not easily infest a host of a different species. *Sarcoptes scabiei* is responsible for sarcoptic mange in a wide range of wild and domestic animals throughout the world.

Sarcoptes scabiei

Morphology: the adult of this species has a round body (Fig. 2.6). The male is about 250 μm in length and is smaller than the mature female which is about 400–430 μm in length. In both sexes, the pretarsi of the first two pairs of legs bear empodial claws and a sucker-like pulvillus borne on a long, stalk-like pretarsus. The sucker-like pulvilli help the mite grip the substrate as it moves. The third and fourth pairs of legs in the female and the third pair of legs in the male end in long setae and lack stalked pulvilli. The dorsal surface of the body of *S. scabiei* is covered with transverse ridges but also bears a central patch of triangular scales. The dorsal setae are strong and spine-like. The anus is terminal and only slightly dorsal.

Life cycle: mating probably takes place at the skin surface, following which the female creates a permanent winding burrow, parallel to the skin surface using her chelicerae and the claw-like empodium on the front two pairs of legs. This burrow may be up to 1 cm in length. Burrowing may proceed at up to 5 mm/day.

Maturation of the eggs takes 3 or 4 days, following which the female starts to oviposit 1–3 eggs/day, over a reproductive life of about 2 months. The eggs, which are oval and about half the length of the adult, are laid singly at the ends of outpockets, which branch off along the length of these tunnels. Three to four days after oviposition, the six-legged larva hatches from the egg and crawls towards the skin surface. Two to three days later the larva moults to become a protonymph. During this time the larva and nymph find shelter and food in the hair follicles. The protonymph moults to become a deutonymph and again a few days later to become an

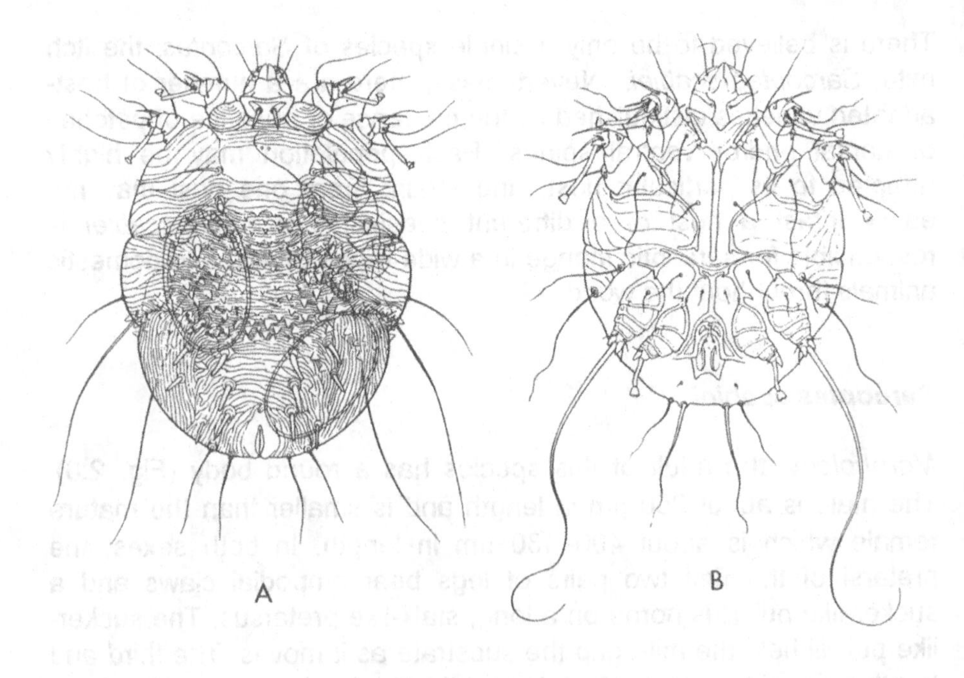

Fig. 2.6 Adults of *Sarcoptes scabiei*: (A) female, dorsal view; (B) male, ventral view (reproduced from Varma, 1993).

adult. Both sexes of adult then start to feed and burrow at the skin surface, creating small pockets of up 1 mm in length in the skin. Despite their short legs, adults are highly mobile, capable of moving at up to 2.5 cm/min. The total egg to adult life cycle takes between 17 and 21 days but may be as short as 14 days.

Pathology: clinical problems caused by *S. scabiei* are are thought to be rare in wild animals but may become common when these animals are kept in captivity. However, periodic outbreaks of sarcoptic mange are common in urban populations of foxes in northern Europe and may cause high levels of mortality.

Sarcoptic mange may affect dogs, pigs, sheep, goats, humans and cattle, but is relatively rare in cats and horses. It occurs in housed farm animals or those in poor condition, usually at the end of winter or early spring. The preferred site of infestation depends on host, with the mites generally being more common on the sparsely haired parts of the body such as the ears, face or muzzle in the dog, the ears and back in the pig, and the neck and tail of cattle. However, with high infestations the mites may spread all over the body of the infested host.

The burrowing and feeding activity of *S. scabiei* cause intense itching, inflammation, hair loss and the formation of crusts of dried exudate. The intense pruritus leads to excoriation, resulting in exudation and even haemorrhage on the skin surface. The cutaneous response reflects inflammation produced by keratinocyte damage and the development of cutaneous hypersensitivity (type I) to mite antigens (e.g. faecal antigens).

Sarcoptic mange is highly contagious and the spread of *S. scabiei* is usually by close physical contact. As a result, single cases are rarely seen in groups of animals kept together. In the early stages infestation may not be apparent. The incubation period, from initial infestation to the development of clinical disease, is usually about 2–3 weeks in the pig and 1–2 weeks in the dog. Infestation may also occur by indirect transfer, since the mites have been shown to be capable of surviving off the host for short periods. The length of time that *S. scabiei* can survive off the host depends on environmental conditions but may be between 2 and 3 weeks.

(b) *Notoedres*

More than 20 species of *Notoedres* have been described, the majority of which are ectoparasites of tropical bats. *Notoedres cati* is of veterinary importance in many parts of the world, although it is rare in northern Europe. Other common species are *Notoedres muris*, which occurs worldwide on rats, and *Notoedres musculi*, which is found on the house mouse in Europe.

Notoedres cati

Morphology: the dorsal striations of the idiosoma of *N. cati* tend to be in concentric rings (thumb-print like) (Fig. 2.7A). There are no projecting dorsal scales. The dorsal setae are simple and not spine-like. It is also considerably smaller than *S. scabiei*; females are about 225 μm in length and males about 150 μm. The anal opening is distinctly dorsal and not posterior.

Life cycle: this species is very similar in behaviour to *S. scabiei*.

Pathology: largely infesting domestic cats, *N. cati* has a far narrower host range than *S. scabiei*. However, it may occasionally also be found on dogs and rabbits. The burrowing activity of the female damages keratinocytes, leading to cytokine release (especially IL-1) leading to cutaneous inflammation and the clinical signs. As with other mite infestations a hypersensitivity may also be involved in the clinical manifestation of dermatitis. The mites occur in clumps in the skin and are usually initially found around the head and ears. As an infestation develops the mites may spread over the body of the host. *Notoedres cati* is highly contagious and transmission from host to host is by the spread of larvae or nymphs. *Notoedres* infestation results in crust and scale formation on the tips of the ears and over the face and neck. The dermatitis causes intense scratching and hair loss. If untreated, the affected animal can become severely debilitated and notoedric mange may be fatal in 4–6 months.

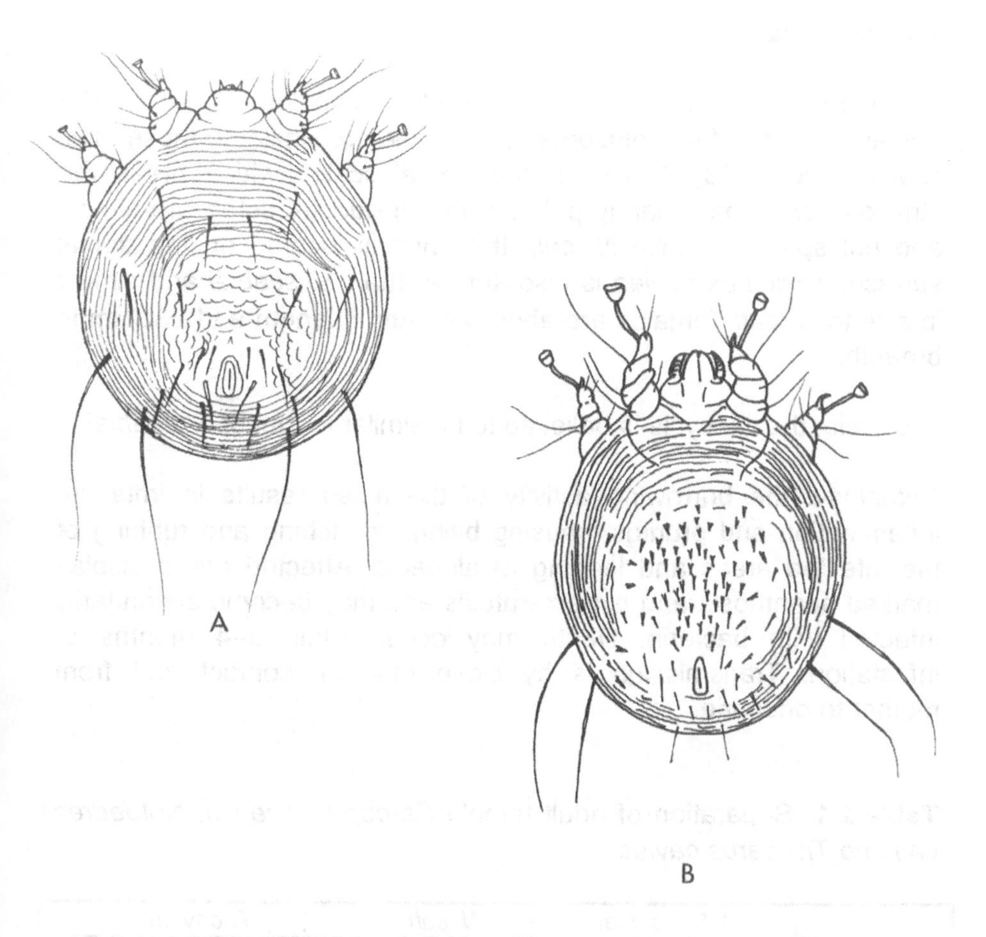

Fig. 2.7 Adult females (dorsal view) of (A) *Notoedres cati* (reproduced from Urquhart *et al.*, 1987) and (B) *Trixicarus caviae*.

(c) *Trixicarus*

Mites described as *Trixicarus* were first reported on rats (*Rattus norvegicus* and *Rattus rattus*) in Europe; these were described as *Trixicarus diversus*. Subsequently, in 1972, a second species of *Trixicarus*, described as *T. caviae*, was identified on guinea-pigs (*Cavia porcellus*) in the UK. This mite has now also been identified in other parts of Europe and was first detected in the USA in 1980 where it is now widespread.

Trixicarus caviae

Morphology: *Trixicarus caviae* superficially resembles *S. scabiei*. The dorsal striations of the idiosoma of *T. caviae* are similar to those of *S. scabiei* (Fig. 2.7B). However, the dorsal scales which break the striations are more sharply pointed and the dorsal setae are simple and not spine-like. Like *N. cati,* the anus is located on the dorsal surface. *Trixicarus caviae* is also smaller than *S. scabiei* and similar in size to *N. cati;* females are about 240 μm in length and 230 μm in breadth.

Life cycle: the life cycle is believed to be similar to that of *S. scabiei*.

Pathology: the burrowing activity of the mites results in irritation, inflammation and pruritus, causing biting, scratching and rubbing of the infested areas and leading to alopecia. Affected areas display marked acanthosis and hyperkeratosis and may become secondarily infected with bacteria. Death may occur within 3–4 months of infestation. Transmission is by close physical contact and from mother to offspring.

Table 2.1 Separation of adult female *Sarcoptes scabiei*, *Notoedres cati* and *Trixicarus caviae*

	S. scabiei	*N. cati*	*T. caviae*
Length (μm)	400–430	225–250	230–240
Anus position	Terminal	Dorsal	Dorsal
Dorsal setae	Some stout dorsal spines	All dorsal setae simple (not spine-like)	All dorsal setae simple (not spine-like)
Dorsal scales	Many, pointed	Few, rounded	Many, pointed

2.7.2 Psoroptidae

The Psoroptidae are oval-bodied, non-burrowing, astigmatid mites. They feed superficially and do not need to burrow into the skin. Some feed on skin scales while others suck tissue fluid. They are generally larger than the burrowing sarcoptid mites, at between 1 and

2 mm in length. The legs are longer than those of the burrowing mites and the third and fourth pairs of legs are usually visible from above. There are two nymphal stages in the life cycle. At the hind end of the male are a pair of prominent circular copulatory suckers which engage with copulatory tubercles of the female deutonymph. Three genera of Psoroptidae are of veterinary importance: *Psoroptes*, *Chorioptes* and *Otodectes*.

(a) *Psoroptes*

Species of the genus *Psoroptes* cause various forms of psoroptic mange. For many years it has been believed that there are a number of very closely related species of *Psoroptes* which specialize on particular hosts: *Psoroptes cuniculi* on rabbits, *Psoroptes ovis* on sheep, *Psoroptes equi* or *Psoroptes natalensis* on horses and *Psoroptes bovis* on cattle. However, more recently the integrity of the various species of *Psoroptes* mites has been called into doubt. Although it is difficult to infest one species of host with mites removed from another, it is possible, and experiments have shown that it is also possible to cross *P. ovis* and *P. cuniculi*. Hence, the view that there are fewer species of *Psoroptes* than was though previously (perhaps only two in temperate habitats, *P. ovis* and *P. natalensis*), with a large number of strains adapted to live on different hosts and in different locations and varying in their degree of virulence, is becoming more widely accepted. *Psoroptes natalensis* is a body mite of domestic cattle, buffalo and horses, which is thought to have originated in South Africa but which is now present in Europe.

Psoroptes ovis

Morphology: all life-cycle stages of *P. ovis* may be recognized by trumpet-shaped, sucker-like pulvilli attached to three jointed pretarsi. The jointed pretarsi are highly diagnostic features since in most other mites the pretarsi are unjointed. Adult female *P. ovis* are large mites, about 750 μm long. They have jointed pretarsi and pulvilli on the first, second and fourth pairs of legs and long, whip-like setae on the third pair (Fig. 2.8A). In contrast, the smaller adult males, which are recognizable by their copulatory suckers and paired posterior lobes, have pulvilli on the first three pairs of legs and setae on the fourth pair (Fig. 2.7B). The legs of adult females are approximately the same length, whereas in males the fourth pair is extremely short.

Life cycle: the eggs of *P. ovis* are relatively large, about 250 μm in length, and oval. The hexapod larva which ecloses from the egg is about 330 μm long and has sucker-like pulvilli on the first two pairs of legs and two long setae on each of the third pair. The larva moults into a protonymph with pulvilli on the first, second and fourth pairs of legs and long setae on the third pair. The protonymph moults into a deutonymph (sometimes also referred to as the tritonymph), which is similar in shape to the protonymph, except that females have posterior copulatory protuberances. The deutonymph moults to become an adult.

Adult males attach to female deutonymphs and remain *in copula* until the females moult for the final time, at which point insemination occurs. The function of this behaviour is unknown, but protection of unmated females by males to ensure that mating occurs

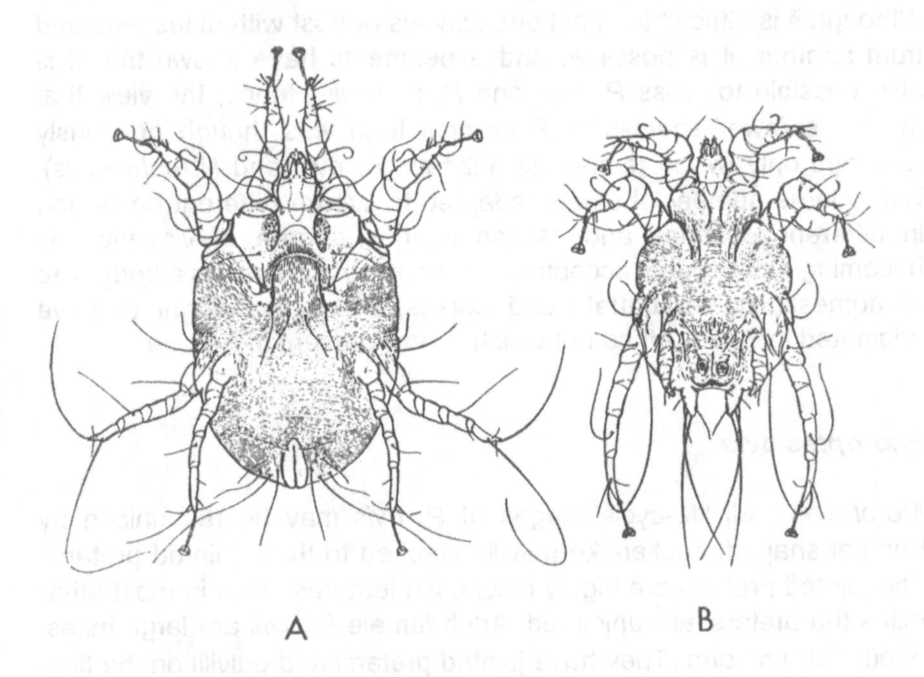

Fig. 2.8 Adult *Psoroptes ovis:* (A) female, ventral view; (B) male, dorsal view (reproduced from Baker *et al.*, 1956).

at the appropriate time so that they ensure paternity of subsequent eggs is common in arthropods.

Eggs are produced at a rate of about 1–3 per day. A female may live for between 40 and 60 days, during which time it may lay 40–100 eggs. The rate of egg production appears to be inversely related to ambient temperature. The minimum duration of the life cycle of *P. ovis,* from egg to egg, is about 14 days.

Pathology: *Psoroptes ovis* is most well known and economically important as an ectoparasite of sheep, where it causes a condition known as sheep scab. It also causes a pruritic dermatosis of cattle. The way in which *P. ovis* feeds at the skin surface is unclear. It was thought for many years that the mites pierce the skin of the infested sheep during feeding. However, this is no longer considered to be the case. As with many other cutaneous mite infestations, the clinical signs of sheep scab are thought to involve a hypersensitivity reaction (type I) by the host to antigenic material produced by the mites, in particular components of mite faeces. This hypersensitivity causes inflammation, surface exudation, scale and crust formation, with excoriation due to self-trauma (scratching). The mouthparts of the mites are adapted for sucking and the mites are believed to feed on the superficial lipid emulsion of lymph, skin cells, skin secretions and bacteria at the skin surface.

The serous exudate produced in response to the mites dries on the skin to form a dry, yellow crust, surrounded by a border of inflamed skin covered in moist crust. Mites are found on the moist skin at the edge of the lesion, which extends rapidly and may take as little as 6–8 weeks to cover three-quarters of the host's skin. Eventually the crust lifts off as the new fleece grows.

Infestation in sheep leads to severe pruritus, wool loss, restlessness, biting and scratching of infested areas, weight loss and, in growing animals, reduced weight gain. When handled, infested sheep may demonstrate a 'nibble reflex', characterized by lip smacking and protrusion of the tongue; others may show fits lasting 5–10 minutes. In sheep, lesions may occur on any part of the body, but are particularly obvious on the neck, shoulders, back and flanks. In severe cases the skin may be excoriated, lichenified and secondarily infected, with numerous thick-walled abscesses of between 5 and 20 mm in diameter. Sheep scab can affect sheep of all ages but may be particularly severe in young lambs.

Mites are usually more active in winter and the oviposition rate is higher at lower temperatures. In summer the disease progresses more slowly, lesions are not obvious and can be missed. The disease can become latent in summer, apparently disappearing, with mites taking refuge in protected sites such as the axilla, eyes, ears or even folds in the skin. Mite activity and associated disease may then flare up again in autumn as the fleece thickens. The short life cycle can contribute to a very rapid build-up of *P. ovis* populations. Scab mites are spread by direct contact between clean and infested animals and can survive for periods of up to 10–14 days off their host (depending on environmental conditions), allowing clean animals to become infested from contaminated housing. Sheep scab is common throughout Europe, Asia and Australasia, although it is thought to have been eliminated from flocks in the US.

Psoroptes ovis infestation in cattle causes lesions initially on the withers, neck and around the root of the tail, which spread all over the body in cases of severe infestation.

Psoroptes cuniculi

Morphology: as discussed previously the specific status of *P. cuniculi* is uncertain and it may be simply a variant of *P. ovis* adapted to an aural environment. Unsurprisingly, therefore, *P. cuniculi* is morphologically almost identical to *P. ovis* and designation is based almost entirely on host and aural location of the infestation on a host. In adult males of *P. cuniculi* the outer opisthosomal setae have been found on average to be 74.0 μm in length, whereas for *P. ovis* these setae are on average 79.3 μm in length. Nevertheless, the usefulness of this character is questionable, since there is considerable variation and overlap in the lengths of the setae between the two groups and the mean length of the setae of mites is known to decrease with the age of a body lesion.

Life cycle: identical to that of *P. ovis*.

Pathology: Psoroptes mites described as *P. cuniculi* are found primarily in rabbits, where it is usually localized in the ears, causing ear mange (psoroptic otocariasis). High populations may cause severe mange, blocking the auditory canal with debris and occasionally spreading over the entire body. Strains of *P. cuniculi* may also be found in the ears of horses, causing irritation and head shaking, and in sheep, associated with haematomas, head shaking and scratching.

(b) *Chorioptes*

There is only one species of *Chorioptes* of veterinary importance in temperate habitats: *Chorioptes bovis.* The names *Chorioptes ovis, Chorioptes equi, Chorioptes caprae* and *Chorioptes cuniculi* used to describe the chorioptic mites found on sheep, horses, goats and rabbits, respectively, are now thought to be synonyms of *Chorioptes bovis.*

Chorioptes bovis

Morphology: adult female *C. bovis* are about 300 μm in length, considerably smaller than *P. ovis. Chorioptes* do not have jointed pretarsi; their pretarsi are shorter than in *Psoroptes* and the sucker-like pulvillus is more cup-shaped, as opposed to trumpet-shaped in *Psoroptes* (Fig. 2.9). In the adult female tarsi I, II and IV have short-stalked pretarsi; tarsi III have a pair of long, terminal whip-like setae. The first and second pairs of legs are stronger than the others and the fourth pair has long, slender tarsi. In the male all legs possess short-stalked pretarsi and pulvilli. However, the fourth pair is extremely short, not extending beyond the body margin. Male *C. bovis* have two broad, flat setae and three normal setae on well-developed posterior lobes.

Life cycle: the life cycle is similar to *P ovis*: egg, hexapod larva, followed by octopod protonymph, deutonymph and adult. *Chorioptes bovis* has mouthparts which do not pierce the skin of the host, but which are adapted for chewing skin debris. The complete life cycle takes about 3 weeks, during which time adult females may produce up to 17 eggs.

Pathology: this mite may be found on a variety of herbivorous mammals, including goats and sheep, but chorioptic mange is the most common form of mange in cattle and horses. *Chorioptes* tends to be confined to certain areas. In horses the mites are found largely on the lower legs, most often on those with feathered fetlocks and especially draft horses. The lesions seen are erythema, crusts, ulceration and alopecia with marked pruritus.

On cattle the mites most commonly cause similar lesions at the base of the tail, perineum and udder. In sheep the mites affect the lower parts of the hind legs and ventral abdomen, producing papules, crusts and ulceration which can lead to temporary ram infertility if the scrotum is affected. Chorioptic mange is considerably less severe than psoroptic mange and, as with psoroptic mange, it is primarily a winter disease. The mites cause irritation and high infestations have been associated with decreased milk production.

(c) *Otodectes*

There is believed to be a single species of importance, *Otodectes cynotis*, known as the ear mite.

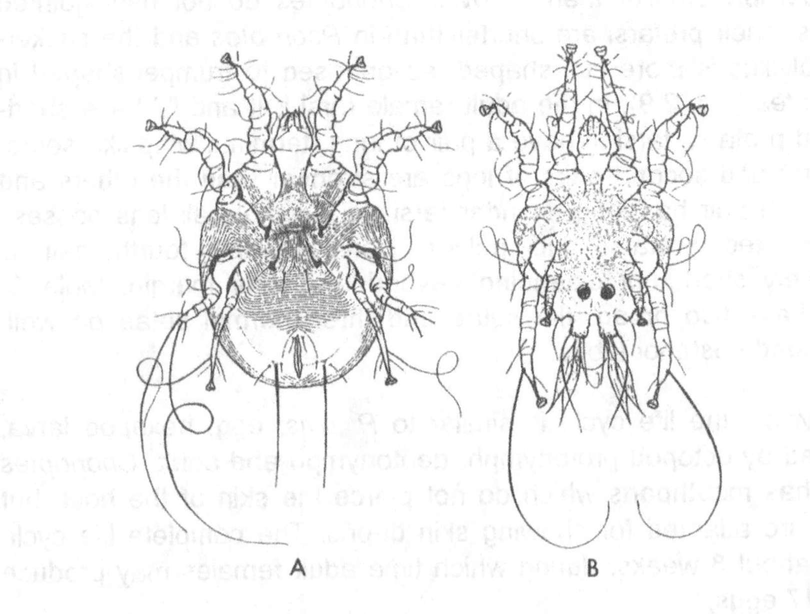

Fig. 2.9 Adult *Chorioptes bovis*: (A) female, ventral view; (B) male, dorsal view (reproduced from Baker *et al.*, 1956).

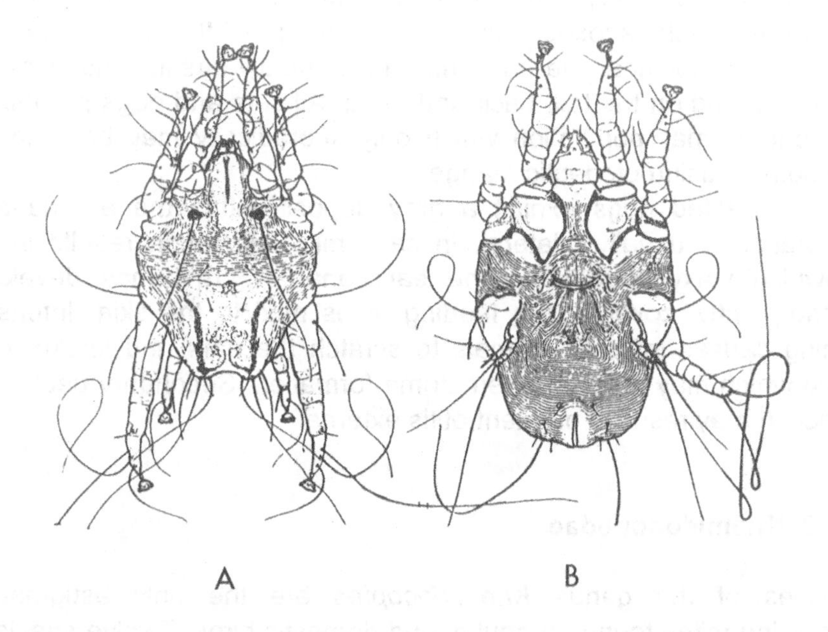

Fig. 2.10 Adult *Otodectes cynotis*: (A) male, dorsal view; (B) female, ventral view (reproduced from Baker *et al.*, 1956).

Otodectes cynotis

Morphology: *Otodectes cynotis* is similar in appearance to *Chorioptes;* it is of similar size and does not have jointed pretarsi (Fig. 2.10). The sucker-like pulvillus is cup-shaped, as opposed to trumpet-shaped in *Psoroptes.* In the adult female, the first two pairs of legs carry short, stalked pretarsi, while the third and fourth pairs of legs have a pair of terminal whip-like setae. The fourth pair is much reduced. The genital opening is transverse. In males all four pairs of legs carry short, stalked pretarsi and pulvilli but the posterior processes are small.

Life cycle: the mites feed on ear debris. The life cycle is similar to that of other psoroptids and takes about 3 weeks. Transfer may occur through direct contact and from infested female hosts to their pups or kittens.

Pathology: it is a very common mange mite of cats, dogs and other carnivores, such as foxes, and is found throughout the world. It exists deep in the ear near the ear-drum but, in heavy infestations, it may also be found on the tail, back and head. All cats and dogs probably harbour a small population which only sporadically may become a problem, causing otodectic mange.

Infested dogs exhibit a grey deposit within the ear canal. Infestation is usually bilateral. In cats, mild infestation results in a brownish waxy exudate in the ear canal. A crust may develop subsequently, covering the feeding mites next to the skin. Intense itching causes infested animals to scratch the ears and shake the head which may result in haematoma formation. Secondary bacterial infection may result in purulent otitis externa.

2.7.3 Knemidocoptidae

Species of the genus *Knemidocoptes* are the only astigmatid burrowing mites found on poultry and domestic birds. Twelve species have been described, of which three are of veterinary importance. The round body, short stubby legs and bird host are generally sufficient for a rough initial identification to the family and genus level.

Knemidocoptes mutans

Morphology: *Knemidocoptes mutans* is somewhat similar in appearance to *Sarcoptes*. Females are rounded and about 400 μm long (Fig. 2.11). The legs are short and stubby and the anus is terminal. The dorsal surface is covered by faint striations. However, mid-dorsally the striations are broken in a plate- or scale-like pattern. The body has no spines or scales. Stalked pulvilli are present on all the legs of larvae and males, but are absent in the nymphal stages and the female. Copulatory suckers are absent in the male.

Life cycle: females are ovoviviparous, giving birth to live larvae. Subsequently, the life cycle is typical of psoroptid mites; the hexapod larva is followed by an octopod protonymph, deutonymph and adult.

Pathology: *Knemidocoptes mutans* burrows into skin of the foot and leg of domestic poultry, causing a crusty white, scaly appearance.

Large populations may cause lameness and deformity of the feet, legs and claws. This is a condition known as 'scaly leg'. The comb and neck may also be affected. As the disease progresses over the course of several months, birds stop feeding and eventually waste away. Mature adult mites may be found beneath the crusts. The condition is more common in birds allowed access to the ground and, therefore, tends to be more prevalent in barnyard and deep-litter systems rather than in caged production facilities. The mites are highly contagious.

Knemidocoptes pilae

Morphology: Knemidocoptes pilae is extremely similar in appearance to *K. mutans.*

Life cycle: similar to that of *K. mutans.*

Pathology: Knemidocoptes pilae is found on budgerigars, parrots and parakeets. It attacks bare or lightly feathered areas, particularly around the beak, causing a condition known as 'scaly face'.

Knemidocoptes laevis gallinae

Morphology: similar in appearance to *K. mutans* except that the pattern of dorsal striations is unbroken (Fig. 2.11).

Life cycle: similar to that of *K. mutans.* Infestation by this mite is especially prevalent in spring and summer and may disappear in autumn.

Pathology: Knemidocoptes laevis gallinae infests poultry, pheasants and geese, where it burrows into the base of the feather shafts, particularly on the back, head and neck, top of the wing and around the vent, causing a condition known as 'depluming itch'. The condition is characterized by intense scratching and feather loss over extended areas of the body. Mites may be found embedded in the tissue at the base of feather quills.

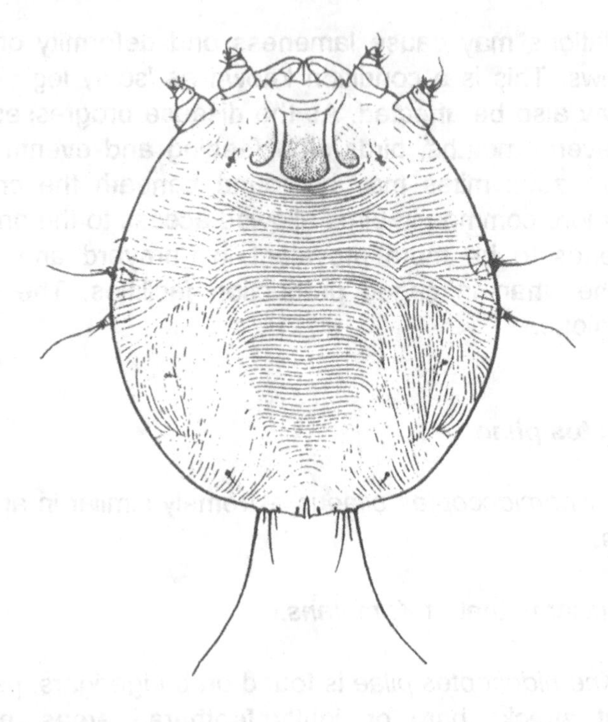

Fig. 2.11 Adult female of *Knemidocoptes laevis gallinae*, dorsal view (reproduced from Hirst, 1922).

2.7.4 Listrophoridae

Members of the family Listrophoridae commonly infest fur-bearing mammals. They are soft-bodied, strongly striated with a distinct dorsal shield, and mouthparts and legs modified for grasping hairs.

Mycoptes musculinus

Morphology: adult females are elongated ventrally, about 300 μm in length, and the propodosomal body striations have spine-like projections (Fig. 2.12). The genital opening is a transverse slit. The anal opening is posterior and ventral. Legs I and II are normal, possessing short-stalked, flap-like pretarsi. Legs III and IV are highly modified for clasping hair. Males are smaller than females, about 190 μm in length with less-pronounced striations and a greatly enlarged fourth pair of legs for grasping the female during copulation (Fig. 2.12B). The posterior of the male is bilobed.

Life cycle: hexapod larvae give rise to octopod nymphs which resemble the adult female. *Mycoptes musculinus* spends its entire life on the hair of the host rather than on the skin, feeding at the base of the hair and gluing its eggs to the hairs.

Pathology: this mite causes myocoptic mange in wild and laboratory mice. It is extremely widespread but is usually of little pathogenic significance. Problems may occcur, however, in crowded laboratory colonies or in animals in poor condition. Infestation causes inflammation, erythema and pruritus, leading to scratching and alopecia.

Chirodiscoides caviae (*Campylochirus caviae*)

Morphology: larger than *M. musculinus*, females of *C. caviae* are about 500 μm and males about 400 μm in length (Fig. 2.13). The

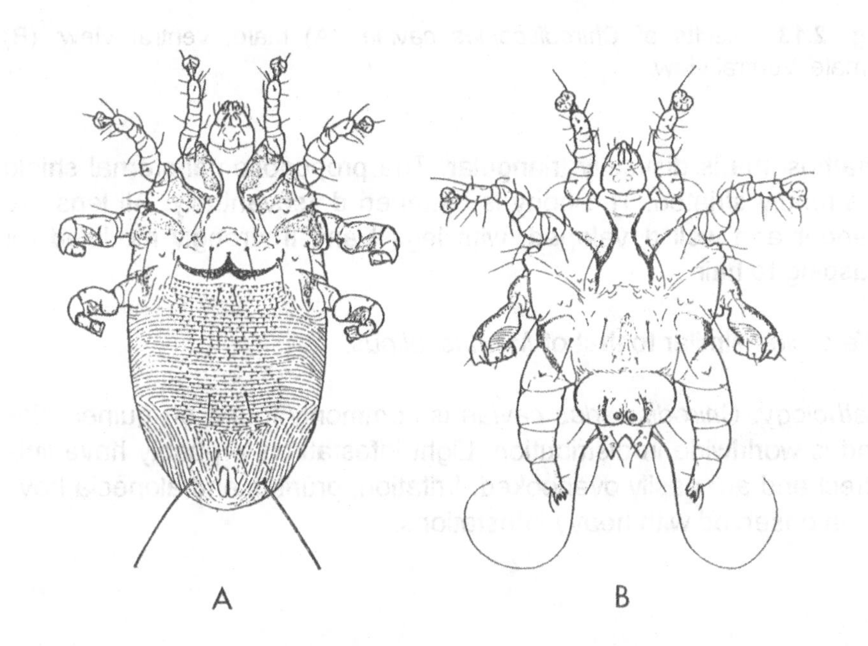

Fig. 2.12 Adults of *Mycoptes musculinus*: (A) female, ventral view; (B) male, ventral view (reproduced from Baker *et al.*, 1956).

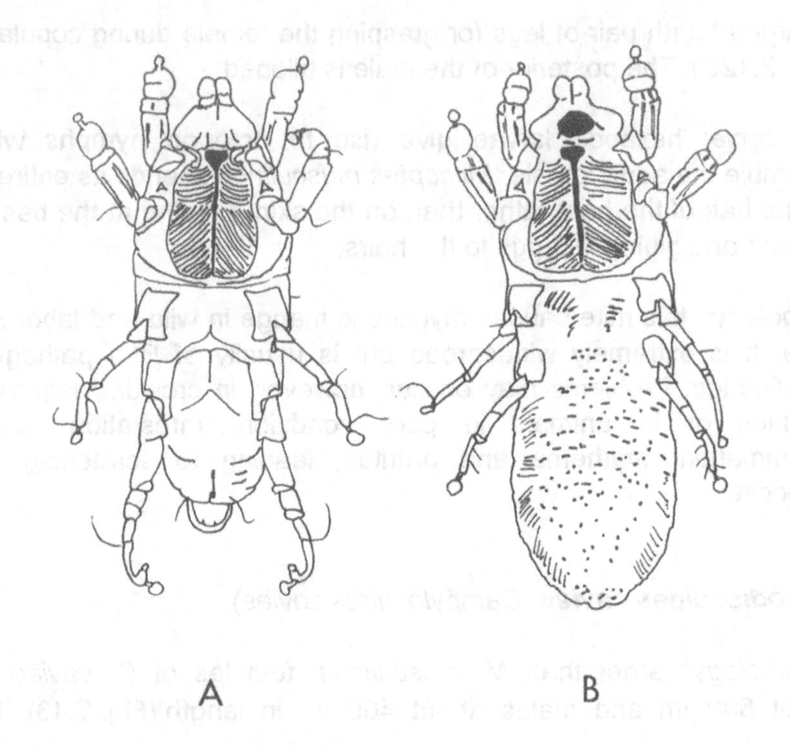

Fig. 2.13 Adults of *Chirodiscoides caviae*: (A) male, ventral view; (B) female, ventral view.

gnathosoma is distinctly triangular. The propodosomal sternal shield is strongly striated. The body is flattened dorsoventrally. All legs are slender and well developed, with legs I and II strongly modified for clasping to hair.

Life cycle: similar to that of *M. musculinus*.

Pathology: *Chirodiscoides caviae* is commonly found on guinea-pigs and is worldwide in distribution. Light infestations probably have little effect and are easily overlooked. Irritation, pruritus and alopecia have been observed with heavy infestations.

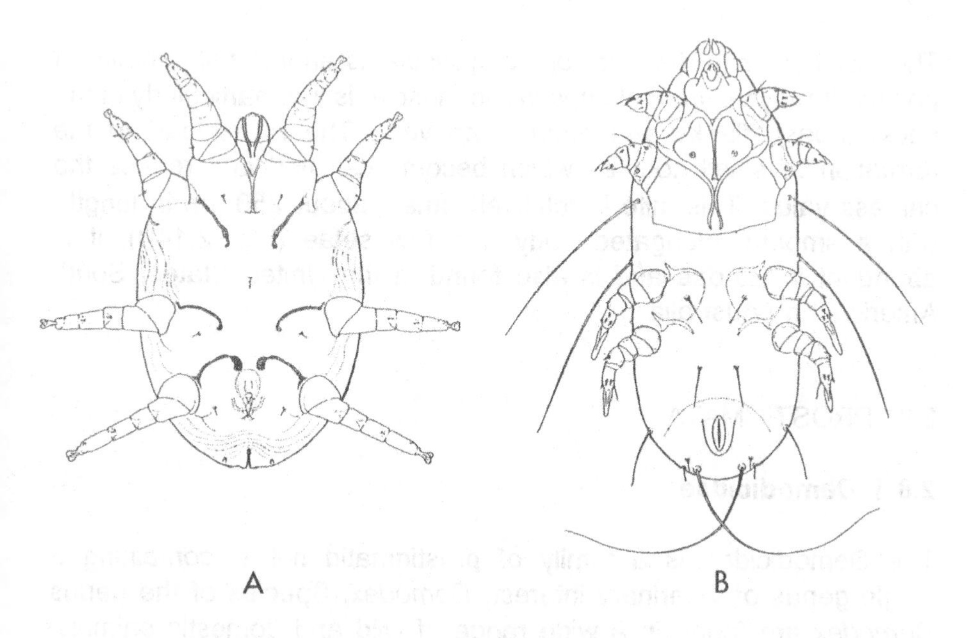

Fig. 2.14 Adults of (A) *Cytodites nudus,* male, ventral view and (B) *Laminosioptes cysticola,* female, ventral view (reproduced from Baker *et al.*, 1956).

2.7.5 Astigmatid mites of minor veterinary interest

(a) Cytoditidae

The cytoditoid mites are respiratory parasites of birds, rodents and bats. Of particular interest is the air-sac mite, *Cytodites nudus,* found in the air passages and lungs of wild birds, poultry and canaries. The mite is oval and about 500 μm long, with a smooth cuticle (Fig. 2.14A). The chelicerae are absent and the palps are fused to form a soft, sucking organ through which fluids are imbibed. Small infestations may have no obvious effect on the animal; large infestations may cause coughing and accumulation of mucus in the trachea and bronchi. Balance may be affected in infested birds. Positive diagnosis is only possible at post-mortem. Infestation may be spread by the host through coughing.

(b) Laminosioptidae

The fowl cyst mite, *Laminosioptes cysticola*, is an internal parasite of poultry. It may occur in clumps in the muscle tissue, particularly in the neck, breast, flanks and around the vent. The mites lead to the formation of small nodules which become calcified and reduce the carcass value. This mite is relatively small, about 250 μm in length, with a smooth, elongated body and few setae (Fig. 2.14B). It is abundant in Europe and is also found in the United States, South America and Australia.

2.8 PROSTIGMATA

2.8.1 Demodicidae

The Demodicidae is a family of prostigmatid mites, containing a single genus of veterinary interest, *Demodex*. Species of the genus *Demodex* are found in a wide range of wild and domestic animals, including humans. They form a group of closely related sibling species, different species being highly specific to particular hosts: *Demodex phylloides* (pig), *Demodex canis,* (dog), *Demodex bovis,* (cattle), *Demodex equi* (horse), *Demodex musculi* (mouse), *Demodex ratti* (rat), *Demodex caviae* (guinea-pig), *Demodex cati* (cat) and *Demodex folliculorum* and *Demodex brevis* on humans.

Morphology: species of *Demodex* are minute mites with an elongated, crocodile-like tapered body, 100–400 μm in length, with four pairs of stumpy legs in the adult (Fig. 2.15). Setae are absent from the legs and body. The legs are located at the front of the body so that the striated opisthosoma forms at least half the body length. Short forms of *Demodex* may be found, although whether these are separate species or phenotypic variants has yet to be established.

Life cycle: they live as commensals, embedded head-down in hair follicles, sebaceous and Meibomian glands of the skin, where they spend their entire lives. Species of *Demodex* are unable to survive off their host. Females lay eggs which give rise to hexapod larvae, in which each short leg ends in a single, three-pronged claw. Unusually, a second hexapod larval stage follows, in which each leg

Fig. 2.15 Adult *Demodex* sp., ventral view (reproduced from Barnes, 1974).

ends in a pair of three-pronged claws. Octopod protonymph, deutonymph and adult stages then follow. All life-cycle stages may be found together in a follicle.

Pathology: for the most part they are non-pathogenic and form a normal part of the skin fauna. As ectoparasites causing significant clinical disease in temperate habitats, they are important primarily in dogs, where they can cause **demodectic mange** or **demodicosis**. Because of the location of the mites within the hair follicles transmission between animals is difficult and it is thought that the normal *Demodex* population is acquired by new-born animals during the first few days of life from the bitch's mammary skin while suckling.

The pathogenesis of canine demodicosis is thought to involve host immunosuppression. Studies have shown that dogs with generalized demodicosis and staphylococcal pyoderma have reduced *in vitro* T-lymphocyte function. This T-cell suppression returns to normal once the staphylococcal pyoderma has been treated. Furthermore, recent evidence suggests that the immune response to *Staphylococcus intermedius* in dogs is humoral with antistaphylococcal IgE production. This type of response may, in theory, reduce the host's cutaneous cell-mediated immunity and allow proliferation of demodex mites and other organisms, such as *Malassezia*. The presence of concurrent demodicosis, staphylococcal pyoderma and malassezia dermatitis may reflect the individual's reduced cutaneous cell-mediated immunity with subsequent parasitic dermatosis.

Demodectic mange has been classified in various ways, depending upon the clinical features seen. These categories are **juvenile demodicosis**, **adult-onset demodicosis**, **localized demodicosis**, **generalized demodicosis** and **pustular demodicosis**. Demodicosis is more common in short-coated breeds of dog, although some long-haired breeds, such as Afghan hound, German shepherd and collie, may be predisposed. However, this is geographically variable and most likely reflects local breed gene pools.

Juvenile demodicosis which occurs between 3 and 15 months of age presents as non-pruritic areas of focal alopecia on head, forelimbs and trunk. This disease is self-limiting. However, if immunosuppressive therapy with glucocorticoids is administered, the dermatosis deteriorates and may become generalized and pustular.

Adult-onset demodicosis is often associated with concurrent staphylococcal pyoderma and is a pustular form. It can be localized or generalized and the clinical features seen are erythema, pustules, crusts and pruritus. The skin often becomes hyperpigmented in chronic cases. The localized form is often confined to the feet. If demodicosis occurs spontaneously in elderly dogs, underlying debilitating diseases, including neoplasia, may be responsible. Immunosuppressive therapy for other diseases may also lead to canine demodicosis.

Although usually benign, infestation with *Demodex bovis* can be important in cattle in the US. The mite damages the hair follicles and can reduce the value of the hide.

2.8.2 Cheyletiellidae

Most species of the family Cheyletiellidae are predatory, feeding on other mites. However, a number of species in the genus *Cheyletiella* are ectoparasites. Three very similar species are of veterinary importance and are common in three types of mammal: *Cheyletiella yasguri* on dogs, *C. blakei* on cats and *C. parasitivorax* on rabbits.

Morphology: all three of the species of veterinary importance are morphologically very similar (Fig. 2.16). Adults are about 400 μm in length. They have blade-like chelicerae which are used for piercing their host, and short, strong, opposable palps with curved palpal claws. The palpal femur possesses a long, serrated dorsal seta. The body tends to be slightly elongated with a 'waist'. The legs are short; tarsal claws are lacking and the empodium is a narrow pad with comb-like pulvilli at the ends of the legs. The peritreme is M-shaped and the stigmata open at the base of the chelicerae. Adults are highly mobile and are able to move about rapidly. The three species can be separated morphologically by the shape of a projection, the **solenidion**, on the genu of the first pair of legs. This is described as globose in *C. parasitivorax*, conical in *C. blakei* and heart-shaped in *C. yasguri* (Fig. 2.16B,C,D). However, this feature can vary in individuals and between life-cycle stages, making identification difficult.

Life-cycle: the mites spend their entire lives on the host, living in the skin debris at the base of hair, fur and feathers. They do not burrow

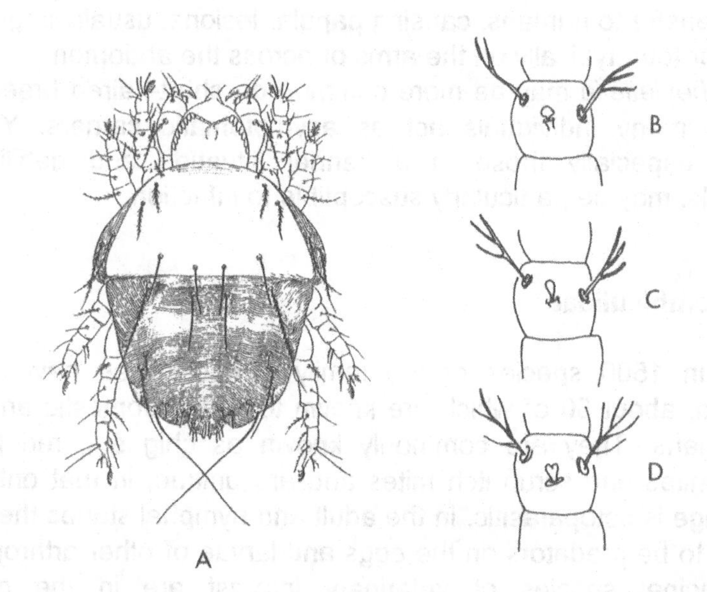

Fig. 2.16 (A) Adult female *Cheyletiella parasitivorax*, dorsal view (reproduced from Baker *et al.*, 1956). Genu of the first pair of legs of adult females of (B) *Cheyletiella parasitivorax*, (C) *Cheyletiella blakei* and (D) *Cheyletiella yasguri*.

but pierce the skin with their stylet-like chelicerae, engorging with lymph. The strong palpal claws are used for grasping fur or feathers during feeding.

The life cycles of the three species are believed to be similar. Eggs are attached to the hairs of the host 2–3 mm above the skin. A prelarva and larva develop within the eggshell. Fully developed octopod nymphs emerge from the egg. There are two nymphal stages before the adult stage is reached. Adults can remain alive for up to 10 days off their host without feeding, in a cool atmosphere. However, immature stages only survive for up to 48 hours off their host.

Pathology: *Cheyletiella* may occasionally cause mild to severe scaling dermatosis with variable pruritus in dogs, cats and rabbits. The skin scales shed into the coat and the presence of mites among this debris gives rise to the common name 'walking dandruff'. In heavily infested dogs excessive shedding of hair, inflammation, and hyperaesthesia of the dorsal skin have been reported. The mites can readily transfer to humans, causing papular lesions, usually in groups of three or four, typically on the arms or across the abdomen.

Cheyletiella may be more common on short-haired breeds of dog and many individuals act as asymptomatic carriers. Young animals, especially those in a kennel situation and debilitated individuals, may be particularly susceptible to infection.

2.8.3 Trombiculidae

More than 1500 species of the family Trombiculidae have been described, about 50 of which are known to attack domestic animals and humans. They are commonly known as chiggers, red bugs, harvest mites and scrub itch mites and are unique, in that only the larval stage is ectoparasitic. In the adult and nymphal stages they are believed to be predators on the eggs and larvae of other arthropods. The principal species of veterinary interest are in the genus *Trombicula*.

The genus *Trombicula* is divided into a number of sub-genera of which two contain species of veterinary interest: *Neotrombicula* and *Eutrombicula*. Key species are the harvest mite, *Trombicula (Neotrombicula) autumnalis*, in Europe; the chiggers, *Trombicula*

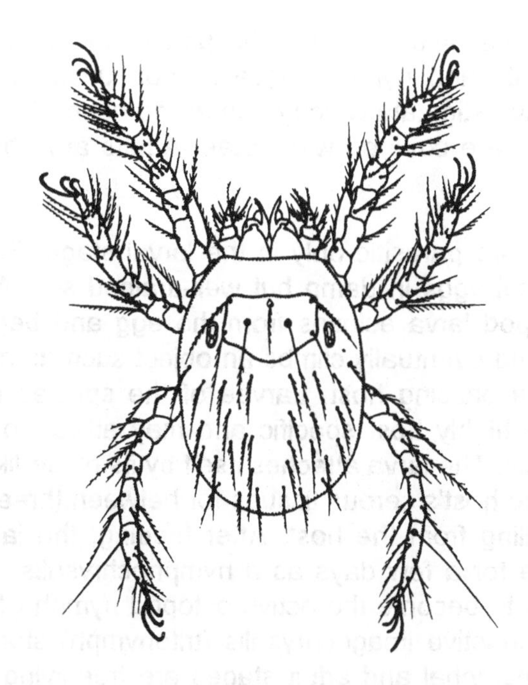

Fig. 2.17 Parasitic larval stage of the harvest mite, *Trombicula (Neotrombicula) autumnalis* (reproduced from Varma, 1993).

(Eutrombicula) alfreddugesi and *Trombicula (Eutrombicula) splendens,* in the New World; and the scrub itch mite, *Trombicula (Eutrombicula) sarcina* in Australasia.

Trombicula (Neotrombicula) autumnalis

Morphology: the hexapod larvae are rounded, red to orange in colour and are about 200 μm long (Fig. 2.17). Larval trombiculids breathe through the cuticle and there are no stigmata. Behind the gnathosoma on the dorsal surface is a dorsal shield known as the **scutum**, bearing a pair of sensillae and five setae. In *T. autumnalis* the scutum is roughly pentangular and has numerous small punctuations. There are two simple eyes on each side of the scutum. The body is covered dorsally with 25–50 relatively long, ciliated, feather-like setae. The chelicerae are flanked by stout, five-segmented palps. The palpal femur and genu each bear a single seta. The palpal tibia has three setae and a thumb-like terminal claw which opposes the palpal tarsus. The palpal claw is three-pronged (trifurcate). Adults and nymphs have a pronounced figure-of-eight shape. They have stigmata which open at the base of the chelicerae and their bodies are covered with setae. Adults are about 1 mm in length.

Life cycle: they are parasitic only in the larval stage. Female adults lay their spherical eggs in damp but well-drained soil. After about a week the hexapod larva ecloses from the egg and begins to crawl about the soil and eventually climbs an object such as a grass stem. Here it awaits a passing host. Larvae of the species of veterinary interest are not highly host specific and may attach to a variety of domestic animals. The larva attaches itself by its blade-like chelicerae and feeds on the host's serous tissues for between three and several days before falling from the host. After feeding, the larva enters a quiescent stage for a few days as a nymphochrysalis (protonymph) before moulting to become the active octopod nymph (deutonymph). After a further inactive imagochrysalis (tritonymph) stage, the adult emerges. The nymphal and adult stages are free-living, mobile and predatory. *Trombicula autumnalis* passes through only one generation per year and its abundance is usually strongly seasonal. In Europe the activity of *T. autumnalis* is most pronounced in late summer and autumn and larvae are most active on dry, sunny days. It will parasitize almost all domestic mammals, including humans, and some ground-nesting birds, and may be particularly abundant in closely cropped chalk grassland, but it may also be found in wooded areas and scrub.

Pathology: harvest mites are commonly found in clusters, on the foot and up the legs of dogs, on the genital area and eyelids of cats, on the face of cattle and horses, and on the heads of birds, after having been picked up from the grass. Infestation causes pruritus, erythema and scratching, though there may be considerable individual variation in response. This variation may reflect the development of a hypersensitivity reaction to the mites, which may result in the development of weals, papules and excoriation.

Trombicula (Eutrombicula) alfreddugesi

The larvae of *T. alfreddugesi*, known as chiggers, are similar in appearance to those of *T. autumnalis.* They are reddish-orange and vary in length between 150 μm when unengorged to 600 μm when fully fed. However, for the larvae of *T. alfreddugesi* the palpal claws are two-pronged (bifurcate), the scutum is approximately rectangular and 22 dorsal setae are present.

The life cycle is similar to that described for *T. autumnalis*. *Trombicula (Eutrombicula) alfreddugesi* is the most important and widespread of the trombiculid mites of veterinary interest in the New World. It is common from eastern Canada through to South America. It is particularly common at the margins of woodland, scrub and grassland but is not highly habitat-specific. In the northern parts of its range it may be most active between July and September, whereas in more southern habitats it may be active all year round. It parasitizes a wide range of mammals, birds, reptiles and amphibians.

Trombicula (Eutrombicula) splendens

Trombicula (Eutrombicula) splendens is morphologically similar and frequently sympatric with *T. alfreddugesi* in North America, although it is not so widely distributed and is generally confined to the east, from Ontario in Canada to the Gulf States, although it may also be abundant in Florida and parts of Georgia. It generally occurs in moister habitats than *T. alfreddugesi*, such as swamps and bogs.

Trombicula (Eutrombicula) sarcina

The scrub itch mite *Trombicula (Eutrombicula) sarcina* is an important parasite of sheep in Queensland and New South Wales of Australia. Its principal host, however, is the grey kangaroo. These mites prefer areas of savannah and grassland scrub. They may be particularly abundant from November to February, after summer rain. The primary site of infestation is on the leg, resulting in intense irritation.

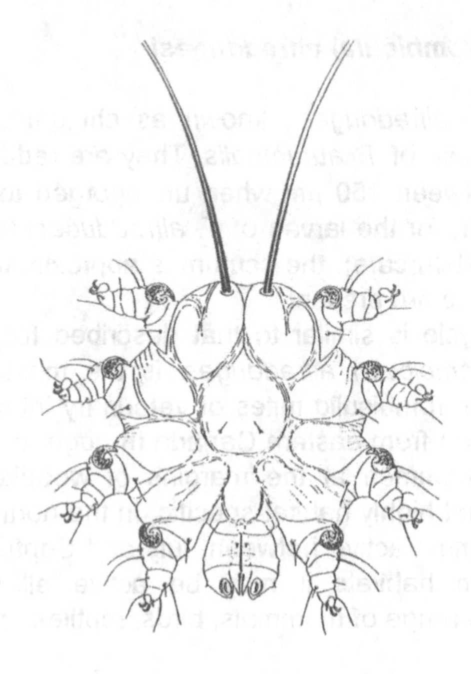

Fig. 2.18 Adult female *Psorergates simplex* (reproduced from Baker *et al.*, 1956).

2.8.4 Psorergatidae

Two species of the family Psorergatidae have been recovered from domestic animals: *P. bos* from cattle and *P. ovis* from sheep. *Psorergates simplex* occurs on mice in the northern hemisphere. Only *P. ovis* is of major veterinary significance.

Psorergates ovis

The sheep itch mite *Psorergates ovis* is an important ectoparasite of sheep in parts of Australia, New Zealand, South Africa and South America. It is thought to have been eradicated from the USA and has not been reported from Europe.

Morphology: the body is almost circular and the legs are arranged more or less equidistantly around the body circumference (Fig. 2.18). Larvae of *P. ovis* have short, stubby legs. The legs become progressively longer during the nymphal stages until, in the adult, the legs are well developed and the mites become mobile. Adults are about 190 μm long and 160 μm wide. The tarsal claws are simple and the empodium is pad-like. The femur of each leg bears a large, inwardly directed curved spine. In the adult female two pairs of long, whip-like setae are present posteriorly; in the male there is only a single pair.

Life cycle: egg, larva, protonymph, deutonymph, tritonymph and adult. The complete life cycle of *P. ovis* takes about 6 weeks.

Pathology: although *P. ovis* does not burrow, it does damage the skin, causing chronic dermatitis with alopecia and scaling. Wool becomes dry, matted and discoloured, with a slightly yellowish colour. The severe pruritus caused by *P. ovis* leads to self-trauma by chewing and rubbing, with resulting wool damage. Infestation is most commonly seen spreading over the lateral thorax, hind quarters and thighs of the host. However, *P. ovis* is very sensitive to desiccation, can only survive for 24–48 hours off the host and only the adults are mobile. As a result, infestation spreads slowly over infested animals and only slowly through a flock. Infestation may take 3–4 years to

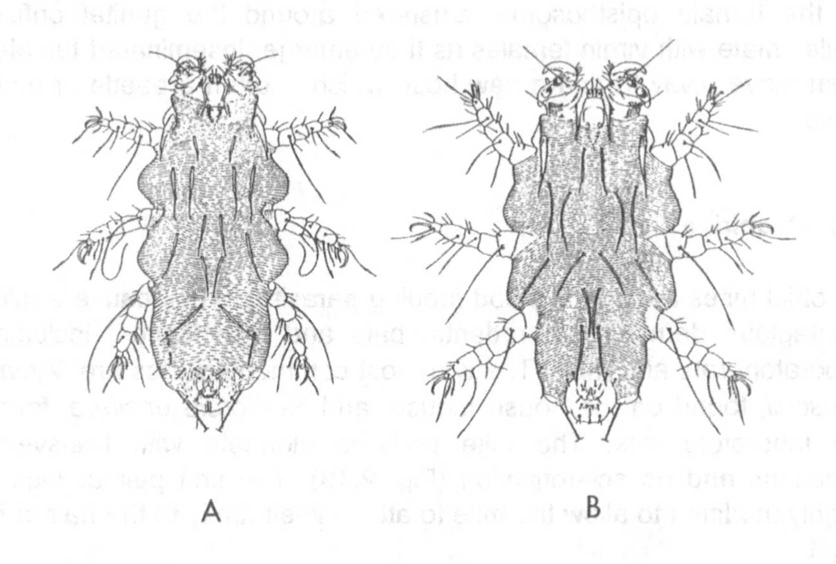

Fig. 2.19 Adults of (A) *Myobia musculi*, female dorsal view and (B) *Radfordia ensifera* male, dorsal view (reproduced from Baker *et al.*, 1956).

become generalized on an infested animal. Mites occur under the superficial layers of the skin and, therefore, skin scrapings are necessary to detect infestations.

2.8.5 Prostigmatid mites of minor veterinary interest

(a) Pyemotidae

Species of Pyemotidae are soft-bodied, predacious mites which infest materials such as grain or hay, sometimes described as forage mites. They feed largely on insect larvae but will attack domestic animals and humans that come into contact with the infested material, causing dermatitis.

They are small mites with elongated bodies. Adults have greatly reduced mouthparts and stylet-like chelicerae. There are several species, of which *Pyemotes tritici* is particularly damaging.

When gravid, the fertilized female becomes enormously swollen with eggs. The eggs hatch within the female and the larval and nymphal stages take place internally, fully formed adults emerging from the female. Up to 200–300 offspring may be produced per female. Males emerge first and remain on the outside of the female opisthosoma, clustered around the genital orifice. Males mate with virgin females as they emerge. Inseminated females then move away to find a new host, which is often a beetle or moth larva.

(b) Myobidae

Myobiid mites are small, blood-feeding parasites. They cause a mild, contagious dermatitis in rodents, bats and insectivores, including laboratory rats and mice. The two most common species are *Myobia musculi*, found on the house mouse, and *Radfordia ensifera*, found on laboratory rats. The mite body is elongate with transverse striations and no sclerotization (Fig. 2.19). The first pair of legs is highly modified to allow the mite to attach itself firmly to the hair of its host.

2.9 MESOSTIGMATA

2.9.1 Macronyssidae

Species of the family Macronyssidae are blood-sucking ectoparasites of birds and mammals. They feed only in the protonymph and adult stages. Only one genus, *Ornithonyssus*, is of general veterinary interest. Species of this genus are ectoparasites of birds. Species of a second genus, *Ophionyssus*, are ectoparasites of reptiles and will not be considered here.

Ornithonyssus sylviarum

The northern feather or northern fowl mite, *Ornithonyssus sylviarum*, is one of the most important ectoparasites of poultry throughout Europe, North America, New Zealand and Australia.

Morphology: adult females are large, between 750 and 1000 μm in length (Fig. 2.20A). They have relatively long legs and can easily be seen with the naked eye. *Ornithonyssus sylviarum* may vary in colour from white to red or black, depending on how recently it has fed. There is a single, relatively narrow dorsal plate which does not entirely cover the dorsal surface and which tapers posteriorly. There are several pairs of long setae on the dorsal plate. The genitoventral shield also tapers posteriorly. The chelicerae are toothless.

Life cycle: the entire life cycle of *O. sylviarum* takes place on the host, with the eggs laid in masses at the base of the feathers, primarily in the vent area. Hexapod larvae eclose from the egg within a day or so of oviposition, giving rise to the protonymph, deutonymph and adult. Only two of the five life-cycle stages, the protonymph and adult, feed on the host; deutonymphs do not feed. The life cycle is short and can be completed within 5 to 12 days. Hence, large populations can develop rapidly on the birds.

Pathology: feathers may become matted and severe scabbing may develop, particularly around the vent. Infested chickens show a grey-black discolouration of the feathers due to the large number of mites present. Heavy infestations may cause decreased egg production, anaemia, loss of condition and death. The mites can survive off the host for 1 to 3 weeks, or occasionally longer, and this ability enhances the chances for mite transmission by the movement of people or materials from infested to uninfested houses. Irritation by bites and allergic reactions of people working with infested chickens have been recorded.

Ornithonyssus bacoti

The tropical rat mite, *Ornithonyssus bacoti* is a blood-sucking ectoparasite of mice, rats, hamsters and small marsupials in both temperate and tropical regions of the world. It is similar in appearance and life cycle to *O. sylviarum* (Fig. 2.20B). The life cycle is rapid and a high population can lead to the death of the host by exsanguination. It is a particularly common parasite in laboratory rodent colonies and can occur worldwide, despite its name. Bites can lead to irritation and dermatitis in animals and people coming into contact with infested animals.

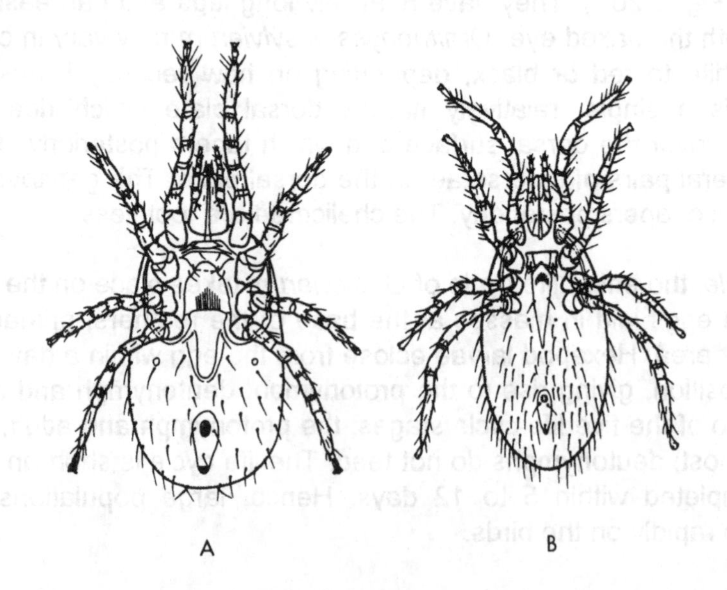

Fig. 2.20 Adult females (A) *Ornithonyssus sylviarum* (northern fowl mite) and (B) *Ornithonyssus bacoti* (tropical rat mite) (reproduced from Varma, 1993).

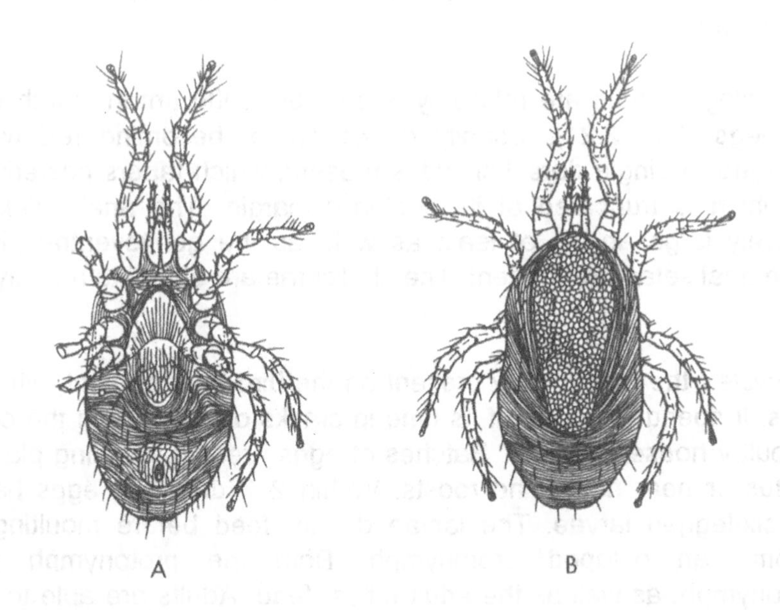

Fig. 2.21 Adult female of the red mite, *Dermanyssus gallinae* (A) ventral view (B) dorsal view (reproduced from Varma, 1993).

Ornithonyssus bursa

The tropical fowl mite, *Ornithonyssus bursa* is found throughout the sub-tropical and tropical regions of the world, where it is an important ectoparasite of poultry and other birds. It is similar in appearance and life cycle to *O. sylviarum*.

2.9.2 Dermanyssidae

Species of the family Dermanyssidae are blood-sucking ectoparasites of birds and mammals.

Dermanyssus gallinae

The red mite or chicken mite, *Dermanyssus gallinae,* feeds off the blood of fowl, pigeons, caged birds and many other wild birds. It occasionally bites mammals, including humans, if the usual hosts are unavailable. It is a common ectoparasite in hen houses, aviaries and pigeon lofts.

Morphology: adults are relatively large, 750–1000 μm in length with long legs (Fig. 2.21). Usually greyish-white, becoming red when engorged. A single dorsal shield is present, which tapers posteriorly, but which is truncated at its posterior margin. The anal shield is relatively large and is at least as wide as the genitoventral plate. Three anal setae are present. The chelicerae are elongate and stylet-like.

Life cycle: the mite is only present on the bird host at night, when it feeds. It spends the rest of its time in cracks or crevices in the cage or poultry house structure. Batches of eggs are laid in hiding places, detritus or near nests and roosts. Within 2–3 days the eggs hatch into six-legged larvae. The larvae do not feed before moulting to become an octopod protonymph. Both the protonymph and deutonymph, as well as the adult mites, feed. Adults are able to live off the host without feeding for up to 34 weeks. In the presence of hosts the life cycle can be completed very quickly (egg to adult in 7 days) allowing large populations to develop rapidly.

Pathology: this mite causes feeding lesions which are most likely to be seen on the breast or legs of the bird. The mites can directly cause anaemia and can lower egg production. Newly hatched chicks may rapidly die as a result of mite activity. *Dermanyssus gallinae* may be an important pest of poultry flocks maintained on the floor in barn or deep litter systems but is less important in caged production facilities. Infestation of pigeons is common.

2.9.3 Mesostigmatid mites of minor veterinary interest

(a) Halarachnidae

The subfamily Halarachninae of the family Halarachnidae are obligatory parasites occurring in the respiratory tract of mammals. The primary species of minor veterinary importance in temperate habitats is *Pneumonyssus caninum,* the dog nasal mite, which occurs in the sinuses and nasal passages of dogs in Australia, South Africa and the USA. Adults are pale yellow, oval mites with few setae. The first pair of legs is equipped with a pair of heavily sclerotized brown claws, while the other three pairs end in a long, stalked pulvillus and two slender claws. Symptoms of infestation are usually mild. However, excessive mucus production or rhinitis has been recorded.

The subfamily Raillietiinae of the family Halarachnidae are obligatory parasites occurring in the external ear of mammals. The primary species of minor veterinary interest is *Raillietia auris,* the cattle ear mite. This is an oval mite, about 1 mm long, which occurs in the ears of domestic cattle in North America, Europe and Australia. The mite feeds on epidermal cells and wax, but does not suck blood. Infestations are usually benign.

(b) Rhinonyssidae

Most species of the family Rhinonyssidae are parasites of the naso-pharynx of birds. The primary species of minor veterinary interest is the canary lung mite *Sternostoma tracheacolum.* It has been recorded from a range of domestic and wild birds, including canaries and budgerigars, and has been found worldwide. It is a yellowish mite, about 0.5 mm in length, usually found in the tracheae, air sacs and bronchi. Females give birth to live larvae. Infestation may result in lesions, inflammation and bronchial haemorrhage. Birds may become listless, waste away and die.

(c) Laelapidae

This family contains a number of blood-feeding species which are parasites of rodents. Five species are common worldwide, *Hirstionyssus isabellinus*, *Haemogamasus pontiger*, *Eulaelaps stabularis*, *Laelaps echidninus* and *Laelaps nuttalli*. Adults have a single dorsal shield.

2.10 FURTHER READING

Baker, E.W. & Wharton, G.W. (1952) *An Introduction to Acarology*, Macmillan, New York.

Baker, E.W., Evans, T.M., Gould, D.J. *et al.* (1956) *A Manual of Parasitic Mites of Medical or Economic Importance*, National Pest Control Association, New York.

Baker, E.W., Camin, J.H., Cunliffe, F. *et al.* (1958) *Guide to the Families of Mites,* Contribution No. 3, Institute of Acarology, University of Maryland.

Barnes, R.D. (1974) *Invertebrate Zoology*, W.B. Saunders, London.

Bronswijk, J.E.M.H. and de Kreek, E.J. (1976) *Cheyletiella* (Acari: Cheyletiellidae) of dog, cat and domesticated rabbit, a review. *Journal of Medical Entomology*, **13**, 315–27.

Burgess, I. (1994) *Sarcoptes scabiei* and scabies. *Advances in Parasitology*, **33**, 235–93.

Desch, C.E. (1984) Biology of biting mites (Mesostigmata), in *Mammalian Diseases and Arachnids* (ed. W.B. Nutting), CRC Press, Florida, Vol. I, pp. 83–109.

Duncan, S. (1957) *Dermanyssus gallinae* (De Geer 1779) attacking man. *Journal of Parasitology*, **43**, 637–43.

Fain, A. (1994) Adaptation, specificity and host–parasite coevolution in mites (Acari). *International Journal for Parasitology*, **24**, 1273–83.

Hirst, S. (1922) *Mites Injurious to Domestic Animals*, Economic Series No. 13, British Museum, Natural History, London.

Kettle, D.S. (1984) *Medical and Veterinary Entomology*, Croom Helm, London.

Krantz, G.W. (1978) *A Manual of Acarology*, Oregon State University Book Stores, Corvallis, Oregon.

Nutting, W.B. (1976) Pathogenesis associated with hair follicle mites (Acari: Demodicidae). *Acarologia*, **17**, 493–506.

Schuster, R. and Murphy, O.W. (1991) *The Acari*, Chapman & Hall, London.

Scott, D.W., Schultz, R.D. and Baker, E.B. (1976) Further studies on the theraputic and immunological aspects of generalized demodectic mange in the dog. *Journal of the American Animal Hospital Association*, **12**, 202–13.

Sinclair, A.N. and Gibson, A.J.F. (1975) Population changes of the itch mite *Psorergates ovis*, after shearing. *New Zealand Veterinary Journal*, **23**, 14.

Sweatman, G.K. (1958) On the life-history and validity of the species in *Psoroptes*, a genus of mange mite. *Canadian Journal of Zoology*, **36**, 905–29.

Urquhart, G.M., Armour, J., Duncan, J.L. *et al.* (1987) *Veterinary Parasitology*, Longman Scientific and Technical, London.
Varma, M.G.R. (1993) Ticks and mites (Acari), in *Medical Insects and Arachnids* (eds R.P. Lane and R.W. Crosskey), Chapman & Hall, London, pp. 597–658.
Walker, A. (1994) *Arthropods of Humans and Domestic Animals*, Chapman & Hall, London.
Woolley, T.A. (1961) A review of the phylogeny of mites. *Annual Review of Entomology*, **6**, 263–84.
Wright, F.C., Riner, J.C. and Guillot, F.S. (1983) Cross mating studies with *Psoroptes ovis* (Hering) and *Psoroptes cuniculi* Delafond (Acarina: Psoroptidae). *Journal of Parasitology*, **69**, 696–700.

3

Ticks (Acari)

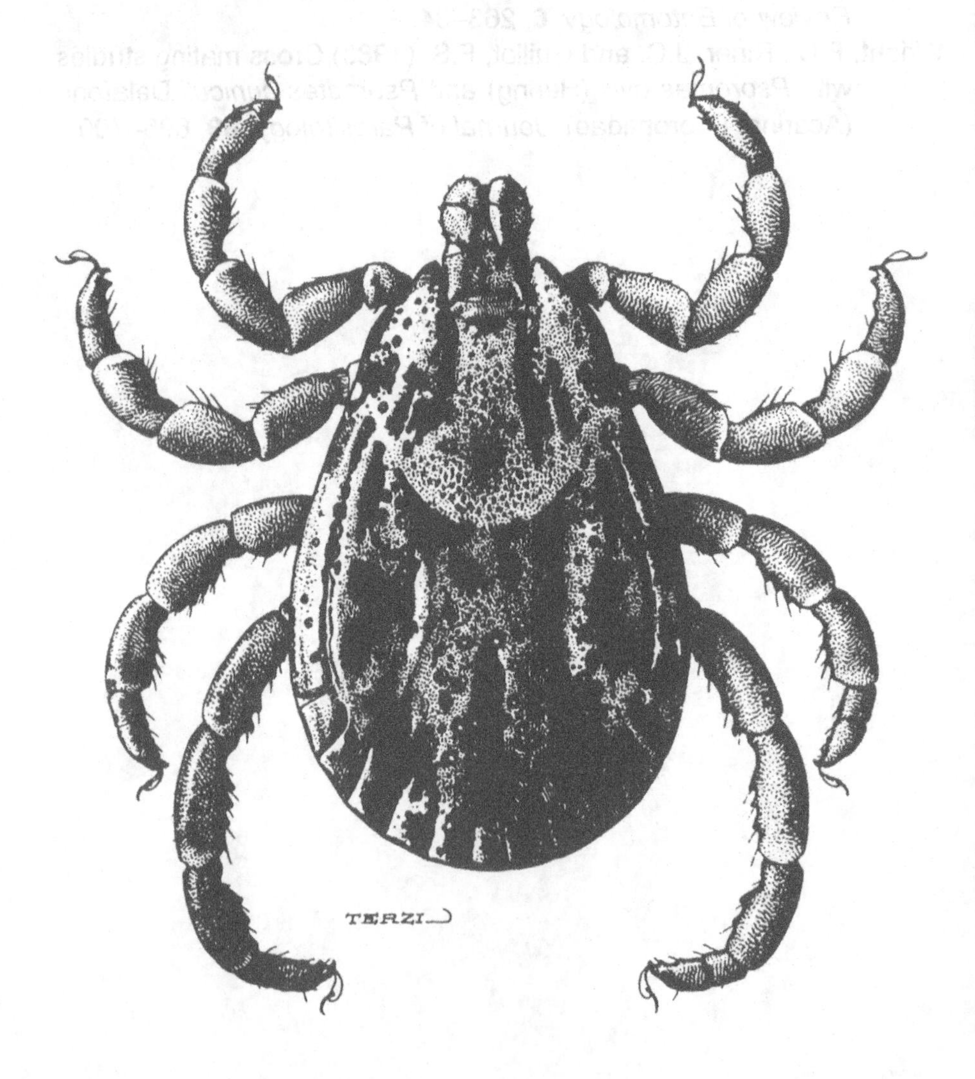

Adult Rocky Mountain wood tick, *Dermacentor andersoni* (from Varma, 1993).

3.1 INTRODUCTION

The ticks are obligate, blood-feeding ectoparasites of vertebrates, particularly mammals and birds. Ticks are **arachnids**, in the sub-class **Acari**, closely related to the mites. They are usually relatively large and long-lived compared to mites, surviving for up to several years. During this time they feed periodically taking large blood meals, often with long intervals between each meal. Tick bites may be directly debilitating to domestic animals, causing mechanical damage, irritation, inflammation and hypersensitivity and, when present in large numbers, feeding may cause anaemia and reduced productivity (see section 3.4). The salivary secretions of some tick species may cause toxicosis and paralysis. However, ticks may also transmit a range of pathogenic viral, rickettsial and bacterial diseases to livestock. Hence, although the ticks are a relatively small order of only about 800 species, they are one of the most important groups of arthropod pests of veterinary interest.

A single family, the **Ixodidae**, known as the **hard ticks**, contains almost all the species of tick of veterinary importance. A second family, the **Argasidae** known as the **soft ticks**, contains a relatively small number of species of veterinary importance. A third family of tick, the **Nuttalliellidae**, contains only a single, little-known species which is found in the nests of swallows in southern Africa.

3.2 MORPHOLOGY

3.2.1 Ixodidae

Ixodid ticks are relatively large, ranging between 2 and 20 mm in length. The body of the unfed tick is flattened dorsoventrally and is similar in structure to that of mites, being divided into only two sections, the anterior **gnathosoma** (or capitulum) and posterior **idiosoma**, which bears the legs. The terminal gnathosoma is always visible when an ixodid tick is viewed from above.

The mouthparts of ticks are structurally similar to those of mites (Fig. 3.1) The gnathosoma carries a pair of four-segmented **palps**, which are simple sensory organs, which help the mite to locate its host. The fourth segment of each palp is reduced and may articulate from the ventral side of the third, forming a pincer-like structure. Between the palps lies a pair of heavily sclerotized, two-

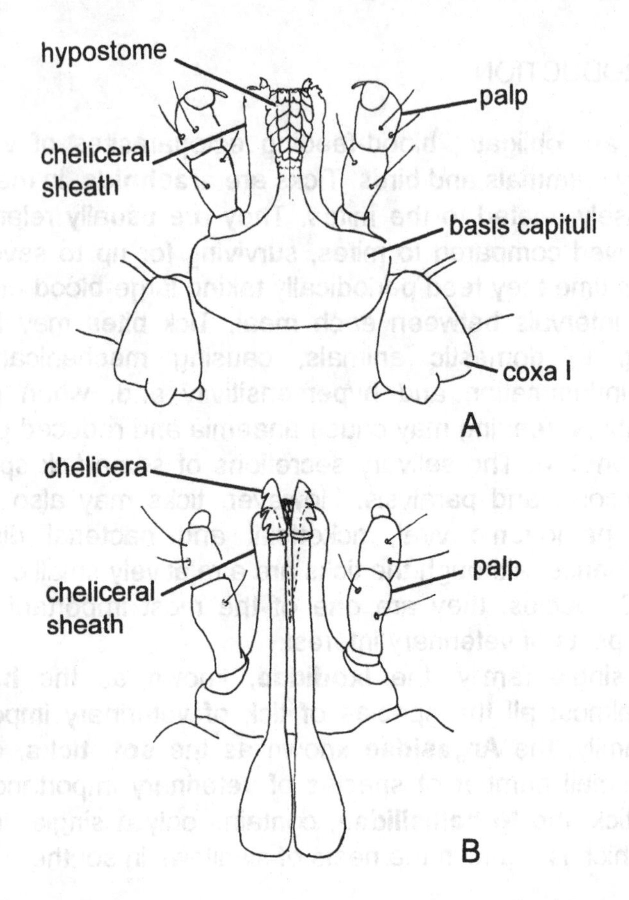

Fig. 3.1 Tick mouthparts: (A) ventral view, showing toothed hypostome; (B) dorsal view, showing the chelicerae behind the cheliceral sheaths (redrawn from Herms and James, 1961).

segmented appendages called **chelicerae**, housed in **cheliceral sheaths.** At the end of each chelicera is a rigid, somewhat triangular, plate bearing a number of sclerotized tooth-like digits. The chelicerae are capable of moving back and forth and the tooth-like digits are used to cut and pierce the skin of the host animal during feeding. The enlarged, fused **coxae** of the palps are known as the **basis capituli**. The basis capituli varies in shape in the different genera, being rectangular, hexagonal or triangular. The lower wall of the basis capituli is extended anteriorly and ventrally to form an unpaired median **hypostome**, like an underlip, which lies below the chelicerae. The hypostome does not move, but in larvae, nymphs and adult females is armed with rows of backwardly directed ventral teeth.

The hypostome is thrust into the hole cut by the chelicerae and the teeth are used to attach the tick securely to its host. As the hypostome is inserted the palps are spread flat on to the surface of the host's skin. Salivary secretions contain anticoagulants and other active compounds which promote lesion development. The large quantity of salivary fluid produced is the principal avenue for disease transmission in the hard ticks. As feeding commences the tick tilts its body to an angle of at least 45° to the skin and this angle may become steeper as feeding proceeds.

Hard ticks remain attached for several days during feeding. For ticks with long mouthparts, attachment by the chelicerae and hypostome is sufficient to anchor the tick in place. However, for ticks with short mouthparts, attachment is maintained during feeding by secretions from the salivary glands which harden around the mouthparts and effectively cement the tick in place. Female ticks show substantial increases in size when they engorge during feeding; some of the larger species of *Amblyomma* can increase from just under 10 mm to over 25 mm in length and increase from about 0.04 g to over 4 g in weight. However, these impresive figures understimate the amount of blood imbibed, because almost as soon as they start to feed ticks begin to digest the blood meal and excrete waste materials.

The immature stages of ticks are very similar morphologically to the adults. The larvae, sometimes known as seed ticks, are six-legged. Adult and nymphal ticks have four pairs of legs (see section 3.3). The region of the idiosoma which carries the legs is called the **podosoma** and the region behind the legs is the **opisthosoma**. Each leg is attached to the body at the coxa. The coxa may be armed with internal and external **ventral spurs**, and their number, size and shape may be important in species identification (Fig. 3.2). The coxa is then followed by the **trochanter**, **femur**, **genu**, **tibia** and **tarsus**. The legs of ticks, like those of mites, usually end in a pair of claws and a well-developed, pad-like **pulvillus**. Located on the tarsi of the first pair of legs is a pit known as **Haller's organ**, which is packed with chemoreceptor setae and used in host location. Chemoreceptors are also present on the palps, chelicerae and scutum.

Male ixodid ticks are usually smaller than females. Males imbibe relatively little blood when they feed and show little increase in size. Ixodid ticks possess a sclerotized dorsal shield or plate on the idiosoma known as a **scutum** (Fig. 3.3). In males, the scutum covers the entire dorsal surface, whereas in females the scutum is relatively small to facilitate the size increase which occurs during feeding. The scutum is difficult to see in a fully engorged female. If the scutum has a pattern of grey and white on a dark background, it is described as **ornate**, if not it is described as **inornate**.

Ticks do not possess antennae, and when eyes are present they are simple and are located dorsally at the sides of the scutum. There are a number of grooves, largely on the ventral side of ticks. The number and presence of these grooves may be important in identification. Uniform rectangular regions on the posterior margins of the body are known as **festoons** and are also separated by grooves (Fig. 3.3).

The larvae lack stigmata and a tracheal system, and water loss and respiration take place directly through the integument. Adult and nymphal ticks have a pair of respiratory openings, the **stigmata**. The stigmata lead to a complex branching system of tracheae (see section 1.6.4). The stigmata are large and positioned posterior to the coxae of the fourth pair of legs. Each is often surrounded by a stigmatal plate which is circular or oval (Fig. 3.4).

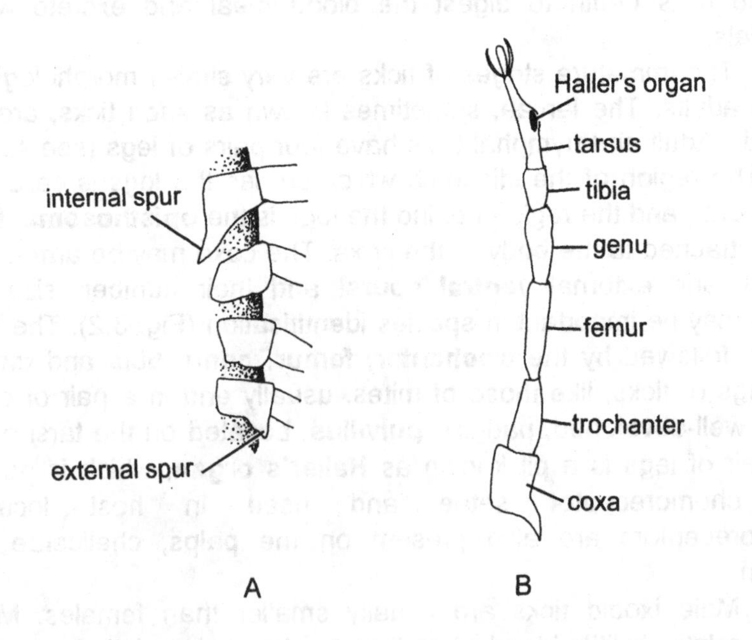

Fig. 3.2 (A) Ventral view of the coxae and (B) segments of the leg of a generalized ixodid tick.

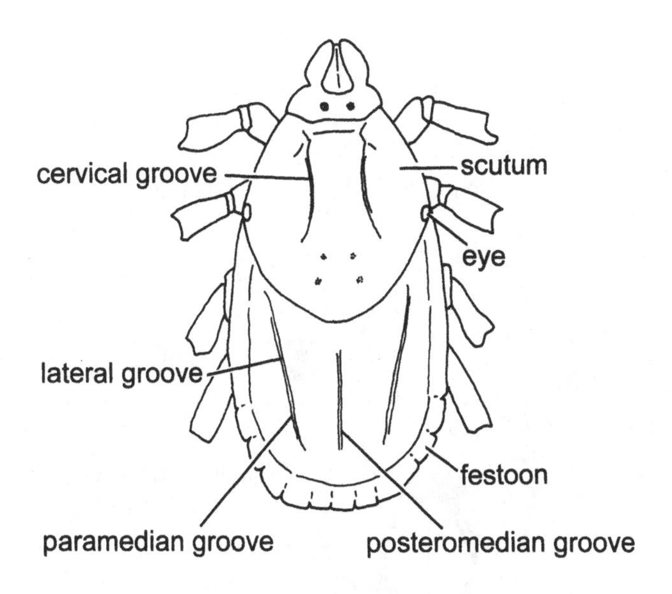

Fig. 3.3 Dorsal view of a generalized female, ixodid tick (redrawn from Varma, 1993).

Sexual differentiation is usually not obvious in the larva and nymph, and immature ixodids generally look like small females but without the genital opening. In adults the genital opening, the **gonopore**, is situated ventrally behind the gnathosoma, usually at the level of the second pair of legs (Fig. 3.4). The **genital apron** is a lightly sclerotized flap originating in front of and covering the genital opening. A pair of **genital grooves** start at the gonopore and extend backwards to the **anal groove**. The anus is also located ventrally, usually being posterior to the fourth pair of legs. The anal groove surrounds the anus. Adult males possess a pair of testes, and a vas deferens extends from each testis to open through the gonopore. Males and females are brought together, usually on the host, by aggregation and sex-recognition pheromones. The male has no external genitalia so, after ejecting a mass of sperm in a spherical spermatophore, the spermatophore is grasped by the chelicerae of the male and implanted into the female gonopore before being pushed into the female genital tract.

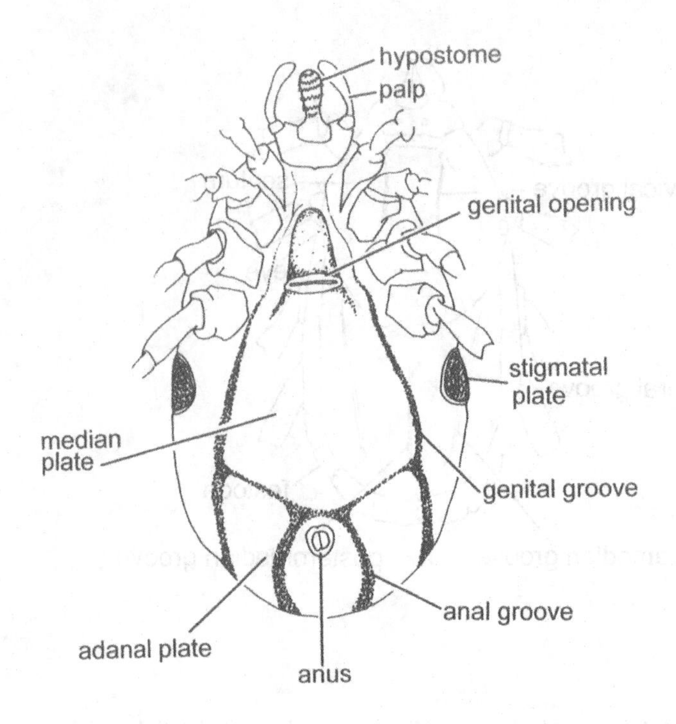

Fig. 3.4 Ventral view of a generalized male, ixodid tick (redrawn from Varma, 1993).

3.2.2 Argasidae

In the Argasidae, the body is leathery and unsclerotized, with a textured surface, which in unfed ticks may be characteristically marked with grooves or folds (Fig. 3.5). The fourth segment of the palp is the same size as the third and the palps may appear somewhat leg-like. In nymphs and adults the gnathosoma is not visible from the dorsal view, being located ventrally in a recess called the **camerostome**. When present the eyes are positioned in lateral folds above the legs. The stigmata of Argasidae are small and placed anterior to the coxae of the fourth pair of legs. The integument is inornate. There is little sexual dimorphism. Pulvilli are usually absent or rudimentary in adults and nymphs, but may be well developed in larvae.

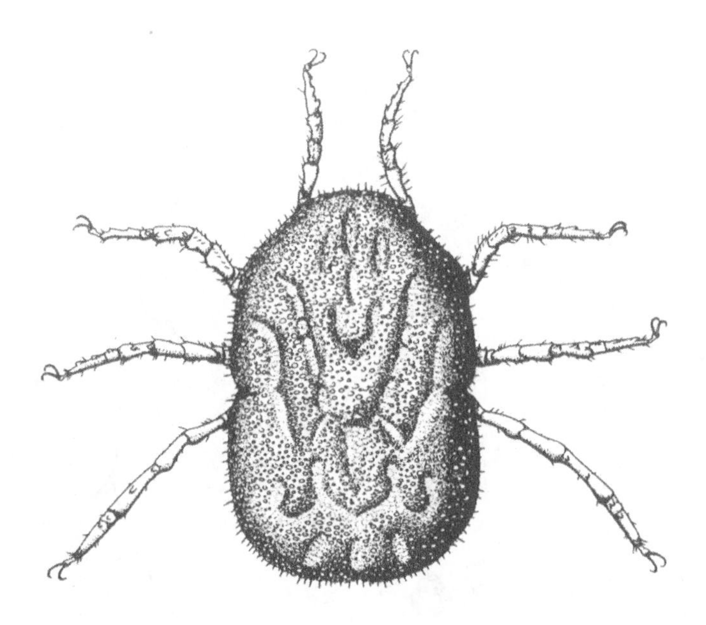

Fig. 3.5 Dorsal view of an argasid tick, *Ornithodoros moubata* (reproduced from Varma, 1993).

3.3 LIFE HISTORY

3.3.1 Ixodidae

The life cycles of ixodid ticks involve four stages: egg, six-legged larva, eight-legged nymph and eight-legged adult (Fig. 3.6). During the passage through these stages ixodid ticks take a number of large blood meals, interspersed by lengthy free-living periods. The time spent on the host may occupy as little as 10% of the life of the tick. They are relatively long lived and each female may produce several thousand eggs.

Most hard ticks are relatively immobile and, rather than actively hunting for their hosts, the majority adopt a 'sit and wait' strategy. However, a few, including some species of *Hyalomma*, actively approach their host. To obtain a blood meal, field-inhabiting ticks show a behaviour known as questing. They climb to the tips of

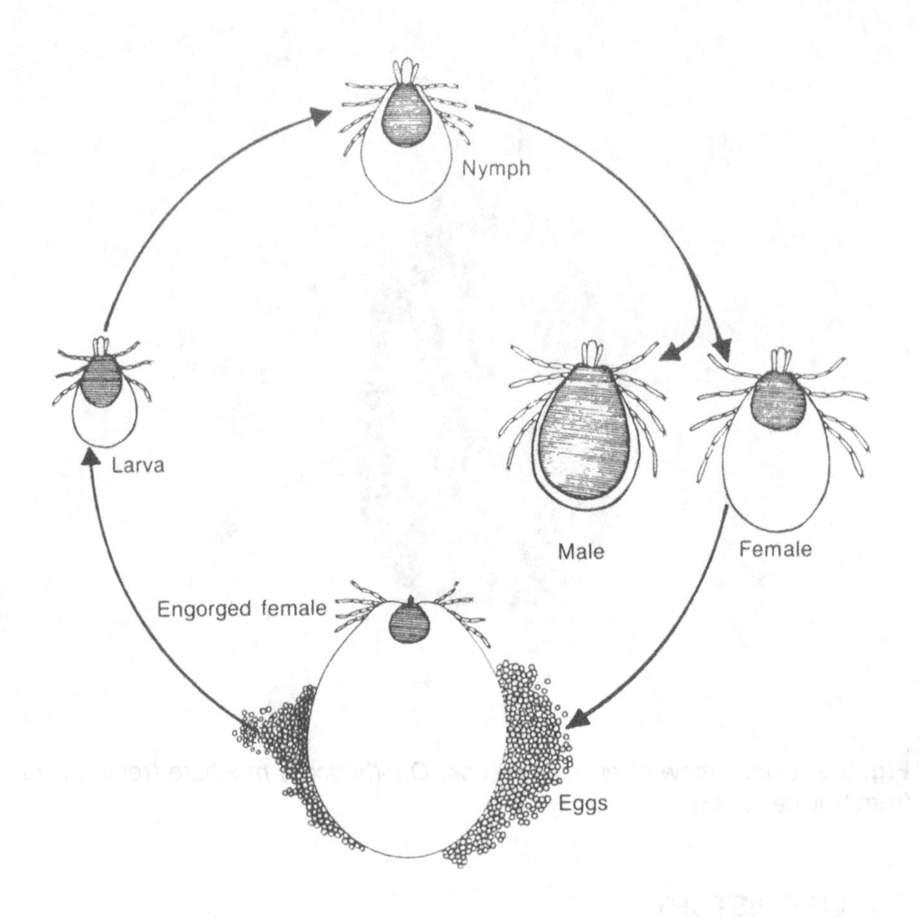

Fig. 3.6 Life cycle of an ixodid tick (reproduced from Urquhart *et al.*, 1987).

vegetation, such as grass, at a height appropriate to their host. At the approach of a potential host the tick begins to wave the first pair oflegs in the direction of the stimulus. The ticks detect the approach of the host using cues such as carbon dioxide and other odours emitted by the host, which they sense using the chemoreceptors, particularly those packed into Haller's organ on the tarsi of the first pair of legs. Vibrations, moisture, heat and passing shadows may also be important cues in host recognition. Once contact is made, as the host animal brushes past, the ticks transfer to the host, and then move over the surface to find their preferred attachment sites, such as the ears. Preferred sites for attachment may be highly specific to the particular species of tick.

The host-location behaviour of ticks represents an intrinsically risky way of finding a host animal and many probably die without ever feeding. To compensate for the risks involved, ticks have developed a variety of complex life cycles and feeding strategies which reflect the nature of the habitat which the various species of tick inhabit and the probability of contact with an appropriate host.

For ixodid ticks, which inhabit forests and pastures where there is a relatively plentiful supply of host animals and in habitats where conditions are suitable for good survival during the off-host phase, a three-host life cycle has been adopted. Larvae begin to quest several days to several weeks after hatching, the precise time depending on temperature and humidity. On finding a suitable host, usually a small mammal or bird, the larvae begin to feed. Blood feeding typically takes between 4 and 6 days. On completion of feeding the larvae drop to the ground where, 2–3 weeks later, they moult to become nymphs. After another interval, again which may vary between several days and several months depending on environmental conditions, nymphs also begin to quest for a second host. The second host need not necessarily be of the same species as the first. After feeding, the nymphs drop to the ground and then moult to become adults. Finally, after a further interval, adults begin to quest. The third host is usually a large mammal. On their final host females mate and then engorge. Following the final blood meal adult females drop to the ground where they lay large batches of several thousand eggs over a period of days or weeks. Oviposition can be delayed for several weeks until conditions are suitable for survival of the eggs. Adult males may remain unattached on the host animal and attempt to mate with as many females as possible. Most ixodid ticks have a three-host feeding strategy.

For a relatively small number of ixodid ticks, about 50 species, which inhabit areas where hosts are scarce and in which lengthy seasonal periods of unfavourable climate occur, two- and one-host feeding strategies have evolved. In two-host ticks, such as *Rhipicephalus bursa*, larvae locate the host, then feed, moult to nymphal stages, feed again and then drop to the ground and moult to become adults. The adults then seek a second and usually larger host on which to feed. Adults then drop to the ground, lay their eggs and die. Where the environment and contact with the host are highly unpredictable, one-host species such as *Boophilus annulatus* and *Dermacentor albipictus* occur. After hatching from the egg, the larvae of these species locate a suitable host, which is usually a large

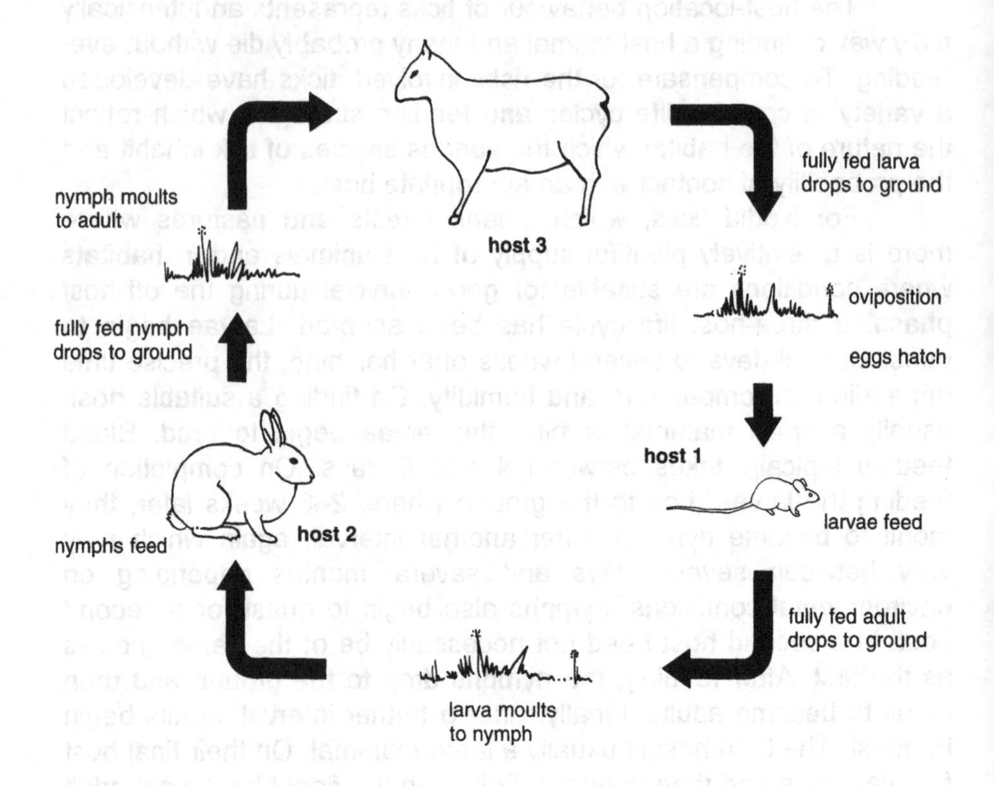

Fig. 3.7 A three-host feeding strategy of an ixodid tick.

herbivorous mammal. They then feed and moult to become a nymph. There is a single nymphal stage, during which they again feed, prior to moulting to become an adult. Finally they feed as adults, drop off their host and females lay their eggs.

In temperate habitats, feeding and generation cycles of hard ticks are closely synchronized with periods of suitable temperature and humidity conditions. Ticks, particularly in the immature stages, are very susceptible to desiccation. The cuticle is covered by a waxy layer which is largely impermeable to water, but when ticks are active water loss through the stigmata is more pronounced. To minimize drying out they start questing when saturated with water and return to the humid ground level when dehydrated. Water may also be imbibed by drinking.

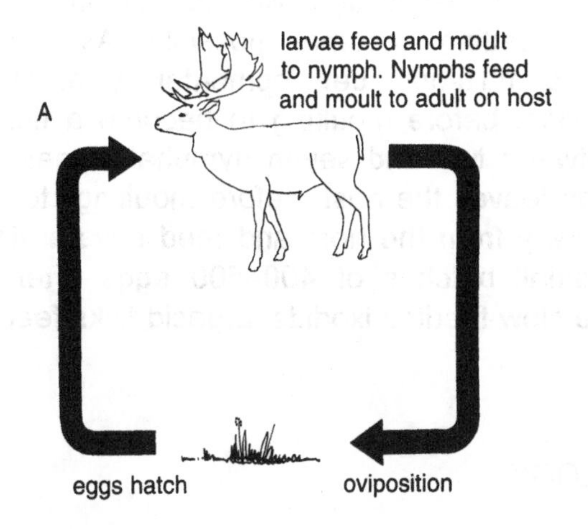

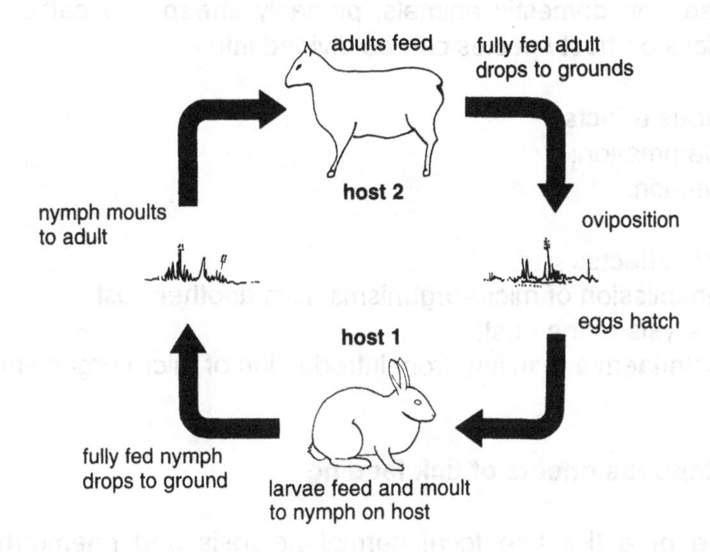

Fig. 3.8 (A) One-host and (B) two-host feeding strategies of ixodid ticks.

3.3.2 Argasidae

Argasid, soft ticks are generally found in deserts or dry conditions. In contrast to the hard ticks, argasid soft ticks tend to live in close proximity to their hosts: in chicken coops, pigsties, pigeon lofts, birds' nests, animal burrows or dens. In these restricted and sheltered habitats the hazards associated with host finding are reduced and more frequent feeding becomes possible. As a result, argasids typically have a multi-host developmental cycle. The single larval stage feeds once, before moulting to become a first-stage nymph. There are between two and seven nymphal stages, each of which feeds and then leaves the host, before moulting to the next stage. Adults mate away from the host and feed several times. The adult female lays small batches of 400–500 eggs after each feed. In contrast to the slow-feeding ixodids, argasid ticks feed for only a few minutes.

3.4 PATHOLOGY

Ticks are primarily parasites of wild animals and only about 10% of species feed on domestic animals, primarily sheep and cattle. The effect of ticks on host species can be divided into:

- Cutaneous effects:
  inflammation;
  infection.

- Systemic effects:
  transmission of micro-organisms from another host;
  paralysis of the host;
  bacteriaemia resulting from introduction of micro-organisms.

3.4.1 Cutaneous effects of tick feeding

At the site of a tick bite focal dermal necrosis and haemorrhage occur, followed by an inflammatory response, often involving eosinophils. Although a hypersensitivity reaction may be involved in the local response, the innate inflammatory response and dermal necrosis is sufficient to damage the hide. Tick-bite wounds can becomeinfected with *Staphylococcus* bacteria, causing local cutaneous abscesses or pyaemia. Heavy tick infestation can result in significant blood loss, reduced productivity, reduced weight-gain and cause restlessness. Tick-bite lesions may also predispose animals to screwworm myiasis (see Chapter 5).

3.4.2 Systemic effects: vectors of disease

Through their blood-feeding habits, ticks are important as vectors of animal disease, transmitting a wide range of pathogenic viruses, rickettsia, bacteria and protozoa. In addition, many of the major diseases transmitted by ticks, such as the tick-borne encephalitides, Lyme borreliosis, relapsing fevers or Rocky Mountain spotted fever, are pathogenic to humans. Wild and domestic animals are particularly important as reservoirs of the organisms causing these diseases through an animal/tick/human cycle of contact.

Ticks are effective vectors because:

- they attach securely to their hosts, allowing them to be transferred to new habitats while on the host;
- the lengthy feeding period allows large numbers of pathogens to be ingested;
- feeding on a number of different hosts allows the transfer of pathogens from host to host;
- many species of tick are long lived;
- females lay large numbers of eggs and therefore tick populations have rapid potential for increase;
- if they fail to find a host they can survive for lengthy periods without feeding;
- ingested pathogens may be passed from larva to nymph and nymph to adult (transstadial) and may also be transferred from adult females to the next generation via the ovaries (transovarial).

Louping ill

This is a tick-borne disease, caused by a flavivirus, which affects the central nervous system of the host, leading to encephalomyelitis. It leads to rapid death in some cases or incoordination, ataxia, torticollis and paralysis followed by coma and death in others. Although it is primarily a problem associated with sheep in the United Kingdom and Ireland, where the primary vector is *Ixodes ricinus,* it may affect cattle, deer, dog and man. It is maintained primarily in a sheep/tick cycle, although ground-nesting birds, such as grouse, and rabbits or hares may also be important reservoirs of the pathogen. However, less than 1 in 1000 ticks usually carry the virus. The morbidity rate amongst lambs can be greater than 50%; however, most are protected against infection for the first 3 months of life by antibodies in colostrum. Lifelong immunity occurs following infection. The disease can be a problem in adult non-immune sheep introduced into affected areas; however, vaccination with an inactivated vaccine is possible.

African swine fever

This is an acute, contagious viral disease caused by an unclassified DNA virus. Originally confined to Africa, the disease spread to Portugal, Spain, Italy and France in the 1960s, causing substantial mortality. It may be transmitted by various species of the argasid genus *Ornithodoros*, in particular *Ornithodorus moubata porcinus*, and may cause high mortality in domestic pigs. Pigs initially have an asymptomatic pyrexia followed by cutaneous blue blotches, depression, anorexia, weakness and death. Vomiting and diarrhoea are variable signs.

Ehrlichiosis (tick-borne fever)

The agent of tick-borne fever in sheep and cattle in western Europe is the rickettsia-like parasite *Ehrlichia* (*Cytocetes*) *phagocytophila*. The infection is characterized by recurrent pyrexia, anorexia and lethargy. Infection in cattle can lead to respiratory distress, reduced milk yield and abortion in the later stages of preganancy. Exposure of naive pregnant ewes to tick-borne fever can also result in outbreaks of abortion, followed by septic metritis. Infection of lambs leads to an increased susceptibility to staphylococcal infections and louping ill. Most ticks carry the parasites, hence infection may be supported by very small tick populations.

A closely related rickettsia-like organism is *Ehrlichia canis*, a parasite of wild and domestic Canidae, which may cause disease in domestic dogs. The organism is spread by *Rhipicephalus sanguineus* and causes anaemia with thrombocytopenia and leucopenia in association with pyrexia.

Rocky Mountain spotted fever

This is a disease caused by *Rickettsia rickettsii* which has been recorded from almost every state in the USA, particularly eastern states, as well as parts of South America. It is transmitted primarily by the Rocky Mountain wood tick, *Dermacentor andersoni*, in western states and the American dog tick, *D. variabilis*, in eastern states. Vectors in the southern states (Texas and Oklahoma), Mexico and South America are *Amblyomma americanum*, *Rhipicephalus sanguineus* and *Amblyomma cajennense*, respectively. It is particularly important as a disease of humans but may have pathogenic significance for dogs, where it may cause fever and lethargy.

Other spotted fevers caused by members of the rickettsia are present in other parts of the world. In South Africa and southern Europe *R. conorii* causes spotted fevers with various local names. The vectors in South Africa responsible for speading this rickettsia are *Rhipicephalus sanguineus* and *Haemaphysalis leachi* and the dog is considered an important reservoir species in southern Europe. The tick-borne *R. australis* causes tick typhus in Queensland, Australia where the ixodid tick *Ixodes holocyclus* is the likely vector.

Q fever

The causative agent of Q fever is the rickettsial parasite *Coxiella burnetii,* which is enzootic in cattle, sheep and goats in many parts of the world. It is transmitted by ixodid ticks. The organism is relatively non-pathogenic in domestic animals but has been associated with anorexia and abortion.

Lyme borreliosis

Lyme disease or borreliosis is a disease caused by a tick-borne infection with the spirochaete, *Borrelia burgdorferi*. It is widespread in the USA and, in recent years, has become increasingly recognized in the UK, continental Europe, the former USSR and China. The disease is transmitted by various species of *Ixodes; I. ricinus* in north-western Europe, *I. dammini* in north-eastern USA, *I. pacificus* in western USA, *Amblyomma americanum* in eastern USA (New

Jersey), *I. persulcatus* in eastern Eurasia, China and Japan. In humans the disease is multisystemic and characterized by a local reaction (skin rash) with flu-like symptoms and subsequent development of polyradiculoneuritis and meningitis. The infection has been detected by serological antibody measurements in numerous animals, including dogs, cats and cattle; however, its significance is still unclear. In the dog clinical signs include pyrexia, lethargy, anorexia, lymphadenopathy and lameness due to arthritis. These signs are usually vague and, since the organism is difficult to culture and identify from tissue samples, a definitive diagnosis is often difficult.

Theileria

Protozoa of the genus *Theileria*, especially *T. parva,* are economically important parasites of domestic livestock, largely cattle, primarily in Afrotropical regions. They cause a range of diseases, including East Coast fever – a highly pathogenic disease of cattle in eastern Central and Southern Africa, causing high mortality in infected stock.

Babesiosis

Protozoan babesias cause pyrexia, severe haemolytic anaemia, haemoglobinuria and death of infected animals, particularly cattle. These diseases may be known commonly as Redwater fever and Texas fever. Many species of *Babesia* have been described, the most important of which are:

- *Babesia bovis,* in Central and South America, Asia, Europe and Australia, is spread in southern Europe by species of *Rhipicephalus* and in other areas by various species of *Boophilus.*
- *Babesia divergens*, which occurs in northern Europe and is spread by *Ixodes ricinus*. It causes anaemia and death in cattle.
- *Babesia bigemina*, which occurs in southern parts of the former USSR and southern Europe, an important vector of which is *Boophilus microplus.* It was also introduced into southern USA from which it has since been eradicated.

In southern Europe, Asia, South Africa and the USA, *Babesia canis* causes a disease known as malignant jaundice in dogs. In Europe it is carried by *Dermacentor marginatus* and *Rhipicephalus sanguineus*. A vaccine for dogs against *B. canis* is available in France.

Adult female ticks acquire infections during feeding which they transfer transovarially to their progeny. Cattle are then infected by feeding tick larvae. Calves may be protected in their first year by antibodies received in colostrum. Various other species of *Babesia* parasitize sheep (*B. ovis*) and horses (*B. cabali* and *B. equi*).

Bovine anaplasmosis

This is an important infection of cattle caused by the rickettsia-like parasites *Anaplasma marginale* and *A. centrale*, which infect the host's erythrocytes. Although originally tropical and sub-tropical in distribution, they now may be found worldwide. Anaplasmosis is an acute or chronic febrile infectious disease producing anaemia and jaundice which may result in high (30–50%) mortality in infected animals. However, the severity of the disease depends on the age of the host; the probability of death increasing with age. Species of *Boophilus*, *Rhipicephalus*, *Ixodes*, and *Dermacentor* have all been implicated as vectors.

3.4.3 Systemic effects: tick paralysis

The injection of a neurotoxin in the saliva of feeding female ticks can cause a disease known as tick paralysis. The toxin disrupts nerve synapses in the spinal cord and blocks neuromuscular junctions. There are about 40 or so species that may cause tick paralysis, each of which may possess a unique toxin. Tick paralysis is caused primarily by *Dermacentor andersoni* in western North America, *Dermacentor variabilis* in eastern North America and *Ixodes holocyclus* in Australia, *Ixodes ribicundus* and *Rhipicephalus evertsi* in South Africa. Tick paralysis results from the feeding of the female tick and the first signs begin approximately 5 days after attachment. In humans, paralysis of the lower limbs spreads to the rest of the body within a few hours. Tick paralysis is an important cause of death amongst sheep, cattle and goats in various parts of the world, including the USA, South Africa and Australia. Tick paralysis also occurs in poultry associated with larval feeding of some *Argas* spp.

3.4.4 Other systemic effects

Tick-bite wounds may allow the entry of bacteria, especially staphylococci, into the circulation, leading to the development of bacteriaemia and septicaemia. This is particularly important in lambs, where tick infestation resulting in staphylococcal bacteraemia can cause heavy losses. The organism spreads to affect many organ systems, including joints and the meninges of the brain and spinal cord, producing arthritis and meningitis, respectively.

3.5 CLASSIFICATION

In general, mites and ticks are considered to belong to the sub-class **Acari**, in the class **Arachnida**. The Acari is divided by some authors, into two orders, the **Parasitiformes** and the **Acariformes** and this is the form adopted here. The Parasitiformes are usually considered to include the sub-order **Metastigmata,** also called the **Ixodida,** known as the ticks, and the **Mesostigmata** (gamesid mites) which were discussed in Chapter 2. The Parasitiformes possess 1–4 pairs of lateral stigmata posterior to the coxae of the second pairs of legs and the coxae usually articulate freely with the body.

There are about 800 species of Metastigmata, divided into two major families of importance: the **Ixodidae**, known as the hard ticks, of which there are about 650 species, and the **Argasidae**, known as the soft ticks, of which there are about 150 species. Thirteen genera of Ixodidae are recognized, two of which are of major veterinary interest in temperate habitats: ***Ixodes*** and ***Dermacentor.*** *Ixodes* is the largest genus of hard tick, containing 217 species, which are widely distributed, particularly in the Old World. The genus *Dermacentor* is relatively small with about 30 species, most of which are found in the New World. A further four genera, ***Rhipicephalus, Haemaphysalis, Boophilus*** and ***Amblyomma*** contain one or more species of veterinary importance, and the genus ***Hyalomma*** is of minor interest in temperate habitats.

The family Argasidae contains four genera of which two contain species of veterinary importance: ***Argas*** and ***Otobius.*** The genus ***Ornithodoros*** is of minor interest in temperate habitats and the fourth genus, ***Antricola***, infests cave-dwelling bats in Central and North America.

3.6 RECOGNITION OF TICKS OF VETERINARY IMPORTANCE

Tick identification beyond family and genus is extremely difficult. This difficulty is exacerbated by the considerable variation that may exist within species and the problems of identifying immature stages. The guide presented below (modified from Varma, 1993) and the species descriptions given in the following pages are intended as a general guide to the ticks of veterinary interest in temperate habitats. For more detailed descriptions of species, their immature stages and the diseases they transmit, readers are directed to more specialist texts at the end of this chapter.

1 Hypostome with backwardly directed barbs; stigmatal shields present behind coxae of the fourth pair of legs or laterally above the coxae of legs 2 or 3; stigmata without peritremes; tarsi of the first pair of legs with a sensory pit **METASTIGMATA (IXODIDA)**

Gnathosoma projecting anteriorly and visible when specimen seen from above; scutum present, covering the dorsal surface completely (male) or the anterior portion only (female); stigmatal plates large, situated posteriorly to the coxae of the fourth pair of legs **IXODIDAE 2**

Gnathosoma ventral and not visible when adult is viewed from above; scutum absent; dorsal integument leathery; stigmatal plates small, situated anteriorly to the coxae of the fourth pair of legs; eyes, if present, in lateral folds **ARGASIDAE 11**

2 Anal groove surrounding the anus distinct, both anteriorly and posteriorly (Figs 3.10–3.12) ***Ixodes***

Anal groove entirely posterior to the anus **3**

3	Eyes absent	4
	Eyes present	5
4	Palps short and broad, about twice as wide as segment 2 with obvious outer angulation at the base (Fig. 3.9)	***Haemaphysalis***
5	Palps wider than long or, at most, only slightly longer than their width	6
	Palps much longer than wide	10
6	Basis capituli usually hexagonal dorsally (Fig. 3.9); medium-sized or small ticks, usually without colour patterns	7
	Basis capituli rectangular dorsally (Fig. 3.9); large ticks with definite colour patterns (Figs 3.13–3.15)	***Dermacentor***
7	Festoons absent; stigmatal plates round or oval; anal groove faint or obsolete	8
	Festoons present; stigmatal plate with a tail-like protrusion; anal groove distinct	9
8	Palps with dorsal and lateral ridges; male with normal legs	***Boophilus***
9	Basis capituli without pronounced lateral angles (Fig. 3.9); males with ventral plates; males with coxae of fourth pair of legs normal (Fig. 3.17)	***Rhipicephalus***
10	Palps with second segment less than twice as long as third segment (Fig. 3.9); scutum without pattern	***Hyalomma***
	Palps with second segment more than twice as long as third segment (Fig. 3.9); scutum with pattern; male without ventral plates (Fig. 3.18)	***Amblyomma***
11	Body periphery undifferentiated, without a definite suture distinguishing the dorsal from ventral surface	12
	Body surface flattened and usually structurally different from the dorsal surface, with a definite suture distinguishing dorsal and ventral surface (Fig. 3.19)	***Argas***

12 Adult integument is granular; hypostome vestigial; nymphal integument spiny; hypostome well developed (Fig. 3.20) ***Otobius***

Adult and nymph integument leathery; hypostome well developed ***Ornithodoros***

3.7 IXODIDAE

3.7.1 *Ixodes*

Ixodes is the largest genus in the family Ixodidae, with about 250 species. They are small, inornate ticks which do not have eyes or festoons (Fig. 3.10). The mouthparts are long and are longer in the female than male. The fourth segment of the palps is greatly reduced and bears chemoreceptor sensilla. The second segment of the palps may be restricted at the base, creating a gap between the palp and chelicerae (Fig. 3.9). Males have several ventral plates which almost cover the ventral surface. *Ixodes* can be distinguished from other ixodid ticks by the anterior position of the anal groove (Fig. 3.11). In other genera of the Ixodidae the anal groove is either absent or is posterior to the anus.

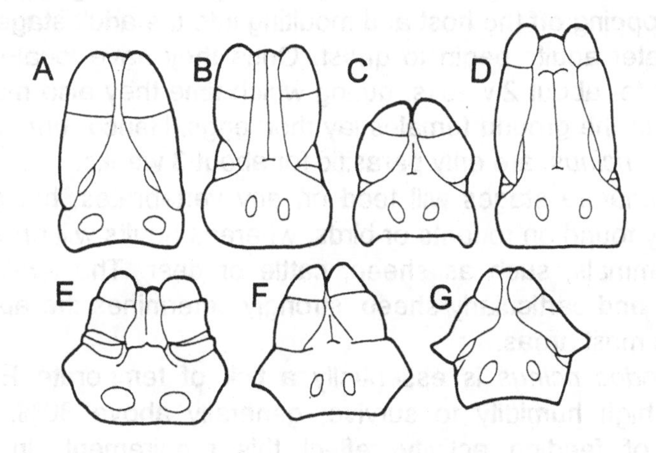

Fig. 3.9 Diagrammatic dorsal view of the gnathosoma of seven genera of ixodid ticks (from Varma, 1993). (A) *Ixodes*, (B) *Hyalomma*, (C) *Dermacentor*, (D) *Amblyomma*, (E) *Boophilus*, (F) *Rhipicephalus* and (G) *Haemaphysalis*.

Ixodes ricinus

The European sheep tick or castor bean tick, *Ixodes ricinus*, is an ectoparasite of major concern to farmers and veterinarians throughout northern and central Europe.

Morphology: adults are red-brown but females appear light grey when engorged. Males are about 2.5–3 mm in length and all four pairs of legs are visible. Unfed females are about 3–4 mm in length and up to 10 mm in length when engorged. The posterior internal angle of the coxa of the first pair of legs bears a spur which overlaps the coxa of the second pair of legs (Fig. 3.11). The tarsi are of moderate length (0.8 mm in the female and 0.5 mm in the male) and tapering (Fig. 3.12).

Life cycle: *Ixodes ricinus* is a three-host tick and its life cycle from egg to adult takes 3 years. Adult females lay 1000–2000 eggs in the soil. These hatch in the summer and larvae feed for the first time the following summer. Larvae are about 1 mm long, are yellowish in colour and have only three pairs of legs. Larvae feed for 3–5 days then drop back on to the vegetation where they digest their blood meal and moult to become nymphs. Nymphs are about 2 mm long, resemble the adults and have four pairs of legs. The nymphs begin to seek a new host after about 12 months, again feeding for 3–5 days, before dropping off the host and moulting into the adult stage. Twelve months later adults begin to quest. Once they have located a host they feed for about 2 weeks, during which time they also mate. After dropping to the ground females lay their eggs. Hence, during their 3-year life, *I. ricinus* are only parasitic for about 3 weeks.

Immature stages will feed on any vertebrates, but are most commonly found on rodents or birds, whereas adults will only feed on large mammals, such as sheep, cattle or deer. The availability of livestock, and particularly sheep, strongly determines the abundance of ticks in most areas.

Ixodes ricinus is essentially a tick of temperate Europe. It requires high humidity to survive, generally above 80%, and the patterns of feeding activity reflect this requirement. In general, feeding activity is most pronounced in April and May. Ticks begin to quest when temperatures rise above a critical threshold of about 7°C.

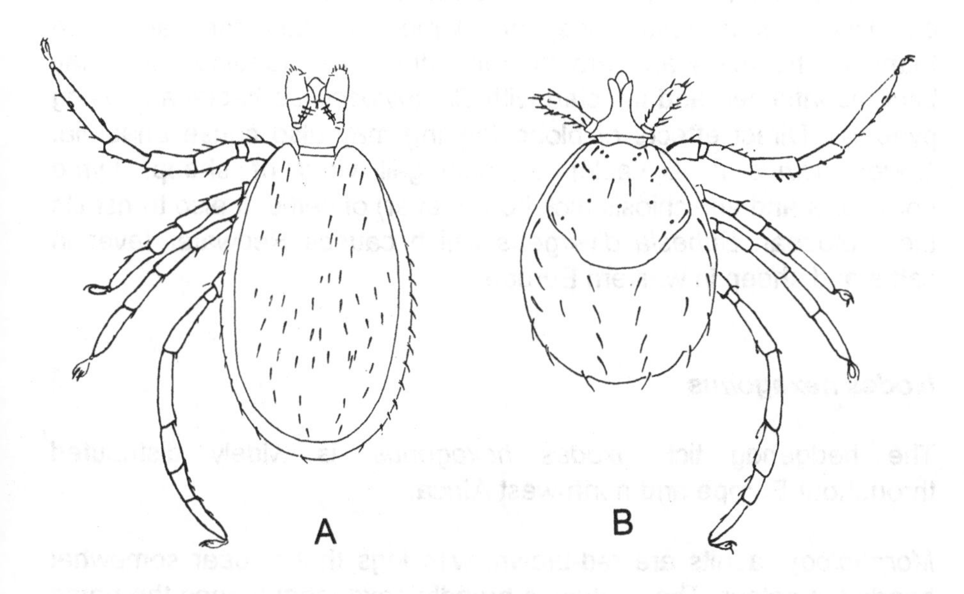

Fig. 3.10 Adult *Ixodes ricinus* in dorsal view, (A) male and (B) female (reproduced from Arthur, 1963).

This is followed by a short period of questing activity, lasting up to 8 weeks. Adults and nymphs start questing earlier than larvae. However, the duration of the active questing phase can be considerably shorter or longer, depending on the nature of the habitat. The high humidity requirements prevents them being active in summer, and ticks that fail to feed in spring desiccate and do not survive over summer. However, the precise pattern of seasonal activity is highly variable and is strongly influenced by life-cycle stage, habitat, climate and host availability. In many parts of its range, a proportion of the tick population feeds in August or September, overwinters as fed ticks and moults the following summer. This is seen particularly in the west of Britain.

As a result of its requirement for high humidity, in general, *I. ricinus* is associated with areas of deciduous woodland containing small mammals and deer. However, in areas with sufficient rainfall large populations may occur in moorland and meadows with rough grazing, where the majority feed on livestock.

Pathology: feeding *I. ricinus* are often found around the ears, eyelids and lips of dogs, cats, sheep and lambs. In cattle they are often found in the axilla and around the udder. The feeding sites may become inflamed and infected with *Staphylococcus* bacteria causing pyaemia. Direct effects of blood feeding may also cause anaemia. *Ixodes ricinus* is a vector of louping-ill virus of sheep, Lyme borreliosis and ehrlichiosis (tick-borne fever) of cattle. It also transmits the protozoan *Babesia divergens* which causes Redwater fever in cattle and sheep in western Europe.

Ixodes hexagonus

The hedgehog tick, *Ixodes hexagonus*, is widely distributed throughout Europe and north-west Africa.

Morphology: adults are red-brown, with legs that appear somewhat banded in colour. The scutum is broadly hexagonal (hence the name *hexagonus*) and, like *I. ricinus,* the coxae of the first pair of legs also bears a spur. However, the spur is smaller and does not overlap the

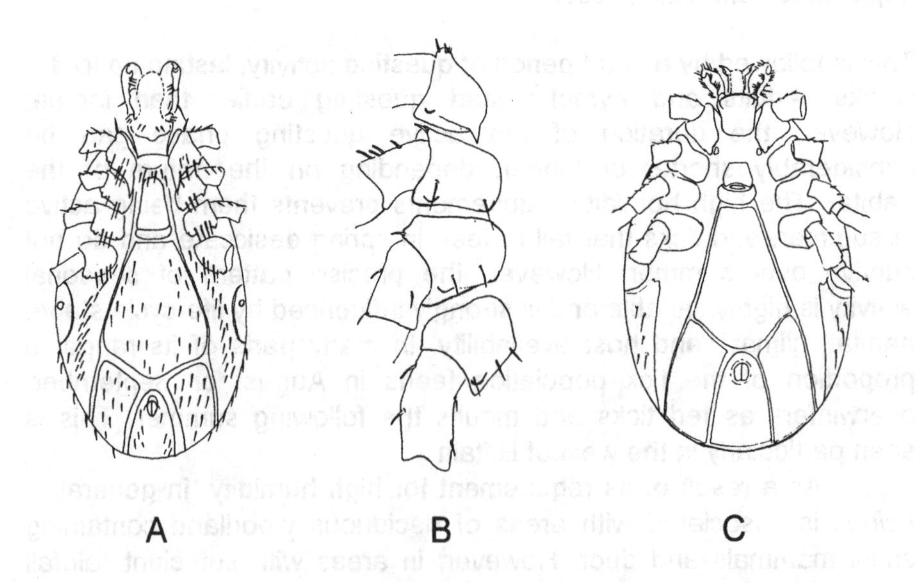

Fig. 3.11 Ventral view of the coxae of adult male (A) *Ixodes ricinus*, (B) *Ixodes hexagonus* and (C) *Ixodes canisuga* (reproduced from Arthur, 1963).

coxa of the second pair of legs (Fig. 3.11). When engorged the female may be up to 8 mm in length. Males are about 3.5–4 mm in length. The tarsi are long (0.8 mm in the female and 0.5 mm in the male) and sharply humped apically (Fig. 3.12).

Life history: this species is a three-host tick adapted to live with hosts which use burrows or nests. It is primarily a parasite of hedgehogs but may also be found on dogs and other small mammals. The life cycle is similar to that of *I. ricinus*: egg, hexapod larva, octopod nymph and adult, occurring over 3years. All life-cycle stages feed on the same host for periods of about 8 days. After dropping to the ground adult females produce 1000–1500 eggs over a period of 19–25 days before death. The ticks may be active from early spring to late autumn, but are probably most active during April and May.

Pathology: this species is occasionally found on dogs and cats, where adult females attach themselves behind the ears, on the jaws, neck and groin. Apart from localized dermatitis and the risk of wound infection, further adverse effects are unrecorded.

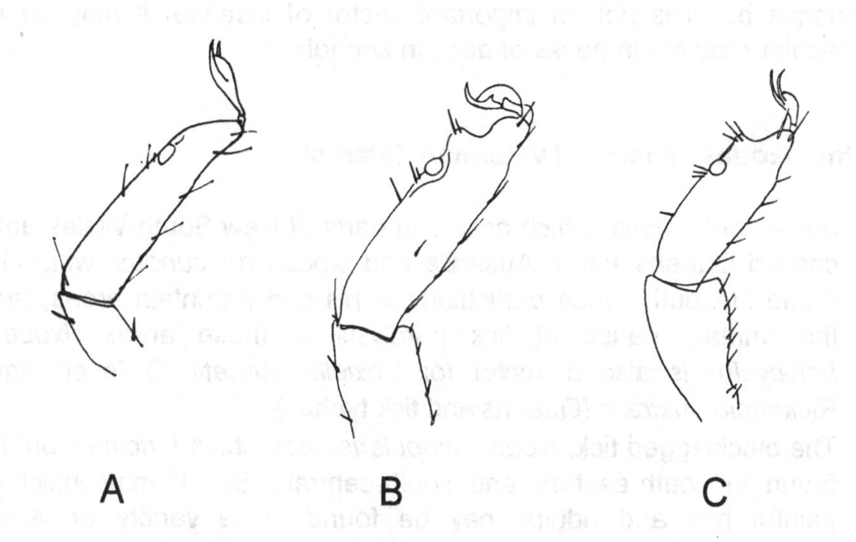

Fig. 3.12 The tarsi of adult male (A) *Ixodes ricinus*, (B) *Ixodes hexagonus* and (C) *Ixodes canisuga* (reproduced from Arthur, 1963).

Ixodes canisuga

The dog tick or *Ixodes canisuga* is found throughout Europe, possibly as far east as Russia.

Morphology: the adults are small and paler in colour than *I. ricinus,* usually being yellowish brown. Females are about 2 mm in length when unfed and up to 8 mm in length when full engorged. Females have no coxal spur and in males the spur is extremely small (Fig. 3.11). The tarsi are of moderate length (0.6 mm in the female and 0.35 mm in the male) and are broad with prominent subapical humps (Fig. 3.12).

Life cycle: this tick is adapted to life in a lair or a den. It may feed on a variety of hosts, including foxes and dogs. The life cycle is similar to that of *I. ricinus*: egg, hexapod lava, octopod, nymph and adult, and takes 3 years. Mating takes place in the den and adult males are only rarely found on the host. Adult females lay relatively small numbers of eggs, probably about 400.

Pathology: infestation may cause dermatitis, pruritus, alopecia and anaemia but it is not an important vector of disease. It may be a particular problem in packs of dogs in kennels.

Other *Ixodes* species of veterinary interest

- *Ixodes holocyclus,* which occurs in parts of New South Wales and coastal Queensland in Australia and *Ixodes rubicundus,* which is found in South Africa particularly in hill and mountain areas, are the primary cause of tick paralysis in these areas. *Ixodes holocyclus* is also a vector for *Coxiella burnetii* (Q fever) and *Rickettsia australis* (Queensland tick typhus).
- The blacklegged tick, *Ixodes scapularis*, resembles *I. ricinus*, but is found in south-eastern and south-central USA. It may inflict a painful bite and adults may be found on a variety of large mammals.
- *Ixodes dammini* is widely distributed in north-eastern USA and Canada. Immatures are parasites of rodents, whereas white tailed deer are the main hosts of adults. It is an important vector of Lyme borreliosis in eastern USA.

- The taiga tick, *Ixodes persulcatus*, is morphologically very similar *to I. ricinus*, but the female adult has a straight or wavy genital opening rather than arched as in *I. ricinus*. It has a similar life history to *I. ricinus*, although it has a more easterly distribution, being widespread throughout eastern Europe, Russia and as far east as Japan. *Ixodes persulcatus* is a major vector of the human diseases Russian spring–summer encephalitis virus and Lyme borreliosis.

3.7.2 *Dermacentor*

Ticks of the genus *Dermacentor* are medium-sized to large ticks, usually with ornate patterning. The palps and mouthparts are short and the basis capituli is rectangular (Fig. 3.9). Festoons and eyes are present. The coxa of the first pair of legs is divided into two sections in both sexes. Most species of *Dermacentor* are three-host ticks, but a few are one-host ticks. The males lack ventral plates and, in the adult male, the coxa of the fourth pair of legs is greatly enlarged. The genus is small with about 30 species, most of which are found in the New World.

Dermacentor variabilis

The American dog tick, *Dermacentor variabilis,* is particularly abundant in central and eastern states of the United States of America although its range may extend as far north as Canada.

Morphology: *Dermacentor variabilis* is an ornate tick, with a base colour of pale brown and a grey colour pattern (Fig. 3.13). Males are about 3–4 mm in length and females about 4 mm in length when unfed and 15 mm in length when engorged. The mouthparts are short. The basis capituli is short and broad (Fig. 3.9). The coxae of the first pair of legs have well-developed external spurs.

Life cycle: *Dermacentor variabilis* is a three-host tick. Adults are important parasites of wild and domestic carnivores. Adult females feed for 1–2 weeks, during which time mating occurs on the host. Females then drop to the ground where they lay approximately 4000–6500 eggs. After about 4 weeks the eggs hatch and larvae begin to

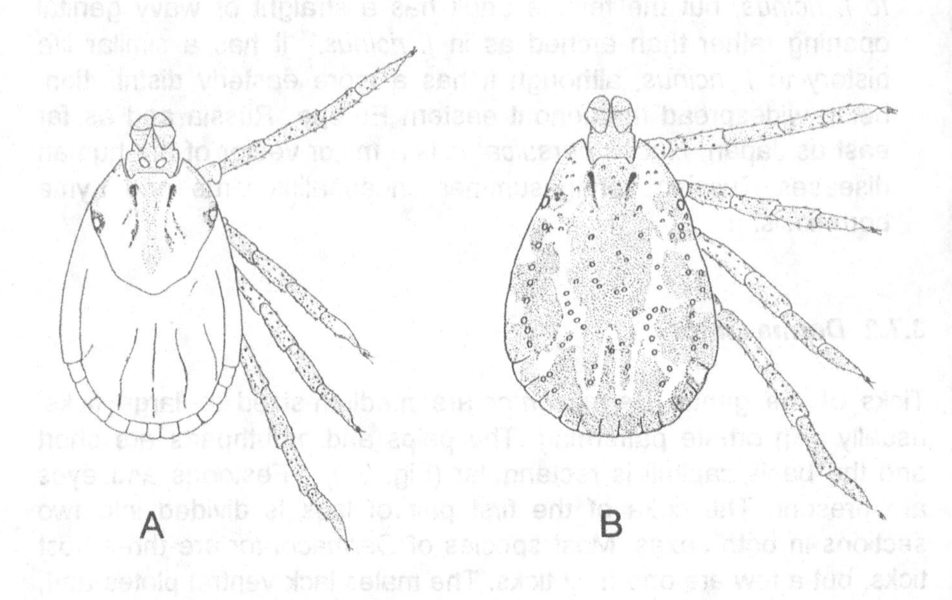

Fig. 3.13 Adult *Dermacentor variabilis*: (A) dorsal view of female, (B) dorsal view of male (reproduced from Arthur, 1962).

quest. Larvae feed for about 4–5 days, before dropping to the ground and moulting. Nymphs feed for about 5–6 days, before leaving the host. The larvae and nymphs usually feed on small rodents. Under favourable conditions, the life cycle from egg to adult may take only 3 months.

Pathology: *Dermacentor variabilis* is a common parasite of dogs and may cause tick paralysis. It occasionally parasitizes cattle and may transmit bovine anaplasmosis. It is also an important vector of *Rickettsia rickettsii* (Rocky Mountain spotted fever) in the USA.

Dermacentor andersoni

The Rocky Mountain wood tick, *Dermacentor andersoni*, is widely distributed throughout the western and central parts of North America from Mexico as far north as British Columbia.

Morphology: *Dermacentor andersoni* is an ornate tick, with a base colour of brown and a grey pattern (Fig. 3.14). Males are about 2–6 mm in length and females about 3–5 mm in length when unfed and 10–11 mm in length when engorged. The mouthparts are short. The basis capituli is short and broad (Fig. 3.9). The legs are patterned in the same manner as the body. The coxae of the first pair of legs have well-developed external and internal spurs.

Life cycle: it is a three-host tick. Adults feed largely on wild and domestic herbivores. Mating takes place on the host, following which females lay up to 6500 eggs over about 3 weeks. The eggs hatch in about 1 month, and the larvae begin to quest. Larvae feed for about 5 days, before dropping to the ground and moulting to the octopod nymphal stage. Larvae and nymphs feed largely on rodents. One- and two-year population cycles may occur. Eggs hatch in early spring and individuals that are successful in finding hosts pass through

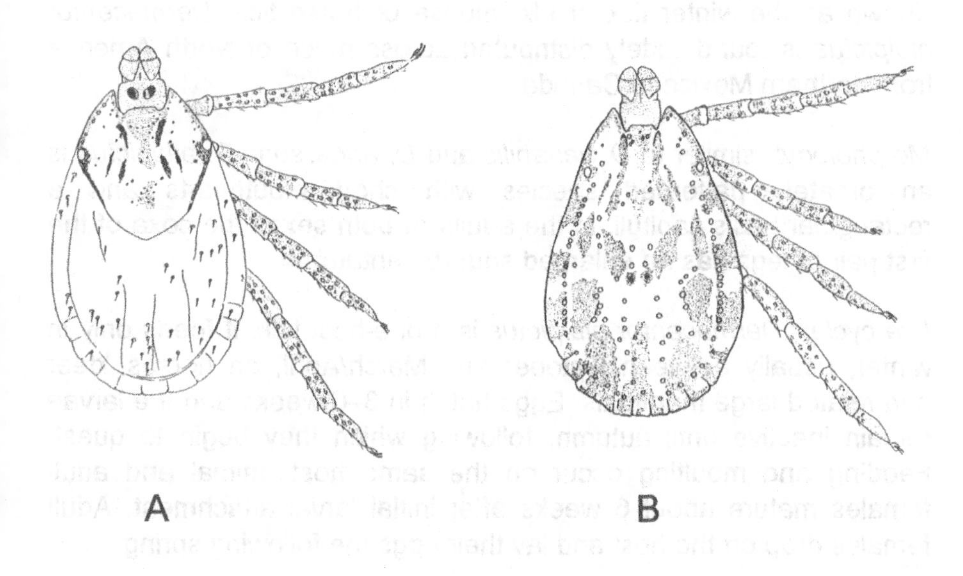

Fig. 3.14 Adult *Dermacentor andersoni*: (A) dorsal view of female; (B) dorsal view of male (reproduced from Arthur, 1962).

their larval stages in spring, their nymphal stages in late summer and then overwinter as adults in a 1-year cycle. Nymphs that fail to feed, overwinter and form a spring-feeding generation of nymphs the following year. Unfed nymphs may survive for up to a year. *Dermacentor andersoni* is most common in areas of scrubby vegetation, since these attract both the small mammals required by the immature stages and the large herbivorous mammals required by the adults.

Pathology: in cattle, *Dermacentor andersoni* may cause tick paralysis, particularly in calves, and may be responsible for the transmission of bovine anaplasmosis, caused by *Anaplasma marginale*. It is also the chief vector of *Rickettsia rickettsii* (Rocky Mountain spotted fever) in western USA.

Dermacentor albipictus

Known as the winter tick or elk, moose or horse tick, *Dermacentor albipictus* is found widely distributed across much of North America from northern Mexico to Canada.

Morphology: similar to *D. variabilis* and *D. andersoni*, *D. albipictus* is an ornately patterned species with short mouthparts and a rectangular basis capituli. In the adults of both sexes the coxa of the first pair of legs has an enlarged spur (bidentate).

Life cycle: *Dermacentor albipictus* is a one-host tick. It feeds only in winter, usually between October and March/April, on horses, deer and related large mammals. Eggs hatch in 3–6 weeks and the larvae remain inactive until autumn, following which they begin to quest. Feeding and moulting occur on the same host animal and adult females mature about 6 weeks after initial larval attachment. Adult females drop off the host and lay their eggs the following spring.

Pathology: Heavy infestations with *D. albipictus* may cause death from anaemia and it may be a vector of *Rickettsia rickettsii* (Rocky Mountain spotted fever).

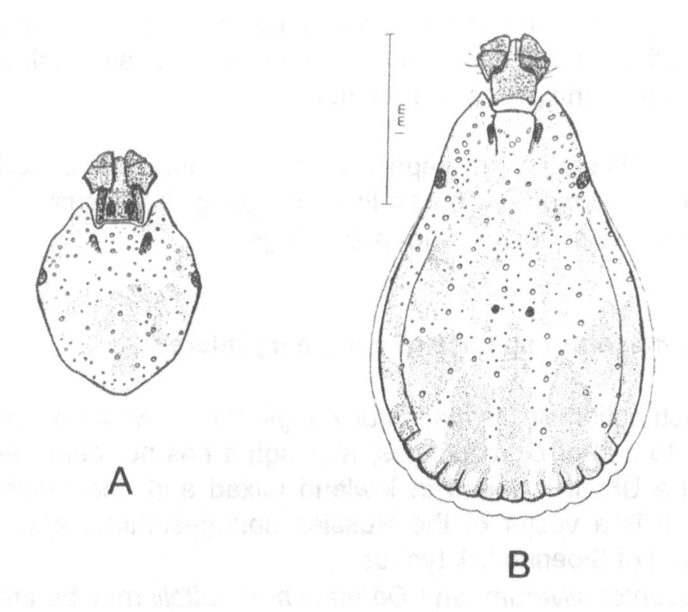

Fig. 3.15 Dorsal view of the gnathosoma and scutum of adult female (A) and male (B) *Dermacentor reticulatus* (reproduced from Arthur, 1962).

Dermacentor reticulatus

This species occurs in many parts of western Europe, extending eastwards to Kazakhstan, particularly in wooded areas.

Morphology: both sexes are ornate, white with variegated brown splashes (Fig. 3.15). In the adult female the ornate scutum is about 1.6 mm in length. In the adults of both sexes the coxa of the first pair of legs has an enlarged spur (bidentate). The other coxae have short internal spurs that become progressively smaller in legs 2–4. The coxae of the fourth pair of legs are large with a narrow tapering external spur.

Life cycle: *Dermacentor reticulatus* is a three-host species. Adults parasitize larger domestic and wild mammals, cattle, horses, sheep, goats and pigs. Copulation takes place on the host. Females feed for 9–15 days. Oviposition may last 6–25 days, during which a female may produce 3000–4500 eggs. Nymphs hatch from the egg after about a month and within about 2 weeks begin questing. The feeding activity of nymphs can last from midsummer to late autumn. Immature stages feed on a variety of small mammals, such as small rodents and carnivores, and occasionally birds.

Pathology: it is particularly important as an ectoparasite of cattle and may be found along their backs in early spring. It transmits *Babesia* infections to cattle, horses, sheep and dogs.

Other *Dermacentor* species of veterinary interest

- The distribution of *Dermacentor marginatus* in western Europe is similar to that of *D. reticulatus,* although it has not been recorded from the UK. It is found in lowland mixed and deciduous forest areas. It is a vector of the Russian spring–summer encephalitis virus and of Siberian tick typhus.
- *Dermacentor silvarum* and *Dermacentor nuttalli* may be important ectoparasites in Eurasia, with *D. silvarum* being more common in the west and *D. nuttalli* in the east.
- *Dermacentor occidentalis*, the Pacific Coast tick, is found in the south-western states of the USA, where adults may be found on larger domestic animals. It may be a vector of bovine anaplasmosis.

3.7.3 *Haemaphysalis*

Ticks of the genus *Haemaphysalis* inhabit humid, well-vegetated habitats in Eurasia and tropical Africa. They are three-host ticks, with the larvae and nymphs feeding on small mammals and birds and adults infesting larger mammals and, importantly, livestock. Most species of the genus are small, with short mouthparts and a rectangular basis capituli (Fig. 3.9). There are about 150 species, found largely in the Old World, with only two species found in the New World.

Haemaphysalis punctata

This species is widely distributed throughout Europe, Scandinavia, North Africa and central Asia.

Morphology: *Haemaphysalis punctata* is a small, inornate tick, with festoons, no eyes and short mouthparts (Fig. 3.16). Sexual dimorphism is not pronounced. The adults of both sexes are about 3 mm in length, the female reaching about 12 mm in length when engorged. The basis capituli is rectangular, about twice as broad as long. The sensory palps are short and broad, with the second segment extending beyond the basis capituli. The males have no ventral shields. The anal groove is posterior to the anus. The coxae of the first pair of legs have a short, blunt internal spur, which is also present on the coxae of the second and third pair of legs and which is enlarged and tapering on the coxae of the fourth pair of legs, particularly in the male where the spur may be as long as the coxa (Fig. 3.16).

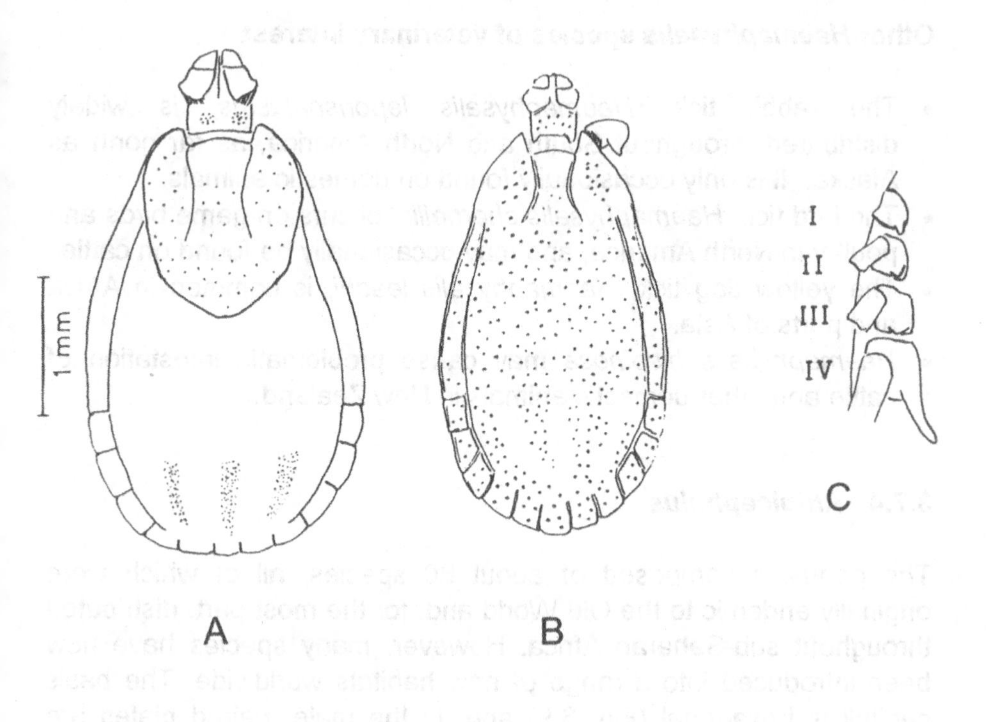

Fig. 3.16 Dorsal view of the gnathosoma and scutum of adult female (A) and male (B) *Haemaphysalis punctata*. (C) Ventral view of the coxae of an adult male (reproduced from Arthur, 1963).

Life cycle: *Haemaphysalis punctata* is a three-host tick, feeding once in each of the larval, nymphal and adult life-cycle stages. After each feed it drops from the host. Engorgement on the host may take 6–30 days to complete and each female lays 3000–5000 eggs. Large mammals, particularly cattle and sheep, are the preferred hosts for adults. Larvae and nymphs are more commonly found on small mammals and birds.

Pathology: in southern England and Wales *H. punctata* may be responsible for the transmission of *Babesia major* in cattle and *B. motasi* in sheep; in other areas of Europe it transmits *B. bigemina* in cattle, *B. motasi* in sheep and *Anaplasma marginale* and *A. centrale* in cattle. It may also cause tick paralysis.

Other *Haemaphysalis* species of veterinary interest

- The rabbit tick, *Haemaphysalis leporispalustris,* is widely distributed throughout South and North America, as far north as Alaska. It is only occasionally found on domestic animals.
- The bird tick, *Haemaphysalis chordeilis,* occurs on game birds and poultry in North America, and may occasionally be found on cattle.
- The yellow dog tick, *Haemaphysalis leachi,* is common in Africa and parts of Asia.
- *Haemaphysalis bispinosa* may cause problematic infestation of cattle and other domestic animals in New Zealand.

3.7.4 *Rhipicephalus*

The genus is composed of about 60 species, all of which were originally endemic to the Old World and, for the most part, distributed throughout sub-Saharan Africa. However, many species have now been introduced into a range of new habitats worldwide. The basis capituli is hexagonal (Fig. 3.9) and, in the male, paired plates are found on each side of the anus. They infest a variety of mammals but seldom birds or reptiles. Most species are three-host ticks but some species of the genus are two-host ticks. The only species of veterinary interest in temperate habitats is *Rhipicephalus sanguineus.*

Rhipicephalus sanguineus

The brown dog tick, *Rhipicephalus sanguineus,* is found worldwide.

Morphology: yellow, reddish or blackish brown ticks with hexagonal basis capituli and short mouthparts. They are usually inornate with eyes and festoons present (Fig. 3.17). Unfed adults may be 3–4.5 mm in length, but size is highly variable. The coxa of the first pair of legs has a large external spur (bifurcate). The legs may become successively larger from the anterior to the posterior pair. The tarsi of the fourth pair of legs possess a marked ventral tarsal hook (Fig. 3.17).

Life cycle: *Rhipicephalus sanguineus* is a three-host tick with a wide host range. However, it is particularly associated with dogs. Eggs are deposited in the cracks of kennel walls or other areas used frequently by dogs. Larvae hatch from the egg within 20–30 days and begin to

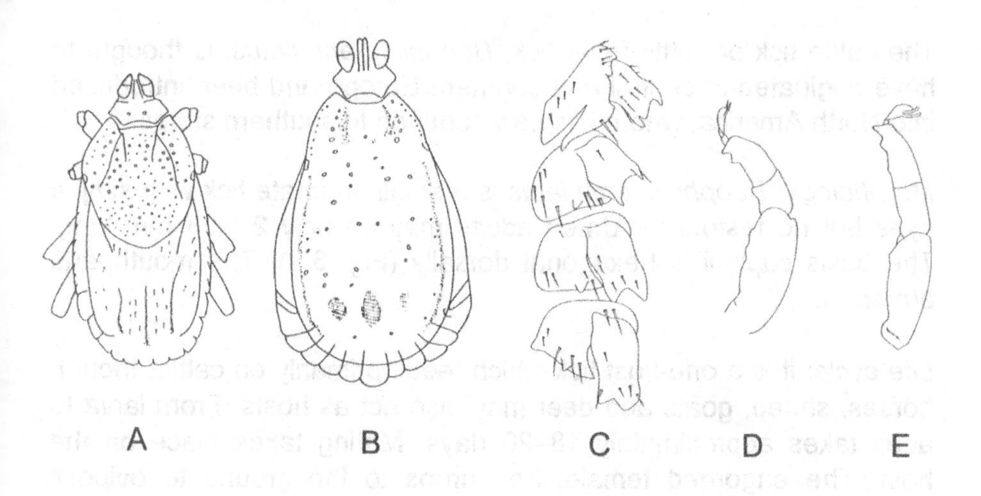

Fig. 3.17 Dorsal view of the gnathosoma and scutum of adult female (A) and male (B) *Rhipicephalus sanguineus*. (C) Ventral view of the coxae and trochanters of an adult male. Tarsi and metatarsi of the fourth pair of legs of an adult male (D) and female (E) (reproduced from Arthur, 1962).

quest shortly after. Larvae, nymphs and adults may infest the same host. It survives protected within the domestic environment; without such protection it would not be able to survive in temperate habitats.

Pathology: on dogs *Rhipicephalus sanguineus* is often found in the ears and between the toes. It is a vector of *Babesia canis*, causing haemolytic anaemia and jaundice in dogs. It transmits East Coast fever among cattle and the virus of Nairobi sheep disease in East Africa. It is also a vector of Rocky Mountain spotted fever in some areas of the USA and Mexico.

3.7.5 *Boophilus*

This small genus contains only five species which are often known as 'blue ticks'. They are one-host ticks and are found predominantly in tropical and sub-tropical habitats. The only species of veterinary interest in temperate habitats is *Boophilus annulatus*.

Boophilus annulatus

The cattle tick or cattle fever tick, *Boophilus annulatus*, is thought to have originated in central and southern Europe and been introduced into North America, where it is now confined to southern states.

Morphology: *Boophilus annulatus* is a small, inornate tick with simple eyes but no festoons. Unfed adults may be only 2 or 3 mm long. The basis capituli is hexagonal dorsally (Fig. 3.9). The mouthparts are short.

Life cycle: it is a one-host tick which feeds primarily on cattle, though horses, sheep, goats and deer may also act as hosts. From larva to adult takes approximately 18–20 days. Mating takes place on the host. The engorged female then drops to the ground to oviposit 2000–3000 eggs. The entire life cycle may be completed in as little as 6 weeks and between two and four generations may occur per year, depending on climate.

Pathology: it is an important vector of Texas cattle fever caused by *Babesia bigemina* and *B. bovis*.

Other *Boophilus* species of veterinary interest

- *Boophilus microplus,* the tropical or southern cattle tick, is widely distributed in the southern hemisphere and the southern states of the USA. It is a one-host tick which parasitizes cattle and other herbivores. It is an important vector of bovine babesiosis.

3.7.6 *Amblyomma*

Members of this genus are large, often highly ornate, ticks with long, often banded, legs. Unfed females may be up to 8 mm in length and when engorged may reach 20 mm in length. Eyes and festoons are present. Males lack ventral plates. They have long mouthparts with which they can inflict a deep, painful bite which may become secondarily infected. There are about 100 species of *Amblyomma*, largely distributed in tropical and sub-tropical areas of Africa. However, one important species is found in temperate North America.

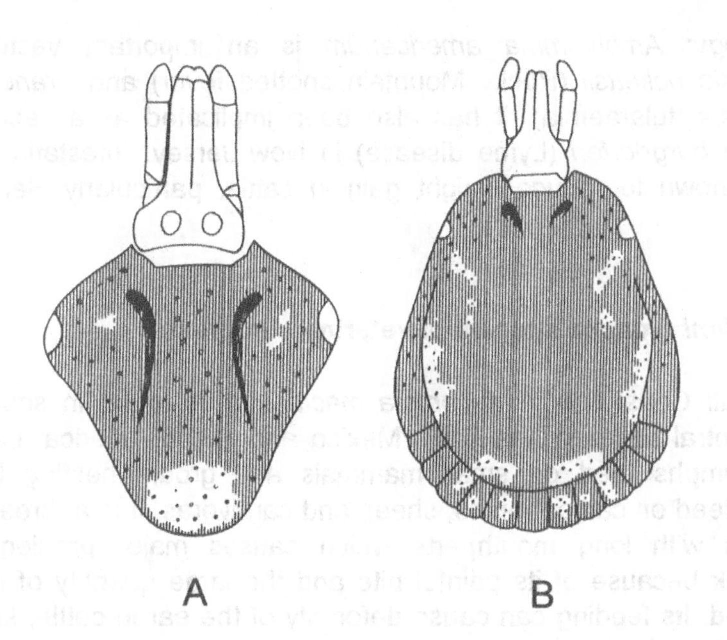

Fig. 3.18 Dorsal view of the gnathosoma and scutum of adult female (A) and male (B) *Amblyomma americanum* (reproduced from Arthur, 1962).

Amblyomma americanum

The lone star tick, *Amblyomma americanum*, so called because of a single white spot on the scutum of the female, is widely distributed throughout central and eastern USA.

Morphology: the female is reddish brown in colour. On the scutum are two deep parallel cervical grooves and a large pale spot at its posterior margin (Fig. 3.18). The male is small with two pale symmetrical spots near the hind margin of the body, a pale stripe at each side and a short oblique pale stripe behind each eye. In both sexes, coxa I has a long external spur and a short internal spur.

Life cycle: it is a three-host tick. Larvae and nymphs feed on rodents, rabbits and ground-inhabiting birds. Adults feed on larger mammals such as deer, cattle, horses and sheep. Feeding larvae, nymphs and adults are active between early spring and late summer in distinct periods corresponding with the feeding activity of each stage. There is usually a single generation per year. It is particularly common in wooded areas.

Pathology: *Amblyomma americanum* is an important vector of *Rickettsia rickettsii* (Rocky Mountain spotted fever) and *Francisella tularensis* (tularaemia). It has also been implicated as a vector of *Borrelia burgdorferi* (Lyme disease) in New Jersey. Infestation has been shown to reduce weight gain in cattle, particularly Hereford steers.

Other *Amblyomma* species of veterinary interest

The Gulf Coast tick, *Amblyomma maculatum*, is found in southern and central states of the USA, Mexico and South America. Larvae and nymphs feed on small mammals and ground-nesting birds. Adults feed on cattle, horses, sheep and carnivores. It is a three-host species with long mouthparts which causes major problems in livestock because of its painful bite and the large quantity of blood removed. Its feeding can cause deformity of the ear in cattle, known as 'gotch ear'.

3.7.7 *Hyalomma*

This is a genus of minor veterinary importance in temperate habitats. Species of this genus are medium-sized or large ticks, with eyes and long mouthparts. The males have ventral plates on each side of the anus. About 20 species are found, usually in semi-desert lowlands of central Asia, southern Europe and North Africa. They can survive exceptionally cold and dry conditions. Adults of a number of species parasitize domestic mammals. *Hyalomma aegyptium* may be introduced on tortoises.

3.8 ARGASIDAE

3.8.1 *Argas*

Species of the genus are usually dorsoventrally flattened, with definite margins which can be seen even when the tick is engorged. The cuticle is wrinkled and leathery. Most species are nocturnal and are parasites of birds, bats, reptiles or, occasionally, small insectivorous mammals. Species of this genus are usually found in dry, arid habitats.

Argas persicus

The fowl tick, *Argas persicus,* is of considerable veterinary importance as a parasite of poultry and wild birds. It originated in the Palaearctic but has been introduced with chickens into most parts of the world and is now found throughout Europe, Asia and North America. However, in North America a number of very closely related species *Argas sanches*, *Argas radiatus* and *Argas miniatus* may also be present.

Morphology: the unfed adult is pale yellow, turning reddish brown when fed. The female is about 8 mm in length and the male about 5 mm. The margin of the body appears to be composed of irregular quadrangular plates or cells and the hypostome is notched at the tip (Fig. 3.19).

Life cycle: it is nocturnal and breeds and shelters in cracks and crevices in poultry houses. Females deposit batches of 25–100 eggs in cracks and crevices in the structure of a poultry house. Up to 700 eggs may be produced by a single female at intervals, each oviposition preceded by a blood meal. After hatching, larvae locate a host and remain attached and feed for several days. After feeding they detach, leave the host and shelter in the poultry house structure. Several days later they moult to become first-stage nymphs. They then proceed through two or three nymphal stages, interspersed with frequent nightly feeds, before moulting to the adult stage. Under favourable conditions the life cycle can be completed in about 30 days.

Pathology: heavy infestations can kill birds through anaemia. It is a

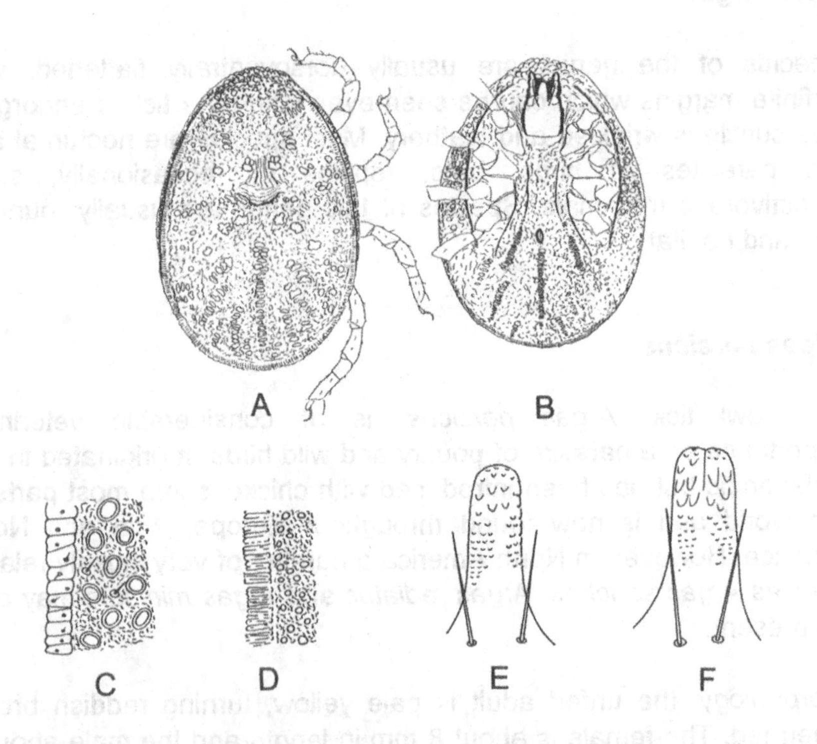

Fig. 3.19 Female *Argas reflexus:* (A) dorsal and (B) ventral view (reproduced from Arthur, 1962). Margin of *Argas reflexus* (C) and *Argas persicus* (D). Hypostome of female *Argas reflexus* (E) and *Argas persicus* (F) (reproduced from Arthur, 1963).

vector of *Borrelia anserina* and *Aegyptianella pullorum* among poultry, as well as avian spirochaetosis. This tick can undergo rapid increases in abundance, passing through 1–10 generations per year, particularly where birds are present all year round.

Argas reflexus

This species is known as the pigeon tick because of its close association with this host. It is abundant in the Middle- and Near-East, whence it has spread into Europe and most of Asia.

Morphology: the adult *Argas reflexus* is between 6 and 11 mm in length and may be distinguished from the fowl tick, *Argas persicus,* by its body margin, which is composed of irregular grooves, and the hypostome, which is not notched apically (Fig. 3.19). It is reddish brown in colour with paler legs.

Life cycle: the life cycle is similar to that of *A. persicus.* It is nocturnal and during the day lives in crevices in the pigeon house or nest material. It can withstand prolonged periods of starvation.

Pathology: heavy infestations may cause death from anaemia. It may also transmit fowl spirochaetosis.

3.8.2 *Otobius*

This small genus contains only two species: *Otobius megnini* and *Otobius lagophilus*.

Otobius megnini

The spinose ear tick is found through western and south-western North America and Canada, where it originated. It has subsequently been introduced to southern Africa and India.

Morphology: the adult body is rounded posteriorly and slightly attenuated anteriorly. Adult females range in size from 5 to 8 mm in length; males are slightly smaller. They have no lateral sutural line and no distinct margin to the body. Nymphs have spines (Fig. 3.20). In adults the hypostome is much redùced and the integument is granular.

Life cycle: *Otobius megnini* uses a single host, primarily cattle or horses, but it may also attack sheep or domestic cats and dogs as well as wild herbivores and canines. It is often associated with stables and animal shelters. Females lay their eggs in cracks and crevices in the walls of animal shelters, under stones or the bark of trees. Larvae attach to a passing host and crawl deep into the outer ear. Here they feed and moult to first- and second-stage nymphs. From 5 weeks to several months may be spent on the host during the larval and nymphal stages. The second-stage nymphs leave the ear and seek the shelter of a suitably protected site where they moult to become adult. Adults do not feed.

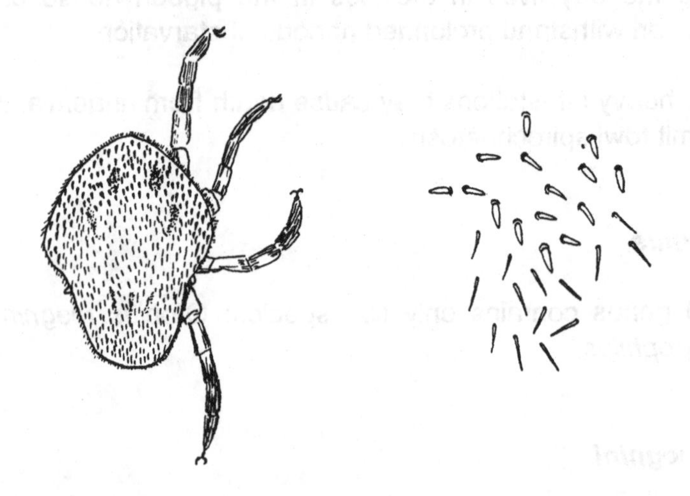

Fig. 3.20 Dorsal view of nymphal *Otobius megnini* and part of the integument showing hairs and spines (reproduced from Arthur, 1962).

Pathology: the larvae and nymphs feed in the external ear canal of the host, producing a severe otitis externa. Affected animals present palpation around the ears and may even develop convulsions.

Otobius lagophilus

Otobius lagophilus is an ectoparasite of rabbits in western USA and Canada.

3.8.3 *Ornithodoros*

This genus includes about 90 species, almost all of which are found in tropical and sub-tropical habitats in both the Old and New World and are of only minor importance as parasites of domestic animals in temperate habitats. They are nocturnal and the mouthparts are well developed. The integument has a wrinkled pattern which runs continuously over the dorsal and ventral surfaces. There is no distinct lateral margin to the body, which appears sac-like. Species of this genus are found largely in habitats such as dens, caves, nests and burrows, and so are not normally a problem for most domestic animals. Three species, *O. hermsi* (of the Rocky Mountains and Pacific Coast of North America), *O. parkeri* and *O. turicata* (in western states of the USA) attack rodents and are vectors of various species of *Borrelia*. *Ornithodoros coriaceus* of south-western North America feeds on cattle and deer, inflicting a painful bite. In parts of Africa *O. savignyi* and the *O. moubata* species complex are important ectoparasites of cattle and humans.

3.9 FURTHER READING

Arthur, D.R. (1962) *Ticks and Disease*, International Series of Monographs on Pure and Applied Biology, Pergamon Press, London, Vol. 9.

Arthur, D. R. (1963) *British Ticks*, Butterworths, London.

Baker, E.W., Camin, J.H., Cunliffe, F. *et al.* (1958) *Guide to the Families of Mites*, Institute of Acarology, Contribution No. 3.

Bennett, C.E. (1995) Ticks and Lyme disease. *Advances in Parasitology*, **36**, 343–405.

Herms, W.B. and James, M.T. (1961) *Medical Entomology*, MacMillan, New York.
Hoogstraal, H. (1966) Ticks in relation to human diseases caused by viruses. *Annual Review of Entomology*, **11**, 261–308.
Hoogstraal, H. (1967) Ticks in relation to human diseases caused by *Rickettsia* species. *Annual Review of Entomology*, **12**, 377–420.
Klomper, J.S.H., Black, W.C., Keirans, J.E. and Oliver, J.H. Jr (1996) Evolution of ticks. *Annual Review of Entomology*, **41**, 141–62.
Lane, R.S., Piesman, J. and Burgdorfer, W. (1991) Lyme borreliosis: relation of its causative agent to its vectors and hosts in North America and Europe. *Annual Review of Entomology*, **36**, 587–609.
Muller, G.H., Kirk, R.W. and Scott, D.W. (1989) *Small Animal Dermatology*, W.B. Sauders, Philadelphia.
Murnaghan, M.F. and O'Rourke, F.J. (1978) Tick paralysis in *Arthropod Venoms* (ed. S. Bettini), Springer-Verlag, New York, pp. 419–64,
Needham, G.R. and Teal, P.D. (1991) Off-host physiological ecology of ixodid ticks. *Annual Review of Entomology*, **36**, 659–81.
Oliver, J.H. Jr (1989) Biology and systematics of ticks (Acari: Ixodidae). *Annual Review of Ecology and Systematics*, **20**, 397–430.
Radostits, O.M., Blood, D.C. and Gay, C.C. (1994) *Veterinary Medicine – a Textbook of the Diseases of Cattle, Sheep, Pigs and Horses*, Ballière Tindall, London.
Sonenshine, D.E. (1986) Tick pheromones. *Current Topics in Vector Research*, **2**, 225–63.
Urquhart, G.M., Armour, J., Duncan, J.L. *et al.* (1987) *Veterinary Parasitology*, Longman Scientific and Technical, London.
Varma, M.G.R. (1993) Ticks and Mites (Acari), in *Medical Insects and Arachnids* (eds R.P. Lane and R.W. Crosskey), Chapman & Hall, London, pp. 597–658.
Waladde, S.M. and Rice, M.J. (1982) The sensory basis of tick feeding behaviour. *Current Themes in Tropical Science*, **1**, 71–118.
Wikel, S.K. (1996) Host immunity to ticks. *Annual Review of Entomology*, **41**, 1–22.

4

Adult flies (Diptera)

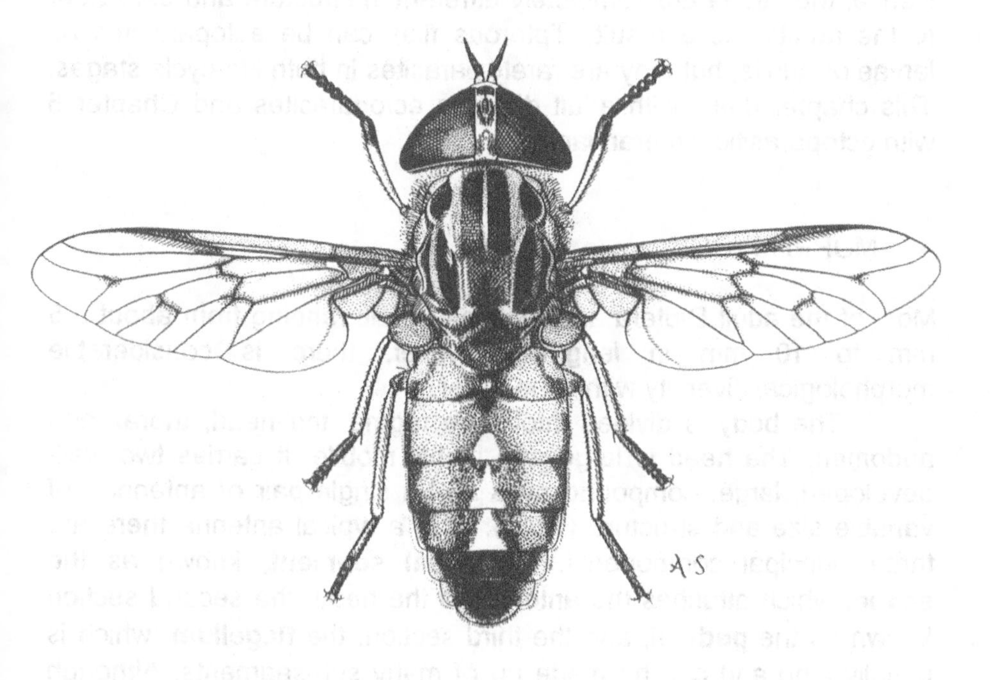

Adult *Tabanus fraternus* (Diptera: Tabanidae) (from Chainey, 1993).

4.1 INTRODUCTION

The Diptera are the true flies. The word 'Diptera' is derived from the ancient Greek *di pteron*, meaning two winged. In most of the orders of winged insect, adults have two pairs of wings. However, the Diptera have only one pair, the hind pair of wings having been reduced considerably to become small, club-like organs called **halteres**.

The Diptera is one of the largest orders in the class Insecta, with over 120,000 described species. All these species have a complex life cycle with complete metamorphosis (see section 1.7.2). Hence, the larvae are completely different in structure and behaviour to the adults. As a result, dipterous flies can be ectoparasites as larvae or adults, but they are rarely parasites in both life-cycle stages. This chapter deals with adult dipteran ectoparasites and Chapter 5 with ectoparasitic dipteran larvae.

4.2 MORPHOLOGY

Most of the adult Diptera are relatively small, ranging from about 0.5 mm to 10 mm in length. However, there is considerable morphological diversity within the order.

The body is divided into three tagma, the head, thorax and abdomen. The head is large and highly mobile. It carries two well-developed, large, compound eyes and a single pair of antennae of variable size and structure (Fig. 4.1). In a typical antenna there are three principal components: the **basal segment**, known as the **scape**, which attaches the antenna to the head, the second section known as the **pedicel**, and the third section, the **flagellum**, which is usually long and can be made up of many sub-segments. Although most have this basic design, the antennae can take on a wide variety of forms and are of considerable importance in the identification and taxonomy of Diptera.

In many dipteran species, especially in male flies, the eyes often meet in the front. This is described as the **holoptic** condition. In, other flies, particularly females, a strip of the head called the **frons** separates the eyes, producing what is known as the **dichoptic** condition. The top of the head commonly bears three simple eyes, known as **ocelli**, arranged in a triangle, although they are absent in

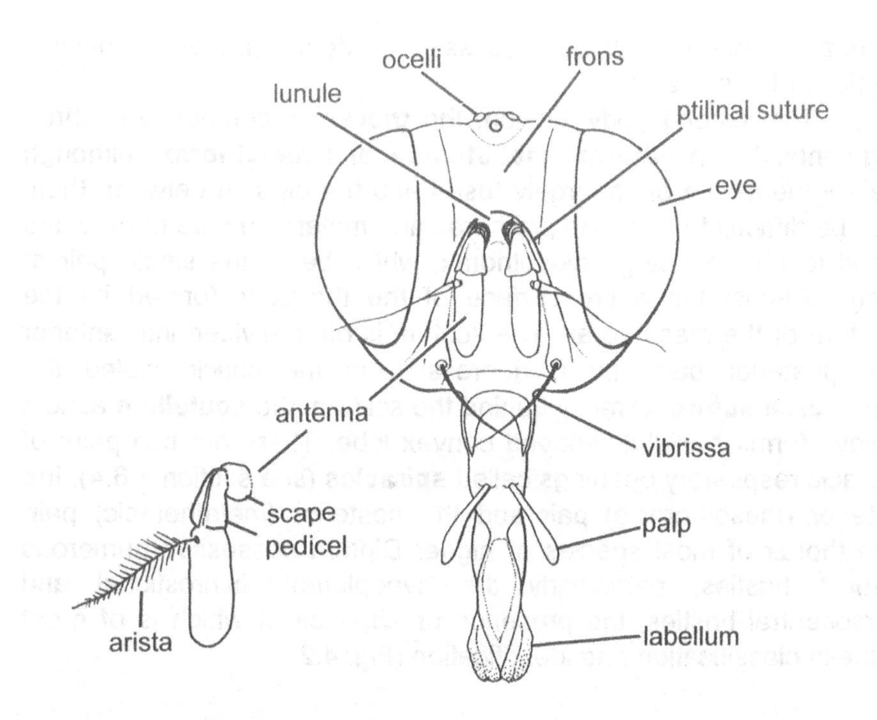

Fig. 4.1 The principal features of the dichoptic head of a typical adult calypterate cyclorrhaphous dipteran (reproduced from Crosskey, 1993a).

some families. In the more advanced dipteran families, there is a conspicuous groove, or **suture**, which marks the position of the **ptilinum**. The ptilinum is an expandable sac which these insects inflate with haemolymph and use to burst out of the pupal case, known as the **puparium.** The ptilinum may then be used to help the newly emerged adult to tunnel out of the ground, where the pupa has remained buried and hidden during pupation. Once the adult fly has emerged and inflated its wings, the sac is withdrawn into the head and the suture is closed.

The fly mouthparts are suspended below the head. The flies are all primarily liquid feeders. The basic feeding apparatus consists of a pair of **maxillae,** a pair of **mandibles**, the **labium** with terminal **labella, hypopharynx** and the **labrum**. However, there is great variation in their feeding habits and in the form of their mouthparts. A small number of flies of veterinary importance have poorly developed mouthparts; in other species there are no functional mouthparts and the adults do not feed at all. The mouthparts of the various groups of

Diptera of interest will be discussed in detail in their respective sections of this chapter.

The second body section, the **thorax**, is composed of three segments, the **prothorax**, **mesothorax** and **metathorax**, although the segments are often largely fused and the division between them may be difficult to see. The prothorax and metathorax are narrow and fused to the very large mesothorax, which bears the single pair of wings. Almost the entire surface of the thorax is formed by the **scutum** of the mesothorax. The scutum is often divided into anterior and posterior parts by a depression in the cuticle called the **transverse suture**. Directly behind the scutum, the **scutellum** almost always forms a well-developed convex lobe. There are two pairs of thoracic respiratory openings called **spiracles** (see section 1.6.4), the anterior (mesothoracic) pair and the posterior (metathoracic) pair. The thorax of most species of higher Diptera possesses numerous distinct bristles, particularly the hypopleural, achrostichal and dorsocentral bristles, the presence or absence of which is of great value in classification and identification (Fig. 4.2).

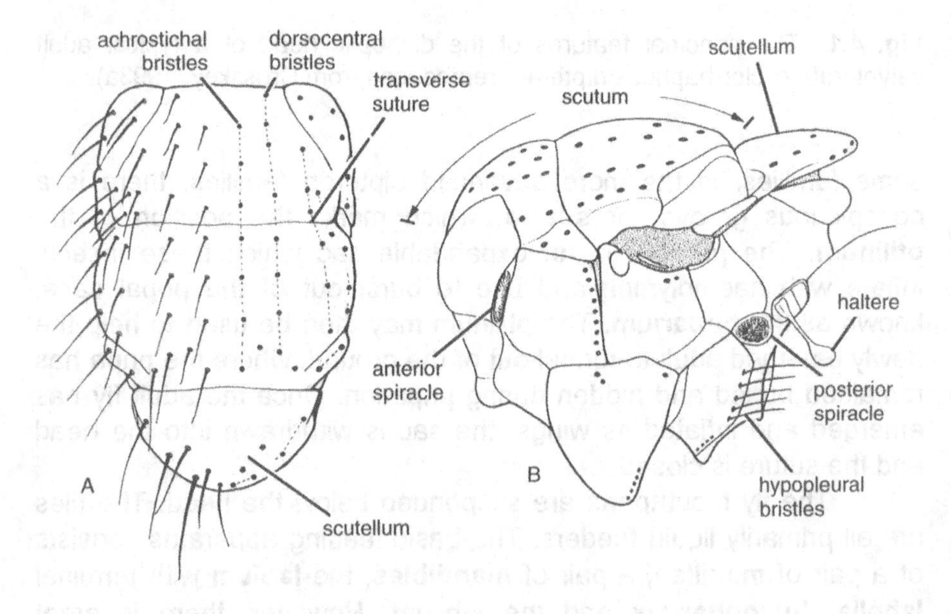

Fig. 4.2 The principal features of the generalized thorax of an adult calypterate cyclorrhaphous dipteran; (A) dorsal view (B) lateral view (reproduced from Crosskey, 1993a).

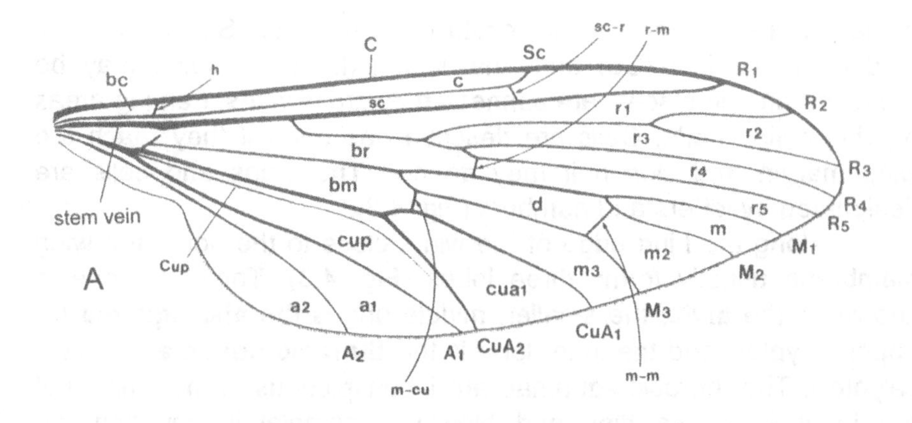

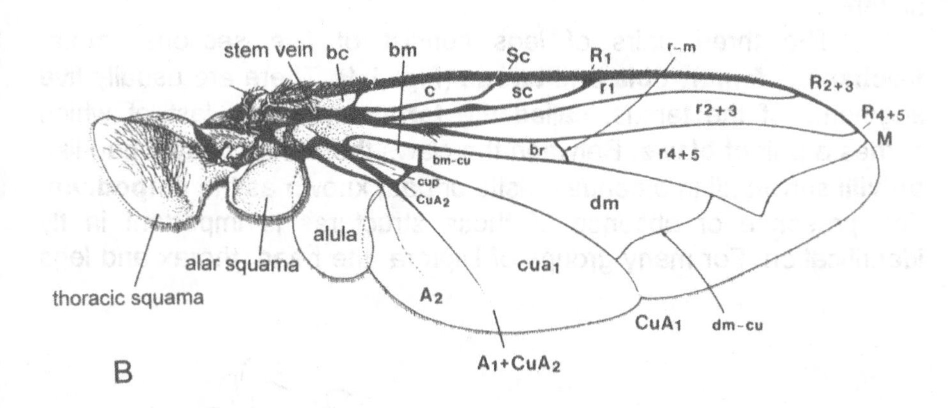

Fig. 4.3 The veins and cells of the wings of adult Diptera: (A) typical groundplan and (B) a typical calypterate dipteran, *Calliphora vicina* (reproduced from Crosskey, 1993a).

The club-like halteres are attached to the metathorax, behind and above the posterior spiracles. In flight the halteres vibrate with the wings but their relatively heavy heads develop inertia and, for a fraction of a second, continue to vibrate in the same direction when the fly changes direction. This produces strain on the cuticle at the base of the halteres which is detected by sensory cells. The halteres, therefore, allow the fly to detect changes in direction and help it to maintain straight and level flight or to judge the angle of turn.

The membranous wings have a remarkably constant species-specific supporting arrangement of hollow, rod-like structures, called **veins**. The venation is extremely important for fly classification. Six primary veins are recognized: costa (C), subcosta (Sc), radius (R), media (M), cubitus (Cu) and anal vein (A). These veins may be branched and, in places, are connected by cross veins, framing areas of wing called **cells**. Cells are described as open if they reach the wing margin and closed if they do not. The veins and cells are designated by letters and numbers (Fig. 4.3).

Along the hind edge of the wing, close to the body, the wing membrane usually forms three lobes (Fig. 4.3). The outer one is known as the **alula**, the smaller, middle one is the **alar squama** (or upper calypter) and the inner lobe is the **thoracic squama** (or lower calypter). The thoracic squamae are inconspicuous in most flies but are large in house flies and blowflies, completely covering the halteres.

The three pairs of legs consist of five sections: **coxa**, **trochanter**, **femur**, **tibia** and **tarsus** (Fig. 4.4). There are usually five segments of the tarsus, called the **tarsomeres**, the last of which carries a pair of **claws**. Between the claws there may be two pad-like **pulvilli** surrounding a central bristle or pad, known as the **empodium**. The presence or absence of these structures is important in fly identification. For many groups of Diptera, the head, thorax and legs

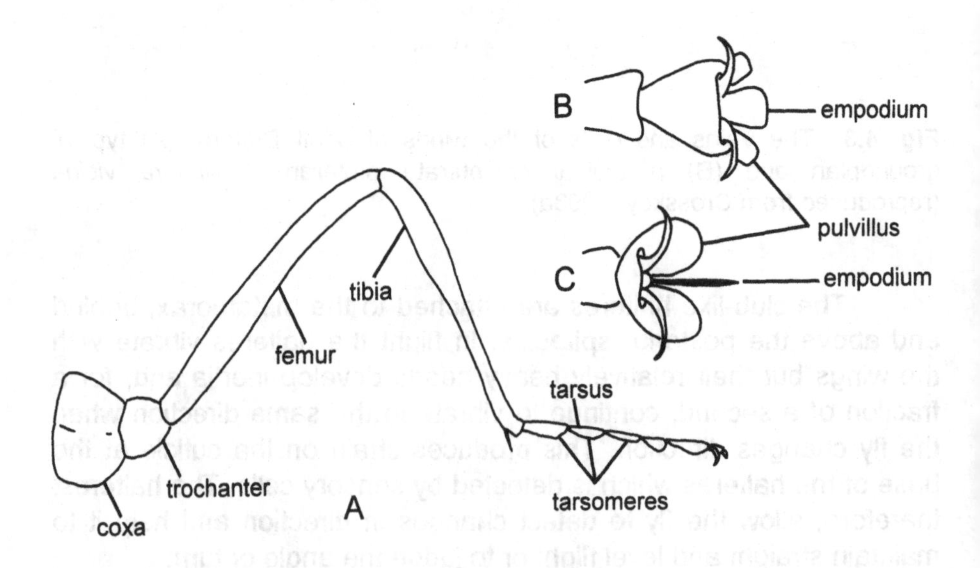

Fig. 4.4 The segments of the leg (A), and the empodium and pulvilli of adult brachyceran (B) and cyclorrhaphous (C) Diptera.

bear a number of distinct bristles, the arrangement of which is also important in fly identification.

The third body section, the **abdomen,** varies a great deal in shape and size among the Diptera. The basic number of segments is 11, but the terminal segments may be greatly modified in association with the genitalia. There are typically seven pairs of spiracles.

4.3 LIFE HISTORY

Most dipterous species are **oviparous**, laying small, oval eggs in discrete batches. Egg laying is referred to as **oviposition**. Embryonic development usually occurs within the egg. However, in a few families, such as the flesh flies (Sarcophagidae), the eggs develop and hatch in the oviduct and the female deposits newly hatched first-instar larvae. This reproductive strategy has been extended in families such as the sheep keds and forest flies (Hippoboscidae) and tsetse flies (Glossinidae), where a single egg is ovulated. The larva ecloses from the egg and is then retained and nourished in the uterus-like oviduct until almost ready to pupariate, when it is **larviposited**. This is known as **adenotrophic viviparity**.

The larvae are soft, legless and segmented, with a head that is well defined in most Nematocera but much reduced in other groups. In general, dipterous larvae need water or high humidity to survive. Those that are not aquatic tend to live in damp or highly humid environments, such as mud, rotting vegetation or faeces. The larva commonly passes through 3–5 stadia before pupation. In almost all Nematocera and Brachycera the last larval stadium sheds its skin and transforms into a **pupa** in which the appendages are externally visible; the adult escapes through a longitudinal slit in the thoracic cuticle. The pupation of Cyclorrhapha involves the formation of a barrel-shaped, shell-like **puparium**, formed from the exoskeleton of the last larval stadium, within which the pupal stage occurs. The adult ecloses from the puparium by pushing off a circular cap.

The adult stage is concerned primarily with mating and egg laying. In many dipteran species adult females require a protein meal to begin the production of yolk and mature their egg batches. These species are described as **anautogenous**. Females that are able to mature their eggs without an initial protein meal are described as

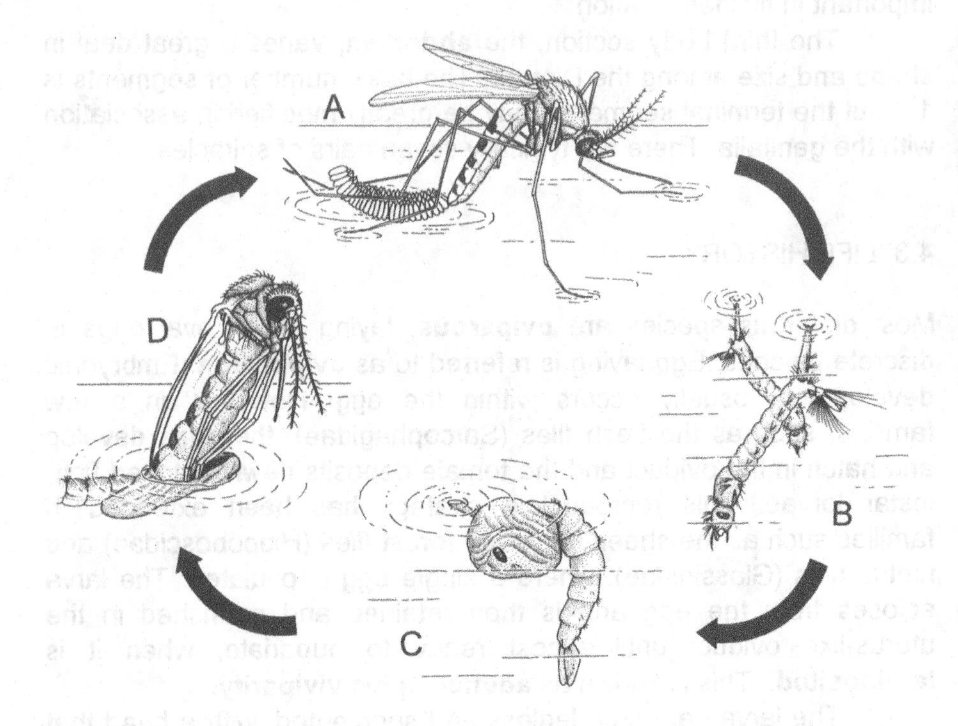

Fig. 4.5 Life cycle of the mosquito *Culex pipiens* (Diptera: Culicidae). (A) Adult ovipositing. (B) Larvae at the water surface. (C) Pupa suspended from the water surface. (D) Adult emerging from its pupal case at the water surface. (Reproduced from Gullan and Cranston, 1994.)

autogenous. In some species, for example species of the black fly (Simuliidae), populations may contain a mixture of autogenous and anautogenous individuals and the proportion of each type may change over the course of a season. This remains a little studied phenomenon. In laboratory cultures adult flies can live for several weeks, sometimes maturing large numbers of egg batches. However, extrapolation from laboratory studies is often misleading and, in the field, the duration of adult life is more commonly in the order of several days, particularly for males.

4.4 PATHOLOGY

Adult flies may feed on blood, sweat, skin secretions, tears, saliva, urine or faeces of the domestic animals to which they are attracted. They may do this either by puncturing the skin directly, in which case they are known as **biting flies**, or by scavenging at the surface of the skin, wounds or body orifices, in which case they may be classified as **non-biting** or **nuisance flies**.

The biting flies may cause particularly acute problems through their blood-feeding habit and in some areas, such as parts of the USA and Canada, populations of mosquitoes and black flies can become so great that livestock die of acute blood loss. Biting flies may act as biological vectors for a range of pathogenic diseases and both biting and non-biting flies may also be mechanical vectors of disease. Mechanical transmission may be exacerbated by the fact that some fly species inflict extremely painful bites and, therefore, are frequently disturbed by the host while blood-feeding. As a result, the flies are forced to move from host to host over a short period, thereby increasing their potential for mechanical disease transmission. The biting activities of blood-feeding flies may also provoke hypersensitivity reactions. In the USA approximately 50% of the estimated annual loss in cattle production from all categories of livestock pests can be attributed to biting flies.

The activity of both biting and non-biting species of fly may be responsible for 'fly-worry' in livestock. This is the disturbance caused by the presence and attempted feeding behaviour of flies. Responses by the host may range from dramatic escape behaviour, in which self-injury can occur, to less sensational movement into shade or simply stamping and tail switching. However, all these changes in behaviour result in reduced time spent feeding and decreased performance. Finally, because many flies are attracted to expired carbon dioxide, large fly populations can occasionally cause death by suffocation after inhalation of the flies by horses, cattle and other animals.

4.5 CLASSIFICATION

The order Diptera is most commonly divided into three sub-orders, **Cyclorrhapha, Brachycera** and **Nematocera** and this is the classification system that will be adopted here. However, recent work has suggested that the sub-order Cyclorrhapha should be replaced as an infraorder known as the Muscomorpha, within an enlarged sub-order Brachycera. This is known as the 'McAlpine classification' but is not, as yet, generally accepted.

4.5.1 Cyclorrhapha

The Cyclorrhapha is a huge sub-order in which, in the adult, the antennae usually have an arista, a feather- or bristle-like structure, borne on the third antennal segment. If the arista is feather-like, it is usually described as plumose. The palps usually have only one segment. The pupa is formed in a puparium from which the adult emerges by pushing off a circular cap.

The sub-order is split into two sections, the **Aschiza** and **Schizophora**. The section Aschiza, containing families such as the hover flies, is of little veterinary importance. Adult Schizophora have a distinct ptilinal suture and **lunule**. The lunule is a small triangular plate (sclerite) situated between the point of attachment of the antennae and the ptilinal suture. The section Schizophora is classified into two sub-sections, the **Calypterae** and **Acalypterae**. Calypterate Diptera have a well-defined groove on the second antennal segment (pedicel) and large squamae. Acalypterate Diptera are a vast assemblage of mainly small flies; they have no groove in the pedicel and small or poorly developed squamae. All the flies of veterinary importance are members of the Calypterae.

The Calypterae are divided into the three super-families **Muscoidea**, **Hippoboscoidea** and **Oestroidea**, all of which contain families of veterinary importance. However, the super-family Oestroidea contains species primarily associated with myiasis, which will be discussed in Chapter 5. The super-families Muscoidea and Hippoboscoidea, each contain two families of veterinary interest – the **Muscidae** and **Fannidae** and the **Hippoboscidae** and **Glossinidae**, respectively.

4.5.2 Brachycera

The Brachycera are small to large, stout-bodied flies with a bulbous face and short antennae, usually composed of three basic segments. Adults have enlarged thoracic squamae like some of the calypterate, cyclorrhaphous Diptera. The palps usually have two segments and are often directed forwards. There is only one family of veterinary importance, the **Tabanidae**.

4.5.3 Nematocera

Flies of the sub-order Nematocera are usually small and slender, with long, narrow wings and long antennae composed of many articulating segments. The palps are usually composed of four or five segments. Following pupation, the adult fly escapes from the pupal case through a dorsal longitudinal slit. Four of the 18 families of Nematocera are of veterinary importance as blood-feeding ectoparasites and disease vectors. They are the **Ceratopogonidae** (biting midges) the **Simuliidae** (black flies), the **Culicidae** (mosquitoes) and the **Psychodidae** (sand flies).

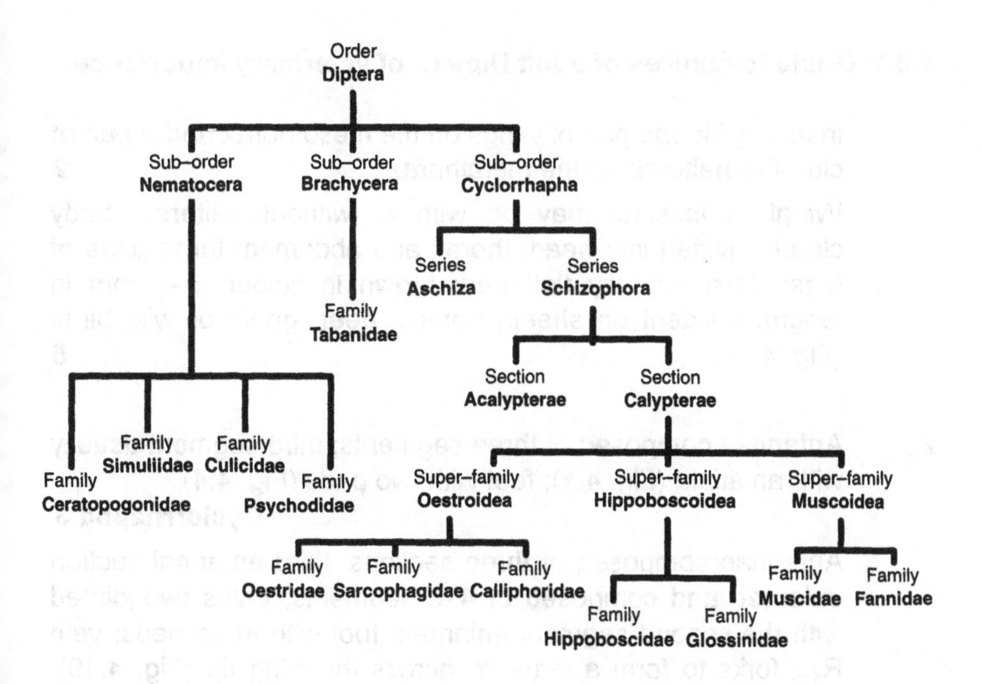

Fig. 4.6 Classification of the families of Diptera of veterinary importance.

4.6 RECOGNITION OF FLIES OF VETERINARY IMPORTANCE

Given that there are at least 125 families and 120,000 species of Diptera, unsurprisingly, their identification can be problematic. Some families are very small and obscure and identification to genus or species may require microscopic examination of the number and position of particular setae, or examination of internal structures, such as the male genitalia. Other families and genera can be easily recognized. This presents something of a problem in an introductory text such as this, since comprehensive keys are voluminous and often difficult to use, but simplified keys leave a large margin for error and misidentification. What follows below, therefore, is simply a generalized recognition guide to aid in the identification of the key families of Diptera that may be found feeding upon or associated with livestock and domestic animals in temperate habitats. Where precise identification is necessary, more specialist keys and the assistance of an expert in the various dipteran groups is recommended.

4.6.1 Guide to families of adult Diptera of veterinary importance

1 Insects with one pair of wings on the mesothorax and a pair of club-like halteres on the metathorax **2**

Wingless insects; may be with or without halteres; body clearly divided into head, thorax and abdomen; three pairs of legs; dorsoventrally flattened; brown in colour; 5–8 mm in length; resident on sheep, horses, deer, goats or wild birds (Fig. 4.14) **5**

2 Antennae composed of three segments; third segment usually with an arista (Fig. 4.1); foot with two pads (Fig. 4.4) **Cyclorrhapha 3**

Antennae composed of three sections; third antennal section enlarged and composed of 4–8 segments; palps two-jointed with the second segment enlarged; foot with three pads; vein R_{4+5} forks to form a large 'Y' across the wing tip (Fig. 4.19); large, stout bodied flies with large eyes **Tabanidae 12**

Antennae long, slender and composed of many articulating segments; palps composed of 4–5 segments; small slender flies with long narrow wings **Nematocera 13**

3 Frons with ptilinal suture (Fig. 4.1) Series **Schizophora 4**

Frons without ptilinal suture Series **Aschiza**

4 Second antennal segment usually with a groove (Fig. 4.1); thoracic transverse suture strong (Fig. 4.2); thoracic squamae usually well developed **Calypterae 5**

Second antennal segment usually without a groove; thoracic transverse suture weak; thoracic squamae often vestigial **Acalypterae**

5 Thorax broad and dorsoventrally flattened; may appear spider or tick-like; often wingless; wings when present with venation abnormal with veins crowded into leading half of wing (Fig. 4.14) **Hippoboscidae**

Wings with veins not crowded together towards the leading edge; thorax not dorsoventrally flattened **6**

6 Proboscis long, forwardly directed and embraced by long palps; arista with feathery short hairs present only on dorsal side; discal medial cell of wings characteristically 'hatchet' shaped (Fig. 4.15); found only in sub-Saharan Africa **Glossinidae**

Discal medial cell of wings widening gradually and more or less regularly from the base **7**

7 Mouthparts small, usually functionless; head bulbous; antennae small; flies more or less covered with soft hair; larval parasites of vertebrates (see Chapter 5) **Oestridae**

Mouthparts usually well developed; antennae not small; flies with strong bristles **8**

8 Hypopleural bristles present (Fig. 4.2) **9**
Hypopleural bristles absent **11**

9 Post-scutellum strongly developed; larval parasitoids of insects **Tachinidae**
Post-scutellum weak or absent **10**

10 Dull grey appearance; three black stripes on the scutum; abdomen usually with chequered or spotted pattern; larval parasites of vertebrates (see Chapter 5) **Sarcophagidae**
Metallic, iridescent appearance (blue-black, violet-blue, green); larval parasites of vertebrates (see Chapter 5) **Calliphoridae**

11 Wings with vein A_1 not reaching the wing edge; strong curved A_2 vein the tip of which approaches A_1 (Fig. 4.13); aristae bare **Fanniidae**
Wings with vein A_1 not reaching the wing edge; A_2 vein not strongly curved (Fig. 4.10); aristae bilaterally plumose to the tip **Muscidae**

12 Antennal flagellum with four segments (Fig. 4.19); wings mottled; proboscis shorter than head ***Haematopota* (Tabanidae)**
Antennal flagellum with five segments (Fig. 4.19); apical spurs on tibiae are small and may be hidden by hair; wings usually with costal region dark and a single dark broad transverse band; proboscis shorter than head ***Chrysops* (Tabanidae)**
Antennal flagellum with five segments (Fig. 4.19); no apical spurs on hind tibiae; wings usually clear but may be dark or banded; proboscis shorter than head ***Tabanus* (Tabanidae)**

13 Small, hairy, moth-like flies; numerous parallel wing veins running to the margin; wings pointed at the tip **Psychodidae 14**

Not like this **15**

14 Palps five-segmented; biting mouthparts at least as long as head; antennal segments almost cylindrical; two longitudinal wing veins between radial and medial forks (Fig. 4.25) **Phlebotominae**

15 Ten or more veins reaching the wing margin **16**

Not more than eight veins reaching the wing margin **17**

16 Wing veins and hind margins of wings covered by scales (Fig. 4.24); conspicuous forward-projecting proboscis **Culicidae**

17 Wings broad; wing veins thickened at the anterior margin; antennae not hairy; thorax humped; antennae usually with 11 rounded segments; palps long with five segments extending beyond the proboscis; first abdominal tergite with a prominent basal scale fringed with hairs (Fig. 4.21) **Simuliidae**

Wings not particularly broad, antennae hairy **18**

18 Front legs often longer than others; median vein not forked; (non-biting midges) **Chironomidae**

Front legs not longer than others; wings with median vein forked; antennae with 14–15 visible segments; palps with five segments; female mouthparts short; legs short and stout; two radial cells and cross vein r-m strongly angled in relation to media; at rest wings close flat over the abdomen (Fig. 4.23) ***Culicoides*** **(Ceratopogonidae)**

4.7 CYCLORRHAPHA

All the cyclorrhaphous calypterate Diptera of veterinary interest are divided into the three superfamilies **Muscoidea**, **Hippoboscoidea** and **Oestroidea**. The Muscoidea and Hippoboscoidea, each contain two families of veterinary interest the **Muscidae** and **Fannidae** and the **Hippoboscidae** and **Glossinidae**, respectively. The super-family Oestroidea contains three families of veterinary interest, Oestridae, Calliphoridae and Sarcophagidae, species of which are primarily associated with myiasis and which will be discussed in Chapter 5.

There are two basic types of mouthpart seen in the cyclorrhaphous Diptera of veterinary interest. Sponging mouthparts are used for feeding on liquid films. Such mouthparts are found in groups such as the house flies, blow flies and face flies. Biting mouthparts are used for puncturing the skin and drinking blood. They occur in groups such as the stable flies, horn flies and tsetse flies.

In the house fly, the proboscis is an elongate feeding tube, composed of a basal **rostrum** bearing the maxillary palps, a median flexible **haustellum**, composed of the **labium** and flap-like **labrum**, and two apical **labella**, which are modified labial palpi (Fig. 4.7). Mandibles and maxillae are absent. The **labrum** and the **hypopharynx** lie within the flexible anterior gutter in the labium. The labella are sponging organs, the inner surface of which are lined by grooves called **pseudotracheae** (Fig. 4.8). The grooves of the

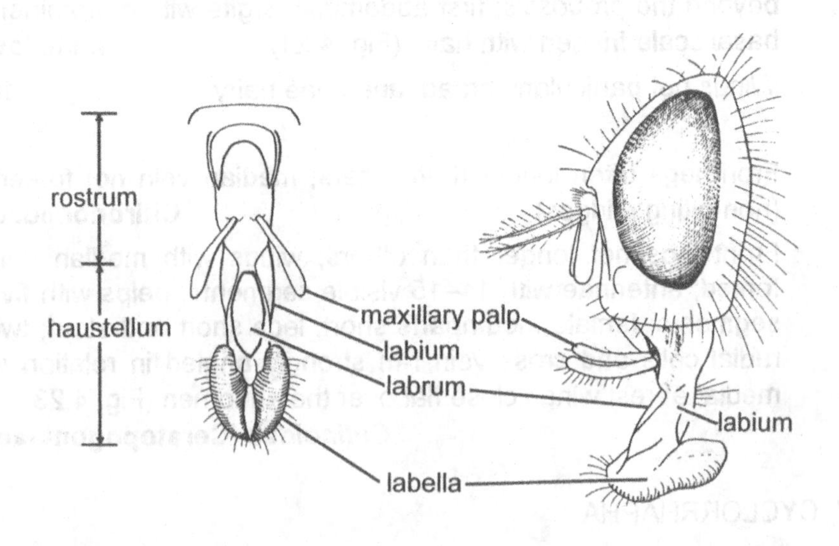

Fig. 4.7 The head and mouthparts of an adult house fly in (A) anterior and (B) lateral views (modified from from Harwood and James, 1979).

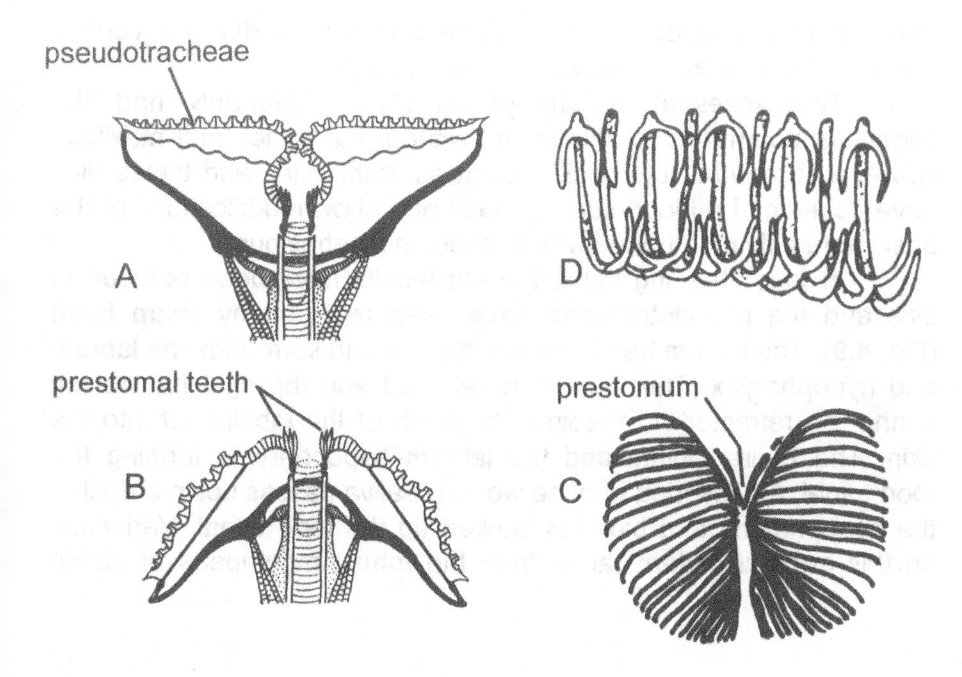

Fig. 4.8 The labella and pseudotracheae of an adult house fly in (A) filtering position and (B) direct feeding position. (C) Surface view of labella in filtering position and (D) pseudotracheal rings. (From Graham-Smith, 1930.)

pseudotracheae are formed by rows of closely packed, curved cuticular rods, which are C-shaped at one end but not the other (Fig. 4.8). Rods are arranged alternately so that the expanded end of one alternates with the non-expanded end of the next rod. Between grooves the rods are joined by a connecting membrane. The grooves lead towards the **oral aperture**, known as the **prestomum**. When feeding, the labella are expanded by blood pressure and opened to expose their inner surface. They are then applied to the liquid film. Liquid flows into and along the grooves by capillary action and then is drawn up the food canal by muscular pumping action. At rest, the inner surfaces of the labella are in close contact and kept moist by secretions from the labial salivary glands.

The house fly proboscis is jointed and can be withdrawn into the head capsule when not in use by the retraction of the rostrum. There are a number of minute teeth surrounding the prestomum, which can be used directly to rasp at the food (Fig. 4.8). These teeth may be well developed and important in the feeding of various species of Muscidae, for example *Hydrotaea irritans*.

The ancestral cyclorrhaphous Diptera probably had the sponging mouthparts as described, without mandibles and maxillae. However, a number of species, such as stable flies and tsetse flies have developed a blood-sucking habit and show modifications of the basic house fly mouthparts which reflect this behaviour.

In blood-feeding Muscidae the labella have been reduced in size and the pseudotracheae have been replaced by sharp teeth (Fig. 4.9). The labium has been lengthened and surrounds the labrum and hypopharynx. The rostrum is reduced and the rigid haustellum cannot be retracted. In feeding, the teeth of the labella cut into the skin. The entire labium and the labrum–hypopharynx, forming the food canal, are inserted into the wound. Saliva passes down a duct in the hypopharynx and blood is sucked up the food canal. Variations on this general pattern range from the robust mouthparts of stable

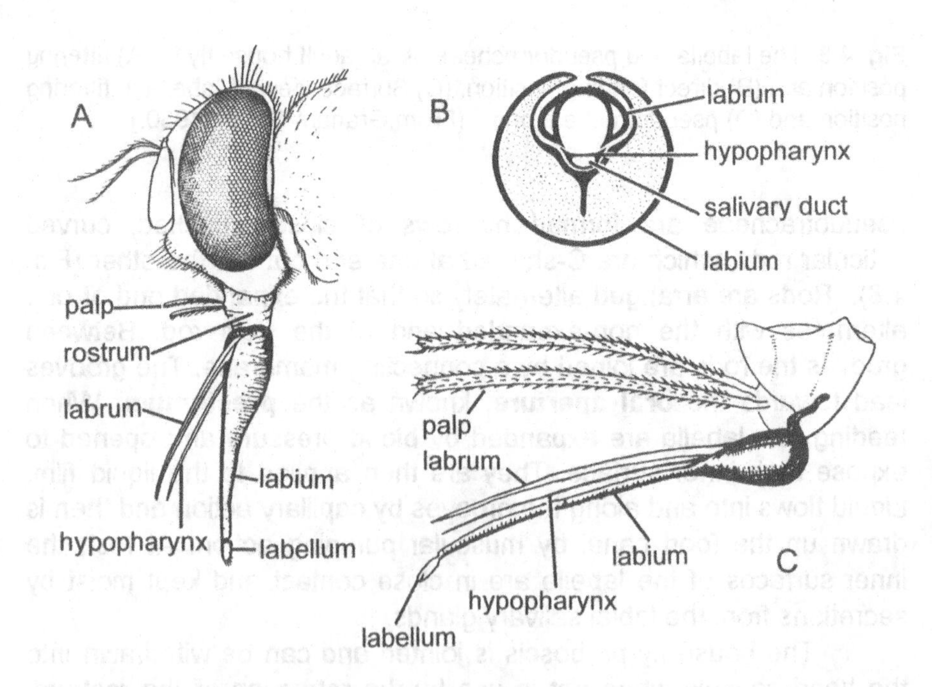

Fig. 4.9 Mouthparts and head of a stable fly: (A) in lateral view and (B) cross-section (reproduced from Harwood and James, 1979). (C) Proboscis and palps of a tsetse fly (reproduced from Newstead, Evans and Potts, 1924).

flies to the delicate mouthparts of tsetse flies.

4.7.1 Muscidae

The Muscidae is the second largest of the calypterate families, with about 4000 species. The family contains a number of sub-families and genera of veterinary importance, notably the genera *Hydrotaea, Musca, Stomoxys* and *Haematobia.* Species of the latter two genera are blood-feeders and together form the sub-family Stomoxyinae.

Musca

The genus *Musca* contains about 60 species, of which the cosmopolitan house fly, *Musca domestica,* and the face fly, *Musca autumnalis,* are of particular importance. In Africa and Oriental regions the bazaar fly, *Musca sorbens*, is widespread, largely replacing *M. domestica*, and in Australia the bush fly, *Musca vetustissima*, which is very closely related to *M. sorbens*, is an important nuisance pest of humans and livestock. All species of *Musca* are liquid feeders and do not have biting mouthparts.

Morphology: female adults of *M. domestica* are 6–8 mm and male adults 5–6 mm in length (Fig. 4.10). The thorax is usually grey with four dark longitudinal stripes. The abdomen has a yellowish-brown background colour with a black median longitudinal stripe. The aristae are bilaterally plumose to the tip. The face fly, *M. autumnalis*, is very similar to *M. domestica* in size and appearance, although the abdomen of the female is darker and that of the male is more orange in colour. Adult *M. vetustissima*, like *M. sorbens*, have two broad longitudinal stripes on the thorax and the first abdominal segment is black.

Life cycle: eggs of the house fly *M. domestica* are laid in animal faeces, manure piles, garbage and other types of decomposing organic material. Each female can produce batches of 100–150 eggs at 3–4 day intervals throughout its life. Each egg is oval and about 1 mm in length. The larvae hatch within 12 hours. The white-coloured larvae pass through three stadia before crawling away from their feeding site, burrowing into the ground and pupariating to form a

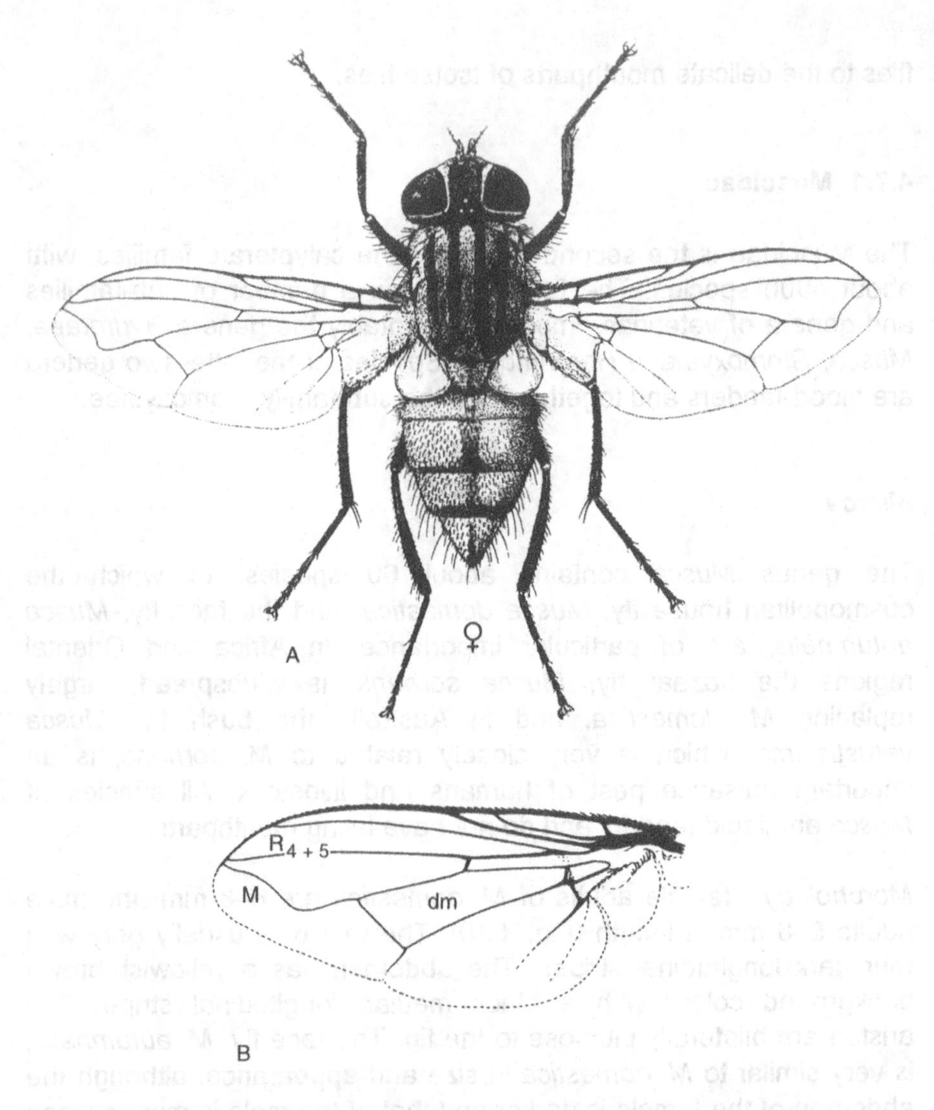

Fig. 4.10 (A) Female house fly, *Musca domestica* and (B) wing venation typical of species of *Musca*, showing the strongly bent vein M ending close to R_{4+5} (reproduced from Croskey and Lane, 1993).

reddish-brown puparium. Under ideal conditions, *M. domestica* can complete its life cycle from egg to adult in as little as 7–10 days. However, low temperatures will greatly extend this developmental period; at 15°C the egg-to-adult life cycle may take 45–50 days.

The face fly, *M. autumnalis*, congregates in large numbers around the faces of cattle. It feeds on secretions from the eyes, nose and mouth as well as from blood in wounds left by other flies, such as tabanids. It lays its eggs just beneath the surface of fresh cattle manure, within about 15 minutes after the dung pats are deposited. The eggs of *M. autumnalis* are about 3 mm in length and possess a short respiratory stalk. Like *M. domestica*, the larvae pass throughout three stadia, before pupariating to form a whitish-coloured puparium. Summer generations require about 2 weeks to complete a life cycle. Face flies prefer bright sunshine and usually do not follow cattle into barns or heavy shade. Adults are strong fliers and can move between widely separated herds. Face flies overwinter as adults, aggregating in overwintering sites such as farm buildings.

Musca vetustissima breeds in excrement, particularly cattle dung. The egg-to-adult life cycle is extremely rapid and may be completed in as few as 7 days at 32–35°C. After mating and protein-feeding, females lay batches of up to 50 eggs in crevices in fresh dung. Adult females live for about 7–14 days, during which time they produce about four egg batches.

Pathology: the house fly, *M. domestica*, is closely associated with humans, livestock, their buildings and organic wastes. Although it may be of only minor direct annoyance to animals, its potential for transmission of diseases and parasites is of significance. However, its pathological importance varies considerably, depending on the precise circumstances in which it occurs. The free availability of livestock or human excrement and low levels of hygiene provide sites in which flies can breed and allow flies to act as vectors as they move from site to site. Pathogens may be carried either on the hairs of the feet and body or regurgitated in the salivary vomit during feeding. *Musca domestica* has been implicated in the spread of mastitis, conjunctivitis and anthrax. In humans, the house fly is important in spread of *Shigella* and other enteric bacteria. Eggs and larvae of various nematodes which affect horses such as *Habronema* spp. also may be carried. The latter, when deposited in wounds, may give rise to skin lesions of habronemiasis, commonly called 'summer sores' in horses. Granulomatous nodules, which contain the nematode larvae, appear on the skin especially around the eye, ventral abdomen, prepuce and lower limbs.

In northern Europe, the face fly, *M. autumnalis*, may often be the most numerous fly worrying cattle in pasture. In Britain this species tends to be found in more southerly areas and, in general, is not found north of Yorkshire. *Musca autumnalis* is also one of the most important livestock pests to invade the United States in recent years. Its introduction into North America from Europe was first detected in 1951 in Nova Scotia. From there it spread southward and, by 1959, many cases were being reported on cattle. It now occurs practically throughout the USA. The annoyance caused by the flies results in cattle seeking shade to escape and contributes to reduced production rates. In the USA, *M. autumnalis* is an important vector of bovine keratoconjunctivitis caused by *Moraxella bovis.* Face flies are also intermediate hosts of *Parafilaria bovicola,* the causative agent of parafilariosis of cattle in northern Europe and elsewhere. Adults are developmental hosts for *Thelazia* (Spirurata: Thelaziidae) nematodes which live in the conjunctival sac of cattle and horses, causing conjunctivitis, keratitis, photophobia and epiphora. This disease is an increasing problem in the USA.

In addition to dung feeding, adults of the bush fly, *M. vetustissima* will persistently attempt to feed at the mouth, eyes and nose. As a result, they are of considerable significance as a nuisance pest in Australia for both livestock and humans. In the absence of native Australian dung beetles capable of disposing of the dung of introduced cattle, the dung of these herbivores is slow to decompose, allowing *M. vetustissima* ample opportunity to breed and reach large and problematic population densities.

Hydrotaea (sweat and head flies)

Species of the genus *Hydrotaea* closely resemble species of *Musca* and are known as the sweat flies. They feed on exudates of the eyes, nose and mouth. They do not bite. The genus contains one particularly important species, *Hydrotaea irritans*, known as the sheep head fly, which is widespread throughout northern Europe but is not believed to be present in North America.

Morphology: adults of *H. irritans* are 4–7 mm in length. The thorax is black with grey patches. The abdomen is olive green and the base of the wings is orange-yellow.

Life cycle: each female produces one or two batches of about 30 eggs in its lifetime. Eggs are laid in decaying vegetation, faeces or carrion. Third-instar larvae may be predatory on other larvae. Adults emerge in spring and there is probably only a single generation per year with peak numbers occurring in midsummer. Final-stage larvae diapause overwinter. Adult flies prefer still conditions and are associated with pastures that border woodlands or plantations.

Pathology: *H. irritans* females are anautogenous and require a protein meal to mature their oocytes. They generally feed on low protein sources such as tears, saliva and sweat. As a result, they need to feed relatively frequently to mature their eggs. However, they are facultative blood-feeders and will ingest blood at the edges of wounds if available. Meal analysis shows that even when flies were caught off sheep, almost 80% had fed on cattle. While this species has the sponging mouthparts typical of most Muscidae, they also have well-developed prestomal teeth, which are used for rasping and can cause skin damage and enlarge existing lesions during feeding. Wounds, such as those incurred by fighting rams, are particularly susceptible to attack; swarms of flies around the head lead to intense irritation, annoyance and can result in self-inflicted wounds. Secondary bacterial infection of wounds is common and this may encourage blowfly strike (see Chapter 5).

Problems caused by head fly in sheep appear to be particularly common in northern England and Scotland. The extent of the problem varies from year to year, depending on the prevailing weather conditions. Incidences of 30–40% of sheep affected have been recorded on farms in fly foci, particularly in Scotland.

In cattle, large numbers of *H. irritans* have been found on the ventral abdomen and udder of cattle and, since the bacteria involved in summer mastitis *Corynebacterium pyogenes*, *Streptococcus dysgalactiae* and *Peptococcus indolicus* have been isolated from these flies, there is a strong presumption that they may transmit the disease.

Stomoxys (stable flies)

The genus contains about 18 species, of which the most common species of importance in temperate habitats is *Stomoxys calcitrans*, the stable fly, which is found worldwide. *Stomoxys niger* and *Stomoxys sitiens* may replace *S. calcitrans* as important pests of livestock in Afrotropical and Oriental regions. Species of this genus have biting mouthparts and both sexes are blood-feeders.

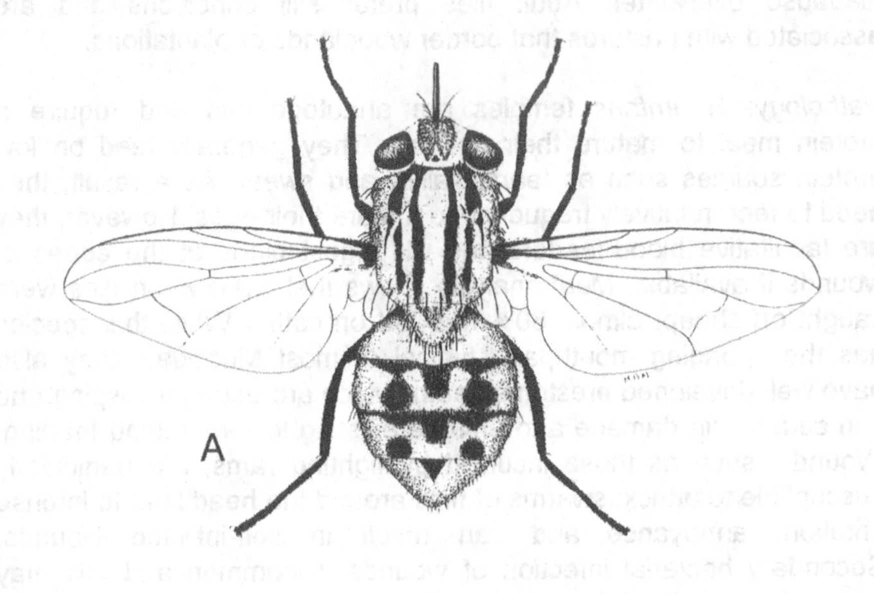

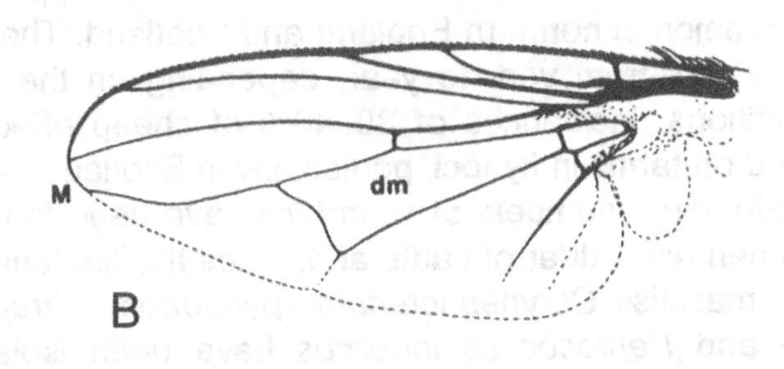

Fig. 4.11 (A) Female stable fly, *Stomoxys calcitrans* (reproduced from Soulsby, 1982) and (B) wing venation typical of species of *Stomoxys*, showing the slight apical forward curve of vein M (reproduced from Crosskey and Lane, 1993).

Morphology: *Stomoxys* are about 7–8 mm in length and are generally grey in colour, with four longitudinal dark stripes on the thorax (Fig. 4.11). The abdomen is shorter and broader than that of *M. domestica* and is also grey with three dark spots on the second and third abdominal segments. The projecting proboscis is sufficiently prominent to distinguish species of this genus from species of *Musca*.

The palps are small and thin and only a quarter to half of the length of the proboscis. The eggs are about 1 mm in length and hatch in 5–10 days, depending on temperature. The cream-coloured, saprophagous larvae pass through three stadia until, when fully developed at about 10 mm in length, they pupariate. The puparia are brown and about 6 mm in length.

Life cycle: *S. calcitrans* may be abundant in and around farm buildings and stables in late summer. Both sexes are blood-feeders and often ingest small blood meals several times a day. After feeding, flies move to a resting site, on structures such as barn walls, fences or trees. Adults prefer strong sunlight but will follow animals into buildings to feed. This habit accounts for the name 'biting house fly' which is also applied to this insect. After the female has had a number of blood meals, it lays its eggs in wet straw, old stable bedding or manure. Each female can lay up to 500 eggs in batches of 25–50. The larvae feed on vegetable matter. In warm weather the average life cycle is 4 weeks, but it can vary from 3 to 7 weeks, depending on temperature.

Pathology: stable flies inflict painful bites and are one of the most annoying and destructive fly pests of cattle, horses, sheep and goats. They are active by day and are persistent and strong fliers, occasionally following potential hosts for considerable distances. They mainly attack the legs and flanks. Loss of blood and disturbance of feeding may result in a 10 to 15% loss of body weight when stable fly populations are high. Reduction in milk production in dairy cattle has been reported to be as high as 40–60% in some cases. Stable flies are known to transmit equine infectious anaemia (retrovirus) and surra among horses, and are suspected of transmitting other diseases such as anthrax among other animals. They may also be mechanical vectors of protozoa and may act as intermediate hosts of the nematode *Habronema*.

***Haematobia* (horn flies)**

There are two common species of *Haematobia* in temperate habitats, *Haematobia irritans*, (sometimes incorrectly referred to as *Lyperosia irritans*) known as the horn fly, which is found in Europe, the USA and Australia, and *Haematobia stimulans* (sometimes referred to as

Haematobosca stimulans) which is found only in Europe. Of these *H. irritans* is the most economically important. It primarily attacks cattle, but horses in adjacent pastures may be also attacked. A third species, *Haematobia exigua* (sometimes referred to as the subspeces *Haematobia irritans exigua*), known as the buffalo fly, is of importance in the Far East and Australia.

Morphology: adult *Haematobia* are grey-black, blood-sucking flies which resemble the stable fly in appearance. Adult *H. stimulans* are slightly smaller that *S. calcitrans*, while *H. irritans* are substantially smaller, at about 3–4 mm in length (Fig. 4.12). In contrast with *S. calcitrans* the palps of the adults are much longer relative to the proboscis and are club-shaped apically. In *H. irritans* the palps are dark greyish whereas in *H. stimulans* they are yellowish in colour.

Life cycle: *H. irritans* adults remain on the host animal day and night, usually congregating on the back, withers and around the head. They will only leave the animal briefly to mate and lay eggs. In contrast, *H. stimulans* is less of a resident on its host and more closely resembles *S. calcitrans* in its behaviour. The reddish-brown eggs are laid almost exclusively in fresh cattle manure. Each female

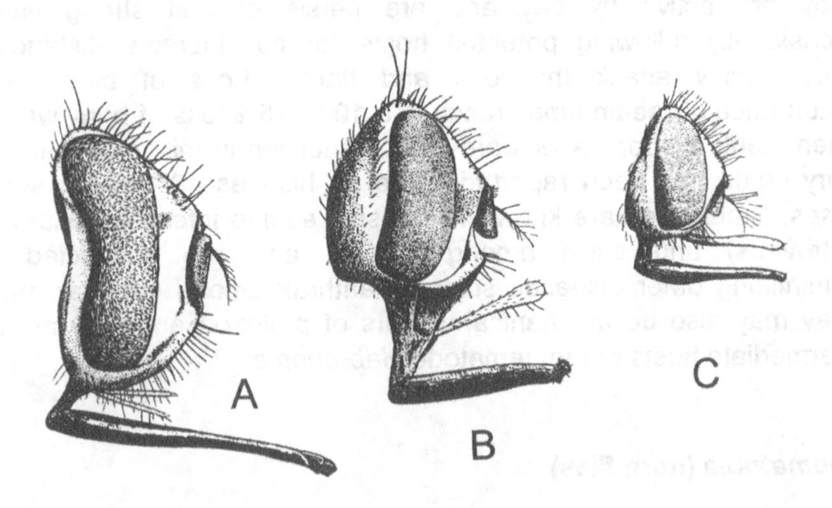

Fig. 4.12 Lateral views of the heads of blood-sucking Muscidae. (A) *Stomoxys calcitrans*, (B) *Haematobia stimulans* and (C) *Haematobia irritans* (reproduced from Edwards, Oldroyd and Smart, 1939).

is capable of producing 300–400 eggs in batches of 20–30 eggs. There are three larval stadia and pupation occurs within the dung pat. The egg-to-adult life cycle may be completed in as little as 10–14 days. Whereas the stable fly usually rests on an animal with its head directed upward, the horn fly faces downward.

Pathology: in North America thousands of *H. irritans* may be found feeding along the back, sides and ventral abdomen of cattle. The loss of blood due to horn flies can be considerable. In addition, during feeding the horn fly withdraws and reinserts its mouth parts many times, resulting in considerable irritation to the host. Horn fly wounds may attract other flies. The feeding activity of horn flies causes disturbance and prevents livestock from feeding normally. Heavy attack by horn fly in the USA has been shown to be able to reduce milk production by up to 25–50%. This species may also transmit *Stephanofilaria stilesi*, a nematode skin parasite of cattle. *Haematobia* spp. may also cause a periorbital and ventral ulcerative dermatitis in horses, the lesions of which may become infected by *Habronema* spp. nematodes.

The pathogenic significance of *H. stimulans* in Europe is not well known. *Haematobia exigua* feeds primarily on buffaloes and cattle, and like *H. irritans* rarely leaves the host unless disturbed. Weight gain and milk production may be affected adversely by infestations, which may reach densities of several thousand flies per host.

4.7.2 Fanniidae

This family contains about 250 species, most of which occur in the Holarctic and temperate Neotropical regions. There is a single genus of importance *Fannia*.

Fannia

The genus contains over 200 species, the most important and cosmopolitan of which is the lesser house fly *Fannia canicularis*. The latrine fly, *Fannia scalaris* and, in North America, *Fannia benjamini* may also be common.

Morphology: species of *Fannia* generally resemble house flies in appearance but are more slender and smaller at about 4–6 mm in length. *Fannia canicularis* is greyish to almost black in colour, possessing three dark longitudinal stripes on the dorsal thorax. The aristae are bare. *Fannia benjamini* has yellow palps, whereas *F. canicularis* and *F. scalaris* have black palps. The larvae are easily recognized by the flattened shape and the branched, fleshy projections from the body (see Chapter 5). The brown-coloured puparium resembles the larva in shape.

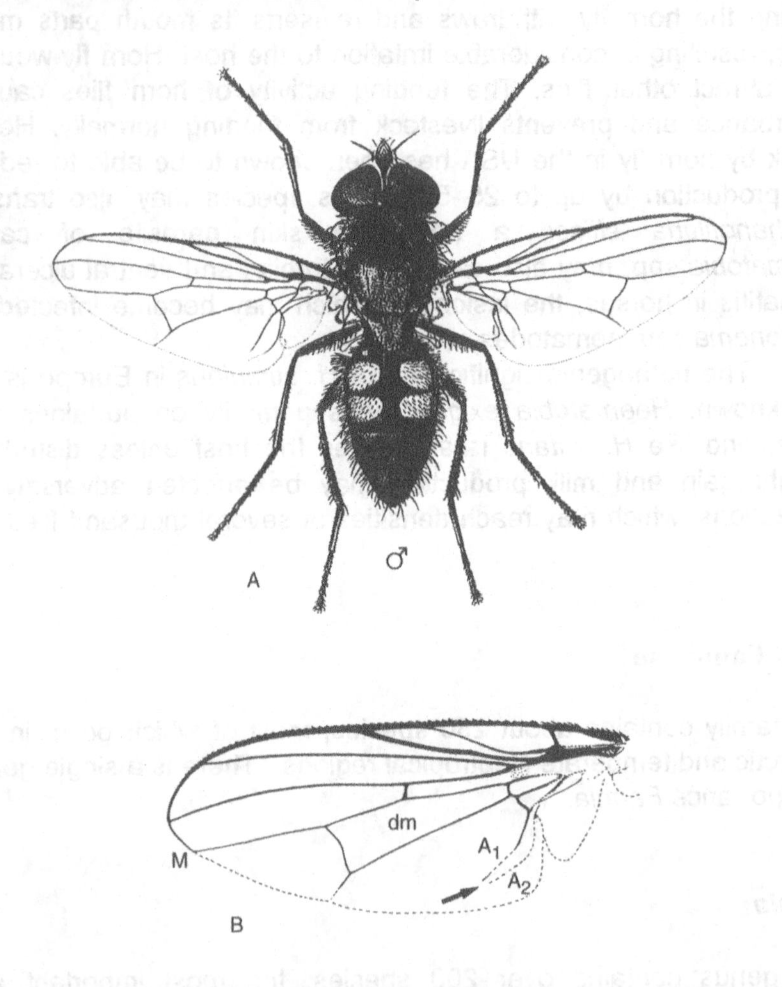

Fig. 4.13 (A) Male lesser house fly, *Fannia canicularis* and (B) wing venation typical of species of *Fannia*, showing the characteristic convergence of the anal veins (reproduced from Crosskey and Lane, 1993).

Life cycle: *Fannia* breed in a wide range of decomposing organic material, particularly the excrement of chickens, humans, horses and cows. However, in contrast to *M. domestica*, the eggs and larvae of most species of *Fannia* are more susceptible to desiccation. Hence, they are more abundant in semi-liquid sites, especially pools of semi-liquid faeces. The complete life cycle requires from 15 to 30 days. Adults are more abundant in the cooler months of spring and autumn, declining in midsummer. Adults of *Fannia* are readily attracted into buildings and adult males are familiar as the flies responsible for the regular triangular flight paths beneath light bulbs or shafts of sunlight from windows in buildings.

Pathology: species of *Fannia* are of interest as nuisance pests of livestock and humans, especially in caged-layer poultry facilities. However, they rarely feed directly from animals.

4.7.3 Hippoboscidae (keds and forest flies)

Adult hippobosciids are unusual, so-called 'degenerate', blood-feeding ectoparasites. They tend to be either permanent ectoparasites or to remain with the host for most of their life. There are about 200 species in the family. Those found on mammals are divided into three main genera: *Melophagus*, *Lipoptena* and *Hippobosca*. The most well-known and widespread species of veterinary importance in this family is *Melophagus ovinus,* the sheep ked. Originally Palaearctic in distribution, sheep keds have now been spread worldwide with domestic sheep.

Morphology: sheep keds, *M. ovinus*, are biting flies which are leathery, dorsoventrally flattened and somewhat tick-like in appearance (Fig. 4.14A). Both sexes are completely wingless, even the halteres are absent. They are brown in colour and 5–8 mm in length. The abdomen is indistinctly segmented and is generally soft and leathery. They have strong claws on their feet to help them cling to wool and hair.

Life cycle: adults are permanent ectoparasites and feed on the blood of sheep and, sometimes, goats. A single egg is ovulated at a time. The egg hatches inside the body of the female and the larva is retained and nourished within the female during its three larval stages, until it is fully developed and ready to pupariate. The female fly then deposits the larva, gluing it to the animal's hair. In 19 to 24 days from the time it was deposited an adult fly emerges, and in another 14 days this fly can produce its first offspring. The cycle may be longer in winter and shorter in summer. Keds that become detached from their host can survive only for periods of up to 4 days. The spread of sheep keds is largely through contact and the movement of keds from ewes to lambs is an important route of infestation.

Pathology: the irritation caused by keds makes sheep restless so they do not feed well, resulting in loss of condition. Heavy infestations may be more common in autumn and winter and ked populations are lowest in summer. Long-woolled breeds may be particularly susceptible to infestation. Shearing may remove a high proportion of the ked population on a sheep. Inflammation leads to pruritus, biting, rubbing, wool loss and a vertical ridging of the skin known as 'cockle'. Faecal staining of the wool reduces its value; heavy infestations may lead to anaemia. The total annual devaluation of sheepskins in the USA due to cockle is estimated at about US$4 million and the overall

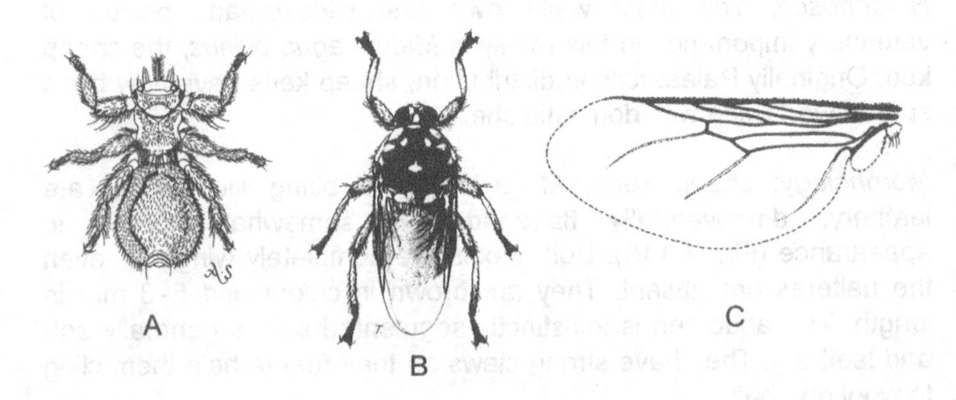

Fig. 4.14 (A) Sheep ked, *Melophagus ovinus;* (B) louse fly, *Hippobosca rufipes;* and (C) wing venation typical of species of *Hippobosca*, showing the characteristic crowding of the veins into the leading half of the wings (reproduced from Smith, 1993).

losses in the USA due to keds are estimated to be about US$40 million per year. Keds may also infest goats.

Hippoboscidae of minor veterinary interest

- The forest fly, *Hippobosca equina*, known in the UK as 'New Forest fly' is found worldwide. Adults are about 10 mm in length. They have wings and a 'shrivelled' body. They primarily attack horses, but may also occasionally be found on cattle. Both sexes of adult are blood-feeders. Like the sheep ked, adult females produce a single, fully grown larva which drops to the ground to pupate immediately. The winged adults emerge and locate a suitable host. Favoured feeding sites are around the perineum and between the hind legs. This species is primarily a nuisance and a cause of disturbance.

- The species *Lipoptena cervi* in Europe and, in North America, *Lipoptena depressa* and *Neolipoptena ferrisi* are common parasites of deer. They are of interest because, like *Hippobosca equina,* the adult is winged on emergence. However, on finding a suitable host the wings are shed. The wingless adults of *L. cervi* can be distinguished from *M. ovinus* by the presence of the halteres.

- Species of the genera *Ornithomya*, *Stenepteryx*, *Lynchia* and *Crataerina* are parasites of wild birds. However, domestic birds, with the exception of domestic pigeons, are not usually parasitized by these flies.

4.7.4 Glossinidae (tsetse flies)

The sole genus in the family Glossinidae is *Glossina*, species of which are known as tsetse flies. Tsetse are a small distinct genus of 22 species, which feed exclusively on the blood of vertebrates. They are entirely restricted to sub-Saharan Africa and, strictly speaking therefore, are outside the geographical remit of this book. As a result, they will be mentioned only briefly here.

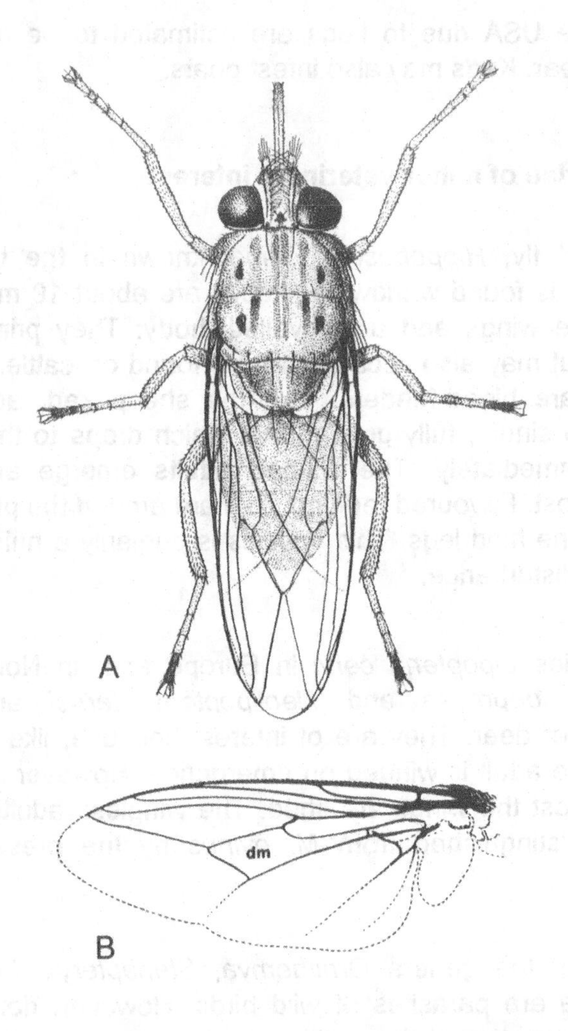

Fig. 4.15 (A) Male tsetse fly, *Glossina longipennis* and (B) wing venation typical of species of *Glossina*, showing the characteristic hatchet shape of the cell dm (reproduced from Jordan, 1993).

Morphology: tsetse are narrow-bodied flies, 6–14 mm in length and yellowish to dark brown in colour (Fig. 4.15). At rest the wings are held, scissor-blade like, overlapping over the abdomen. In many species the abdomen is marked with conspicuous dark lines and patches. The proboscis is long, forwardly directed and embraced by long palps. Each antenna has an elongated third segment with an arista with 17–29 short hairs, present only on the dorsal side. The most characteristic diagnostic feature of the genus is the discal medial cell of the wings which is described as 'hatchet' shaped.

Life cycle: rather like the Hippoboscidae, the form of reproduction of tsetse is a highly specialized form of adenotrophic viviparity. The fertilized egg is retained in the oviduct, where it hatches after about 4 days at 25°C. The larva passes through three stadia, nourished by secretions from the highly modified accessory glands. Larviposition occurs about 9 days after fertilization. At this stage the third-instar larva is fully grown and ready to undergo metamorphosis. Pupariation is initiated within 1–2 hours of larviposition, after the larva has burrowed into the ground. The adult fly emerges after about 30 days at 25°C. Hence, the female supplies all the requirements for the growth and development of the larva, which usually weighs at least as much as the mother at larviposition. Adult tsetse are relatively long-lived, but each female produces only two or three offspring during the course of its life. Both sexes of adult are blood-feeders.

Pathology: tsetse flies are important because they are the primary vectors of a number of species of the parasitic protozoan trypanosomes which cause trypanosomiasis in humans and domestic animals, described as sleeping sickness or Nagana, respectively. The normal hosts of tsetse flies are African wild, large mammals and reptiles, which experience few or no ill effects from the presence of the trypanosomes in their blood, unless subject to stresses such as starvation. These wild animals act as reservoirs of the disease. When humans or domestic animals become infected, however, the pathogenic effects of the trypanosomes are usually fatal unless treated.

4.7.5 Cyclorrhaphous flies of minor veterinary interest

Garbage flies

Muscidae of the genus *Ophyra*, are small (about 5 mm in length), shiny, black flies. They breed in refuse and animal excrement, particularly poultry manure. The genus is distributed worldwide. Species may occur along with *M. domestica* and may be an important nuisance to livestock.

Muscina

Muscina stabulans and *Muscina assimilis* are widespread throughout the USA, Asia and Europe. Adults resemble house flies, although they are slightly larger. Eggs are laid in livestock faeces or other decaying material. As with house flies, species of *Muscina* may be responsible for the mechanical transmission of pathogens.

Bat flies

The Nycteribiidae are spider-like, wingless insects parasitic on bats, closely related to the Hippoboscidae (Fig. 4.16). Adults leave their hosts to deposit their larvae on beams and walls, where they immediately pupate. The head is greatly reduced and when at rest, is folded back into a groove in the upper surface of the thorax.

4.8 BRACHYCERA

There is only a single family of Brachycera of major veterinary

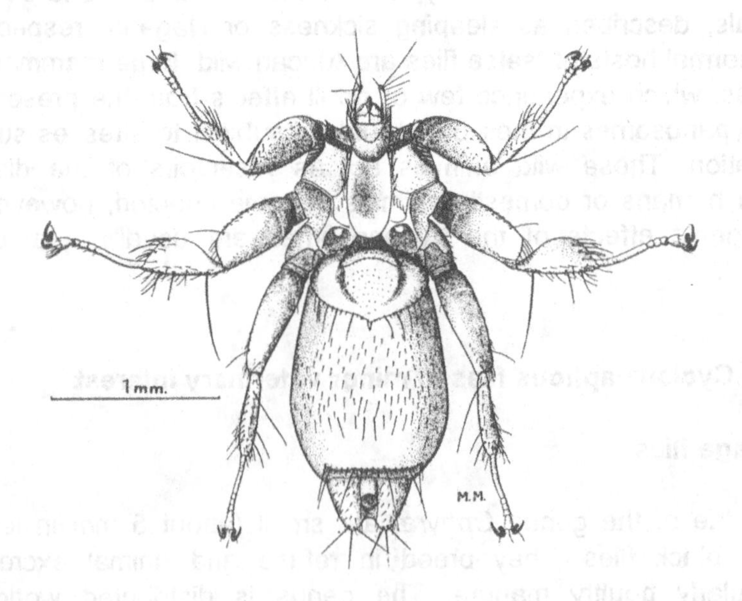

Fig. 4.16 Adult female *Nycteribia pedicularis* (from Edwards, Oldroyd and Smart, 1939).

interest, the Tabanidae, often known as horse flies, deer flies and clegs. Other families of Brachycera include the Stratiomyidae (soldier-flies), Asilidae (robber flies) and Rhagionidae (snipe flies), which feed on decaying vegetation or other small insects.

4.8.1 Tabanidae (horse flies, deer flies and clegs)

The family Tabanidae contains over 4000 species divided into 30 genera, only three of which are of major veterinary importance in temperate habitats: *Tabanus*, *Haematopota* and *Chrysops*. Species of the genus *Tabanus* are found worldwide; the *Haematopota* are largely Palaearctic, Afrotropical and Oriental in distribution; species of the genus *Chrysops* are largely Holarctic and Oriental.

Morphology: all the Tabanidae are large robust flies, up to 25 mm long, with large, broad heads and bulging eyes (Fig. 4.17). The body is generally dark in colour, but this can be very variable, ranging

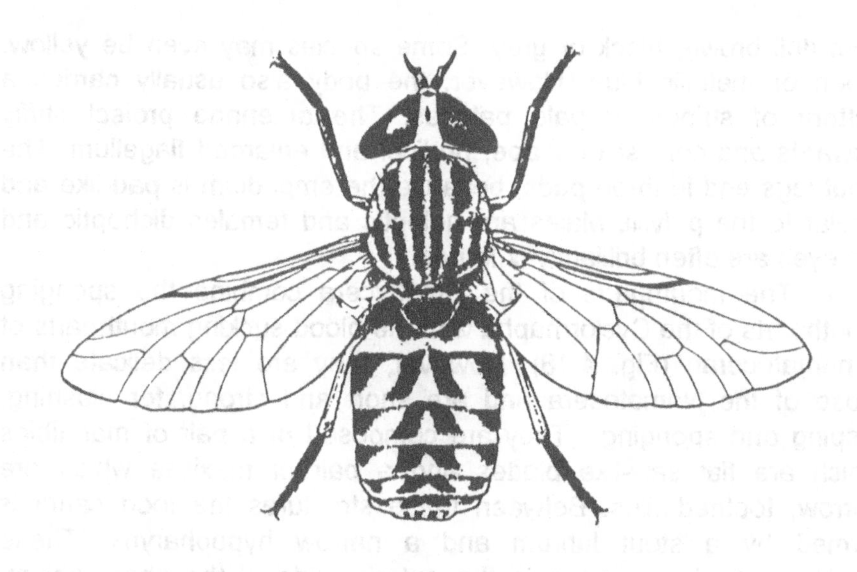

Fig. 4.17 The horse fly, *Tabanus opacus* (reproduced from McAlpine, 1961).

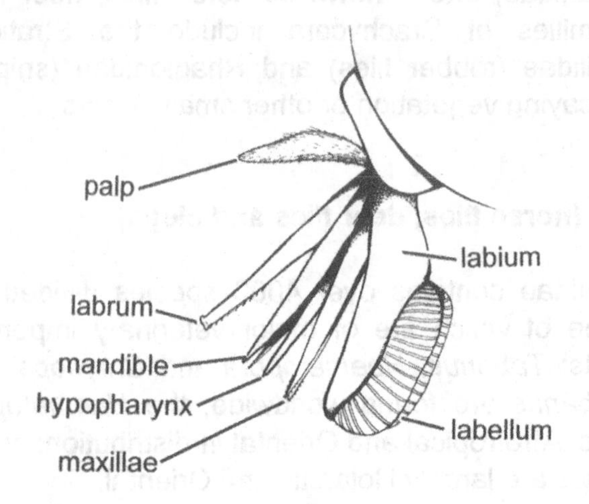

Fig. 4.18 Mouthparts of a tabanid (Diptera: Brachycera) (from Urquhart *et al.*, 1987).

from dull brown, black or grey. Some species may even be yellow, green or metallic blue. However, the body also usually carries a pattern of stripes or pale patches. The antennae project stiffly forwards and consist of scape, pedicel and enlarged flagellum. The stout legs end in three pads, because the empodium is pad-like and similar to the pulvilli. Males are holoptic and females dichoptic and the eyes are often brilliantly coloured.

The mouthparts of the Brachycera combine the sponging mouthparts of the Cyclorrhapha with the blood-sucking mouthparts of a nematoceran (Fig. 4.18). However, they are less delicate than those of the Nematocera and are short and strong, for slashing, rasping and sponging. They are composed of a pair of mandibles which are flat saw-like blades and a pair of maxillae which are narrow, toothed files. Between these structures the food canal is formed by a stout labrum and a narrow hypopharynx. These structures lie in a groove in the anterior side of the short labium, which bears a pair of large, fleshy sensory labella. The maxillary palps are two-segmented.

When a female tabanid feeds, the labella are retracted and the closely associated labrum, mandibles and maxillae penetrate the skin. During this process the mandibles move across each other with a scissor-like action. Saliva, which contains an anticoagulant, is pumped into the wound, before blood is sucked up into the food canal. When feeding ceases and the mouthparts are withdrawn, the labella close together, trapping a small film of blood between them. This is important because pathogens in this blood may be protected and survive for an hour or more. Next time the fly attempts to feed pathogens may escape from the blood from the previous feed to infect the new host. Mechanical transmission of pathogens is made more likely if the fly is disturbed during its first feeding attempt and therefore is unable to obtain a full blood meal. In this case the fly is likely to attempt to feed again quickly, during the period when blood pathogens remain alive.

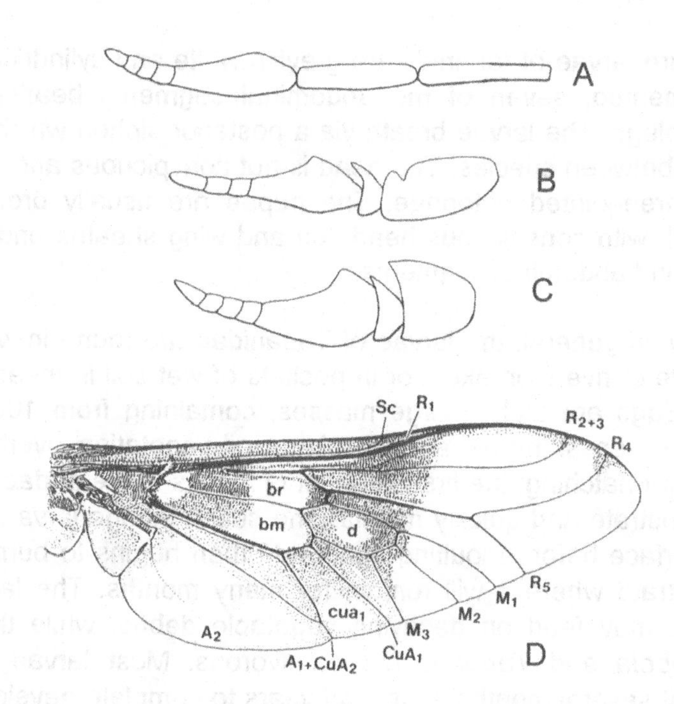

Fig. 4.19 Antennae of (A) *Chrysops* (B) *Haematopota* and (C) *Tabanus*. (D) Wing venation of Tabanidae (reproduced from Chainey, 1993).

The wing venation of Tabanidae is very characteristic with R_{4+5} forked to form a large 'Y' across the wing tip (Fig. 4.19).

- Species of the genus *Tabanus* are the largest of the Tabanidae. Their wings are usually clear. The antennal flagellum is composed of five segments, with the first segment of the flagellum bearing a horn-like projection (Fig. 4.19).
- Species of the genus *Haematopota,* often known as clegs, have speckled or dark-banded wings and eyes marked with zigzag bands. They have only four segments in the antennal flagellum (Fig. 4.19).
- Species of *Chrysops* are often known as deer flies because of their preferred habitat of woodland. They have wings often with a single, broad, dark stripe and spotted eyes. The proboscis is not longer than the head. The antennal flagellum has five segments (Fig. 4.19).

The mature larvae of tabanids are greyish-white and cylindrical; they are segmented, seven of the abdominal segments bearing eight fleshy prolegs. The larvae breath via a posterior siphon which varies in length between species. The head is not conspicuous and bears a pair of three-jointed antennae. The pupae are usually brown and cylindrical, with conspicuous head, leg and wing sheaths and visible thoracic and abdominal segments.

Life cycle: in general, the larvae of Tabanidae are found in wet mud at the side of rivers or lakes, or in pockets of wet soil in meadows or forests. Eggs are laid in large masses, containing from 100 up to 1000 eggs, on stems of aquatic plants or vegetation overhanging water. After hatching, the first-stage larva moves to the surface of the damp substrate and quickly moults. The second-stage larva remains at the surface before moulting again and then begins to burrow into the substrate where it will remain for many months. The larvae of *Chrysops* may feed on decaying vegetable debris, while those of *Haematopota* and *Tabanus* are carnivorous. Most larvae require periods of several months to several years to complete development, during which time they pass through between 6 and 13 stadia. Pupation takes place close to, or within, dryer parts of the substrate. Most temperate species have only a single generation per year and adults live for 2–4 weeks.

Adults are strong fliers and are usually diurnal. Both sexes feed on nectar and, in most species, females are also haematophagous and feed on a wide range of hosts.

Pathology: tabanids feed primarily on large mammals and occasionally birds. The flies stay on animals only long enough to feed. The bites of these flies are deep and painful and may cause considerable disturbance. In the USA studies have shown that 20 to 30 flies feeding over a period of 6 hours may take approximately 100 ml of blood. The activities of flies may also reduce milk yields substantially. In 1976, estimated losses in the USA due to tabanids were US$40 million. Tabanids may also be important mechanical vectors of anthrax, pasteurellosis, equine infectious anaemia (retrovirus), hog cholera (pestivirus), tularaemia, trypanosomiasis and anaplasmosis.

In northern Europe, the dull-grey *Haematopota pluvialis*, sometimes called a cleg, is probably the most well known, being a persistent species with a painful bite. In North America, the most important representative of this genus is *Haematopota americanus*, which occurs from Alaska to New Mexico. In eastern states of North America, *Tabanus atratus* is the widespread black horse fly; *Tabanus lineola* and *Tabanus similis* are also annoying pests. In the west, *Tabanus punctifer* is an important pest as is *Tabanus sulcifrons* in the midwest. *Tabanus quinquevittatus* and *Tabanus nigrovittatus* are well known in North America as 'greenheads'. *Chrysops discalis* is thought to be a vector of *Pasteurella* in the USA.

4.9 NEMATOCERA

Flies of the sub-order Nematocera are usually small, slender and delicate with long, filamentous antennae composed of many articulating segments. The wings are often long and narrow, with many branching veins. The palps are usually pendulous, though not in mosquitoes, and usually are composed of four or five segments. Following pupation, the adult fly escapes from the pupal case through a dorsal longitudinal slit.

The mouthparts or proboscis of Nematocera are considerably modified (Fig. 4.20). The labium, which is more delicate than seen in the Cyclorrhapha or Tabanidae, forms a protective sheath for the other mouthparts, known collectively as the stylets, and ends in two

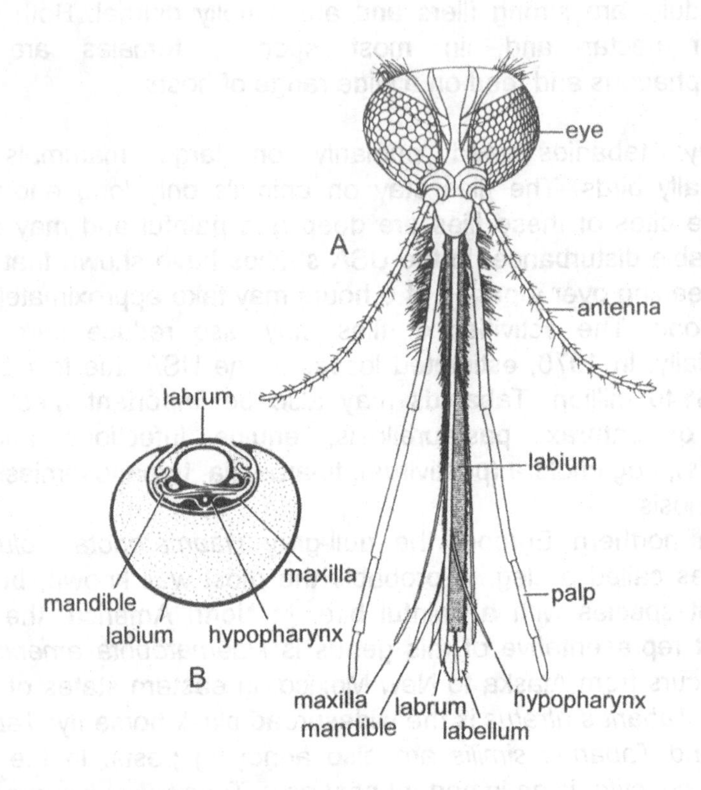

Fig. 4.20 Mouthparts of a mosquito (Diptera: Nematocera). (A) Frontal view (reproduced from Gullan and Cranston, 1994). (B) Transverse section (reproduced from Service, 1993).

small, sensory labella. Inside the labium lies the labrum which is curled inwards to the edges so that it almost forms a complete tube. The gap in the labrum is closed by the very fine paired mandibles to form a food canal. Behind the mandibles lies the slender hypopharynx, bearing the salivary canal, and behind this are the paired maxillae (laciniae) (Fig. 4.20). Both the mandibles and maxillae are finely toothed towards their tips. At the base of the mouthparts are a single pair of sensory maxillary palps.

The structure of these mouthparts is essentially similar in all families of blood-feeding Nematocera (Ceratopogonidae, Simuliidae, Culicidae and Phlebotominae). However, they are greatly elongated in the mosquitoes. When a female mosquito feeds, the labella test the surface of the skin and select a suitable location. The labrum, mandibles, maxillae and hypopharynx are then thrust into the skin, while the labella rest on the surface with the labium bending backwards, like a snooker or pool cue going through a player's hand rest. Saliva is pumped down the middle of the hypopharynx to cause local vasodilation and blood is sucked up the food canal by two muscular pumps. The shorter, more robust mouthparts of the other blood-feeding Nematocera allow them only to puncture the skin surface and then feed from the accumulating pool of blood. Most nematocerous males do not feed on blood and have either poorly developed or non-functional mouthparts.

4.9.1 Simuliidae (black flies)

The members of the nematoceran family Simuliidae are known as black flies or buffalo gnats. There are more than 1500 known species in 19 genera distributed worldwide. Only the single large genus *Simulium* is of significant economic veterinary importance in temperate habitats. They feed on the blood of cattle, horses, sheep, goats, poultry, other livestock, wild mammals and birds.

Morphology: the adults are small, 1–5 mm in length, with squat bodies and a characteristic humped thorax with broad wings (Fig. 4. 21). The antennae are shorter than in most other nematocera, usually with 11 rounded segments. The palps are conspicuously long with five segments and extend beyond the proboscis. Females are dichoptic and males holoptic but with characteristic enlarged ommatidia in the upper part of the eye. The description 'black fly' can be somewhat misleading because, while the adults are often black, the colouration may vary between black, grey or even dark yellow. The wings are short and broad, with a large anal lobe, and have veins which are thickened at the anterior margin of the wing. The first abdominal tergite is modified to form a prominent basal scale, fringed with fine hairs.

Life cycle: females deposit their eggs in sticky masses or irregular strings of 150–600 eggs. Batches are usually laid on partially submerged stones or vegetation in flowing water. The eggs hatch in 6–12 days. The newly hatched larva spins a silken thread from its salivary glands and uses this to assist it to drift downstream in search of a suitable place to settle. Once an appropriate site has been located, on stones or stems close to the surface, the larvae spin a patch of silk and attach themselves to the silk by their posterior suckers. The larvae remain in areas of fast-flowing current, since they require highly oxygenated water to survive. In deoxygenated water, the larvae detach from their silken pads and drift downstream.

The larvae of most species of *Simulium* possess a pair of mouth brushes which are used to filter small food particles, such as bacteria, algae and detritus, from the water. The larvae pass through up to eight stadia over the course of between a few weeks to a year (Fig. 4.22). Some species overwinter as larvae. Pupation takes place underwater in a cocoon that is firmly attached to shallowly covered objects such as rocks. The pupa has a number of filaments attached

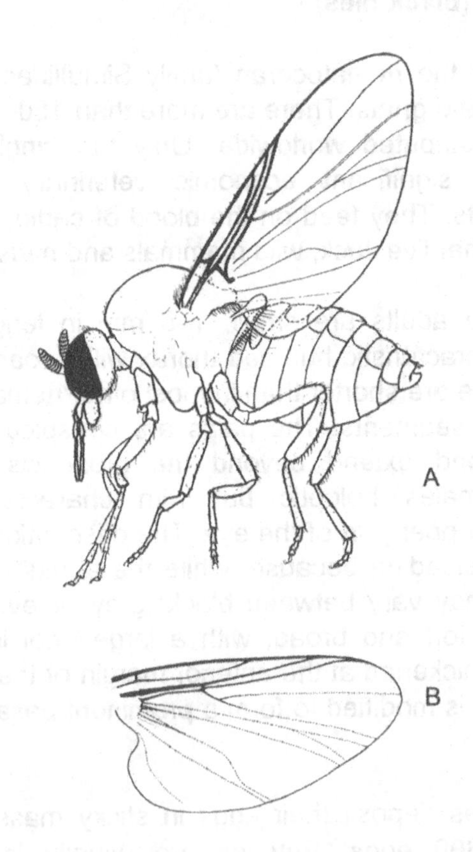

Fig. 4.21 (A) Adult female *Simulium* (reproduced from Soulsby, 1982). (B) Wing venation typical of *Simulium*, showing the large anal lobe and crowding of the veins towards the leading edge (reproduced from Crosskey, 1993b).

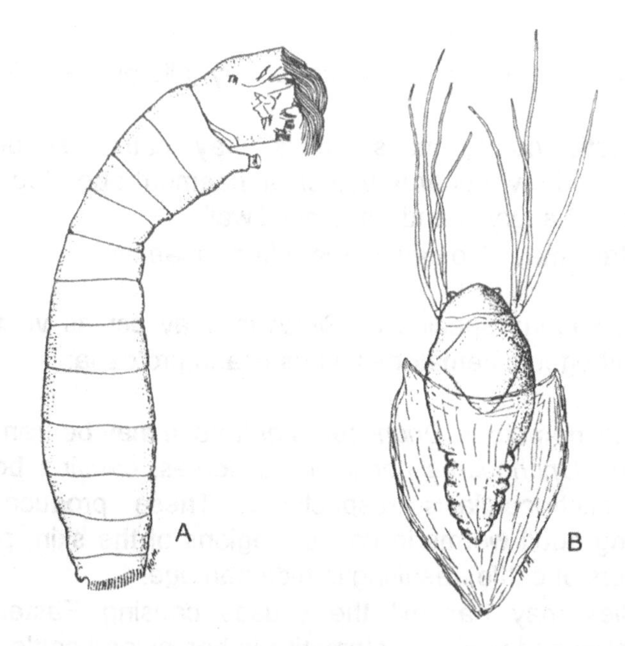

Fig. 4.22 Immature stages of Simuliidae: (A) larvae and (B) pupa within its cocoon (reproduced from Soulsby, 1982).

anteriorly to the dorsal part of the thorax, through which it respires. In the final stages of pupation a film of air is secreted between the developing adult and the pupal cuticle. When the pupal case splits the emerging adult rises to the surface in a bubble of air.

Both males and females feed on plant nectar. However, in most species adult females are anautogenous, requiring a blood meal to obtain the protein necessary to mature their eggs. Adults are predominantly diurnal and are particularly active during morning and evening in warm, cloudy weather. Adults are strong fliers and respond strongly to carbon dioxide and other host-animal odours.

Pathology: simuliids are economically important livestock pests in many parts of the world, particularly areas of North America and eastern Europe. They feed primarily on poultry, cattle and horses, in the latter two species usually feeding on the legs, abdomen, head and ears.

- Even at relatively low population densities the painful bites inflicted by adult female black flies may cause considerable disturbance and reduced productivity.

- Biting stress may be compounded by allergic reactions to fly saliva.
- In domestic cattle, mass attack may cause sudden death characterized by general petechial haemorrhage, together with oedema of the larynx and abdominal wall.
- Mass attack may also cause death from anaemia.

In addition, various species of *Simulium* may act as vectors for a range of pathogenic nematodes, viruses and protozoa:

- The filarial nematode *Onchocerca gutterosa* may be transmitted to cattle and *Onchocerca cervicalis* to horses, causing bovine and equine onchocerciasis respectively. These produce nodules containing adult worms in various regions of the skin, particularly the withers of cattle, resulting in hide damage.
- Black flies may transmit the viruses causing Eastern equine encephalitis and vesicular stomatitis in horses and cattle.
- In North America, black flies may transmit various species of protozoa of the genus *Leucocytozoon* to turkeys and ducks.
- From a medical perspective Simuliidae are particularly important as vectors of the filaroid nematode *Onchocerca volvulus*, which causes river blindness in humans in Africa, Central and South America.

Possibly the most damaging simuliid of temperate latitudes in the new world is *Simulium arcticum* which can be a major livestock pest in western Canada. Populations can reach densities which are high enough to kill cattle. In the USA, *Simulium venustum* and *Simulium vittatum* may be common and widespread pests of livestock, being particularly common in June and July. *Simulium pecuarum,* the southern buffalo gnat, may cause losses in cattle in the Mississippi valley. The turkey gnat, *Simulium meridionale* is common in southern USA and the Mississippi valley, where it may be a significant pest of poultry. *Simulium equinum*, *Simulium erythrocephalum* and *Simulium ornatum* may cause problems in western Europe and *Simulium kurenze* in Russia. Particularly damaging in central and southern Europe is *Simulium colombaschense,* which may cause heavy mortality of livestock.

4.9.2 Ceratopogonidae (biting midges)

The large nematoceran family, the Ceratopogonidae, are known as the biting midges to distinguish them from the non-biting nematoceran midges, the Chironomidae. The family contains more than 60 genera and over 4000 species. However, there is only a single genus of veterinary interest in temperate habitats, *Culicoides*. There are over 1000 species of *Culicoides*, all of which feed on birds or mammals, inflicting a painful bite.

Morphology: all the *Culicoides* are small flies, 1–4 mm in length. The legs are relatively short and stout, particularly the forelegs. The antennae are long and filamentous with 14 to 15 segments in females. Adults may be grey or brownish-black with an iridescent sheen. The thorax is humped over a small head. The wings are

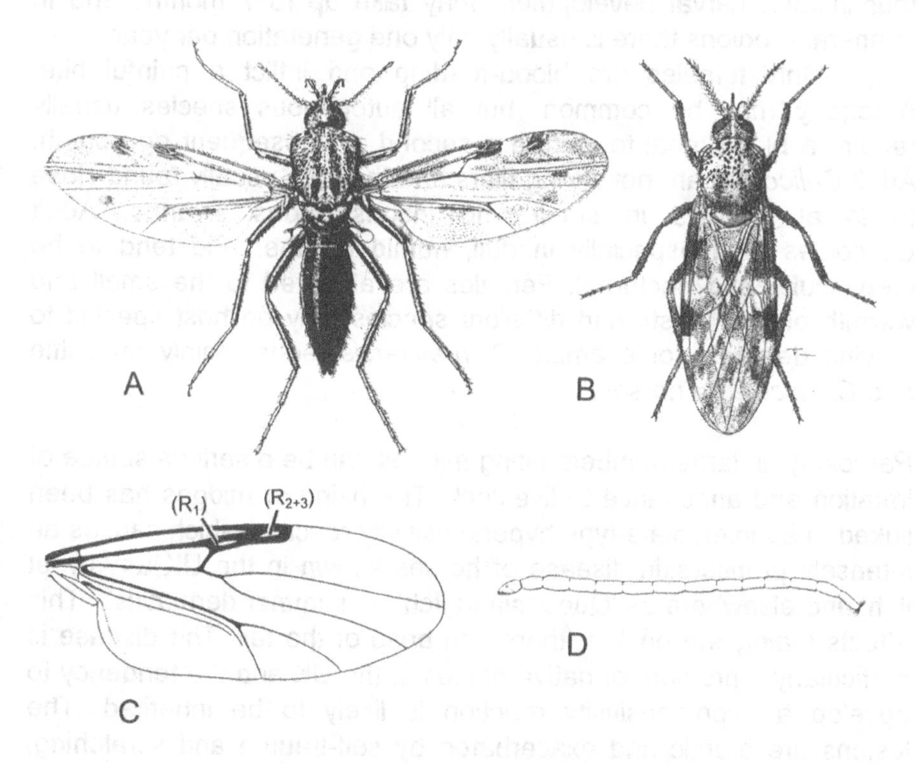

Fig. 4.23 Adult female *Culicoides:* (A) wings spread and (B) living fly at rest. (C) Wing venation typical of species of *Culicoides*, showing the two elongate radial cells. (D) Typical *Culicoides* larva. (Reproduced from Boorman, 1993.)

mottled and hairy. At rest the wings are folded over each other and held flat over the abdomen. Ceratopogonids have a forked media vein (M_1, M_2) and species of the genus *Culicoides* usually have a distinct pattern of radial cells and an r-m cross-vein on their wings.

Life cycle: the eggs are small, dark and cylindrical and are usually deposited in masses of 25–300. The eggs are usually oviposited in damp, marshy ground or decaying vegetation. The larvae are typical of Nematocera, with a well-developed head, 11 body segments and no appendages. The larvae are aquatic and occur in a wide variety of semi-solid habitats, including the edges of lakes and streams, muddy water-filled holes, marshland and swamps. In general, *Culicoides* appear able to exploit a wide range of moist habitats, but individual species tend to utilize only specific breeding sites. The larvae usually burrow into the surface of the substrate, where they pass through four instars. Larval development may take up to 7 months and in temperate regions there is usually only one generation per year.

Only females are blood-feeding and inflict a painful bite. Autogeny may be common, but all autogenous species usually require a blood meal to mature a second or subsequent egg batch. Adult *Culicoides* are not strong fliers and they are usually found close to larval habitats in small and inconspicuous swarms. Adult *Culicoides* feed especially in dull, humid weather and tend to be crepuscular and nocturnal. Females are attracted to the smell and warmth of their hosts and different species may be host specific to varying degrees, for example *C. brevitarsis* feeds mainly on cattle and *C. imicola* on horses.

Pathology: in large numbers biting midges can be a serious source of irritation and annoyance to livestock. The biting of midges has been linked to an immediate-type hypersensitivity reaction which causes an intensely pruritic skin disease of horses known in the UK as 'sweet itch' and elsewhere as 'Queensland itch' or 'summer dermatitis'. This affects mainly the back, withers and base of the tail. The disease is particularly a problem of native ponies in the UK and the tendency to develop a hypersensitivity reaction is likely to be inherited. The lesions are pruritic and exacerbated by self-trauma and scratching, resulting in hair loss, hyperpigmentation and skin thickening. Several species are involved, *C. pulicaris* in Britain, *C. robertsi* in Australia and *C. insignis*, *C. stelifer* and *C. venustus* in the USA.

Species of *Culicoides* may act as mechanical vectors for the filaroid nematodes *Onchocerca reticulata* and *Onchocerca gibsoni* to cattle, *Onchocerca cervicalis* to horses and several species of protozoa (*Haemeproteus, Leucocytozoon*) to poultry and other birds.

At least 20 types of viral pathogens have been isolated from various species of *Culicoides*, including those responsible for the important livestock diseases causing blue-tongue in sheep, African horse sickness and bovine ephemeral fever and, in the USA, eastern equine encephalitis.

- Blue-tongue virus (BTV) exists as a number of distinct serotypes, 24 of which have been recognized to date. These viruses can infect a wide range of ruminant species, but usually only cause severe disease in certain breeds of sheep, particularly the fine-fleeced species, such as Merino and Dorset Horn. In sheep it causes fever, enteritis, upper respiratory tract infection, ulceration of the tongue and lameness. BTV can cause very high mortality, in excess of 25%, and morbidity in excess of 75%. Blue-tongue occurs generally in Africa, the Middle East, Asia, Australia and parts of North America, and serious outbreaks have occurred in the past 50 years in southern Europe. In one such outbreak, between 1956 and 1960, over 180,000 sheep died in Spain and Portugal. In the USA blue-tongue is estimated to cost the livestock industry over US$100 million per year.
- African horse sickness is caused by a retrovirus (AHSV) and is among the most lethal of equine diseases. It frequently causes mortality rates in excess of 90%. It is enzootic in Africa. A series of epizootics in Spain and Portugal from 1987 to the present have resulted in the deaths of over 3000 equines. *Culicoides imicola* is one of the members of the genus able to transmit the virus and occurs widely in Spain, Portugal and southern Greece.
- Eastern equine encephalitis is a viral disease of horses and humans found only in the New World. It is caused by a species of the *Alphavirus* genus which is part of the Togaviridae. The disease is present throughout North and South America as far south as Argentina. The wild reservoir hosts are birds and the primary midge vector is *Culicoides melanura*.
- Bovine ephemeral fever, also known as 3-day sickness, is caused by an arbovirus. It is found throughout Africa, the Oriental region and occasionally causes epizootics in Australia. It affects cattle

causing morbidity, but usually not mortality, resulting in reduced milk yields.

There are a large number of species of *Culicoides* of varying importance as nuisance pests and vectors. Of particular note in Europe and Asia are *Culicoides pulicaris*, *Culicoides obsoletus* (a complex of four separate species), *Culicoides impunctatus* and *Culicoides sibirica*. *Culicoides imicola* is found throughout Africa and southern Europe and is a key vector of African horse sickness. In North America *Culicoides furens* and *Culicoides denningi* inflict painful bites and *Culicoides variipennis* is the primary vector of blue-tongue virus.

4.9.3 Culicidae (mosquitoes)

The Mosquitoes, family Culicidae, are a diverse group of over 3000 species. They occur worldwide from the tropics to the arctic. There are three sub-families: Anophelinae, consisting of *Anopheles* and two other rare genera, *Bironella* and *Chagasia*; Culicinae, composed of nearly all the other genera; and Toxorhynchitinae, which do not feed on blood. There are more than 2500 species of Culicinae, of which the main genera are *Aedes*, containing over 900 species, and *Culex,* with nearly 750 species.

Morphology: mosquitoes are small, slender flies, 2–10 mm in length, with long legs. Adults have scales on the wing veins and margins (Fig. 4.24) and the adult females possess an elongated proboscis which is used in blood-feeding. Male mosquitoes have plumose antennae, whereas those of females have fewer, shorter hairs.

Living anopheline adults can readily be distinguished from culicines, such as *Aedes* and *Culex*, when resting on a flat surface. On landing anopheline mosquitoes rest with the proboscis, head, thorax and abdomen in one straight line at an angle to the surface. The culicine adult rests with its body angled and its abdomen directed towards the surface. The palps of female anopheline mosquitoes are as long and straight as the proboscis, whereas in female culicine mosquitoes the palps are usually only about one-quarter of the length of the proboscis.

Life cycle: anopheline mosquitoes lay batches of eggs on the surface

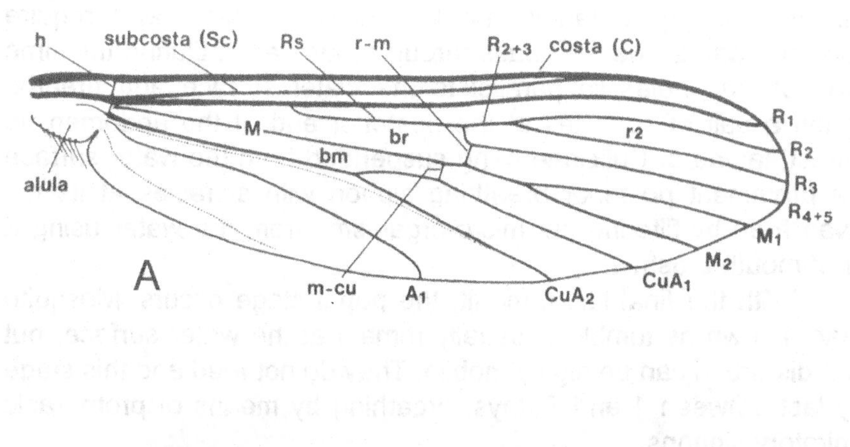

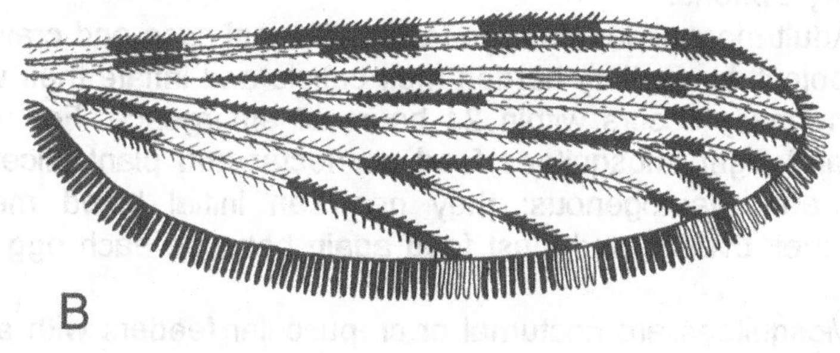

Fig. 4.24 Wing of an adult mosquito: (A) the basic structure with veins and cells; (B) wing of *Anopheles* showing the characteristic patterns produced by the scales on the veins (reproduced from Service, 1993).

of water, usually at night and often deposit more than 200 eggs per oviposition (Fig. 4.5). The eggs possess characteristic lateral floats that prevent them from sinking. Such eggs usually hatch within 2 or 3 days. Species of the genus *Culex* lay their eggs side by side and form them into a raft. Rafts generally contain between 100 and 300 eggs. When the eggs mature they will hatch regardless of the availability of water. Most species of *Aedes* lay their eggs on moist substrates, not on the water itself, where they mature and await adequate water to stimulate hatching. In some cases the eggs may remain viable for up to 3 years.

The larvae of all species are aquatic and occur in a wide variety of habitats, such as the edge of permanent pools, puddles, flooded tree-holes or even, for some species, temporary water-filled containers. Mosquito larvae are known as wrigglers and require between 3 and 20 days to pass through four stadia. During this time larvae of *Anopheles* lie parallel to the water surface and breathe through a pair of spiracles at the posterior end of the abdomen. In contrast, larvae of Culicinae hang suspended from the water surface by a prominent posterior breathing siphon with spiracles at its tip. Larvae feed by filtering out micro-organisms from the water using a pair of mouth brushes.

With the final larval moult, the pupal stage occurs. Mosquito pupae, known as tumblers, usually remain at the water surface, but when disturbed can be highly mobile. They do not feed and this stage may last between 1 and 7 days, breathing by means of prothoracic respiratory siphons.

Adult mosquitoes emerge from the pupal case and crawl to a nearby object where they harden their cuticle and inflate their wings. Mating normally occurs within 24 hours of emergence. For normal activity and flight, mosquitoes feed on nectar and plant juices, but females are anautogenous; they need an initial blood meal to develop their ovaries and must feed again between each egg batch matured.

Mosquitoes are nocturnal or crepuscular feeders with a wide host range. Host location is achieved using a range of olfactory and visual cues, orientation to wind direction and body warmth.

Pathology:

- Mosquito populations can reach large sizes, especially in parts of the southern United States, and the persistent feeding activity of adults females may cause considerable nuisance and reduce the productivity of livestock.
- Mosquito bites can cause hypersensitivity reactions.
- Mosquitoes can be vectors of the dog heartworm, *Dirofilaria immitis* (although this occurs mainly in tropical and sub-tropical regions).
- Mosquitoes can act as vectors of various viral diseases, including arboviruses, such as equine encephalitis (a togavirus), rabbit myxomatosis and infectious equine anaemia (a retrovirus).
- In general, mosquitoes are far more important in terms of human disease than animal disease, since species of the genus *Anopheles* are the sole vectors of malaria which kills over 1 million people each year worldwide. Culicines are important vectors of

arboviruses, causing yellow fever and the filarial nematode *Wuchereria bancrofti* which causes human elephantiasis.

4.9.4 Psychodidae (sand flies)

The large family Psychodidae contains over 600 species, widely distributed in the tropics, sub-tropics and around the Mediterranean. In the sub-family Phlebotominae, the true sand flies, there is a single genus of veterinary importance in the Old World, *Phlebotomus*, and a single genus of veterinary importance in the New World, *Lutzomyia*. Species of both genera may act as vectors of *Leishmania*.

Morphology: phlebotomine sand flies are narrow-bodied, up to 5 mm in length. They are hairy in appearance with large black eyes and long legs. The wings are narrow, long, hairy and held erect over the thorax when at rest. The antennae are long, 16-segmented, filamentous and covered in fine setae. There is no sexual dimorphism in the antennae, but males have conspicuous genital apparatus. The mouthparts are moderately long and the palps are five-segmented. In the Old World genus *Phlebotomus* the longest palpal segment is the fifth, whereas in the New World genus *Lutzomyia* the third palpal segment is usually the longest.

Life cycle: females lay 50–100 eggs per egg batch in small cracks or holes in damp ground, in leaf litter and around the roots of forest trees. The grey, segmented larvae pass through four stadia before pupation. The larvae feed on organic debris, such as faeces and decaying plant material. The larvae and pupae are very sensitive to desiccation.

Only females have functional mouthparts and are blood-feeders. Adult females of most species are anautogenous. Species of the genus *Phlebotomus* feed on mammals and are found in savannah and desert areas. In contrast, species of the genus *Lutzomyia* feed on both mammals and reptiles and are found in damper, forested areas. Adults often accumulate in the burrows of rodents or in other shelters, such as caves, where the microclimate is suitable. The adults remain in these refugia, feeding at night, dawn or dusk, on the occupants or mammals in the close vicinity. Adults have very limited powers of flight, with a range of perhaps only 100–200 m. They move in characteristic short hops of flight and are only able to

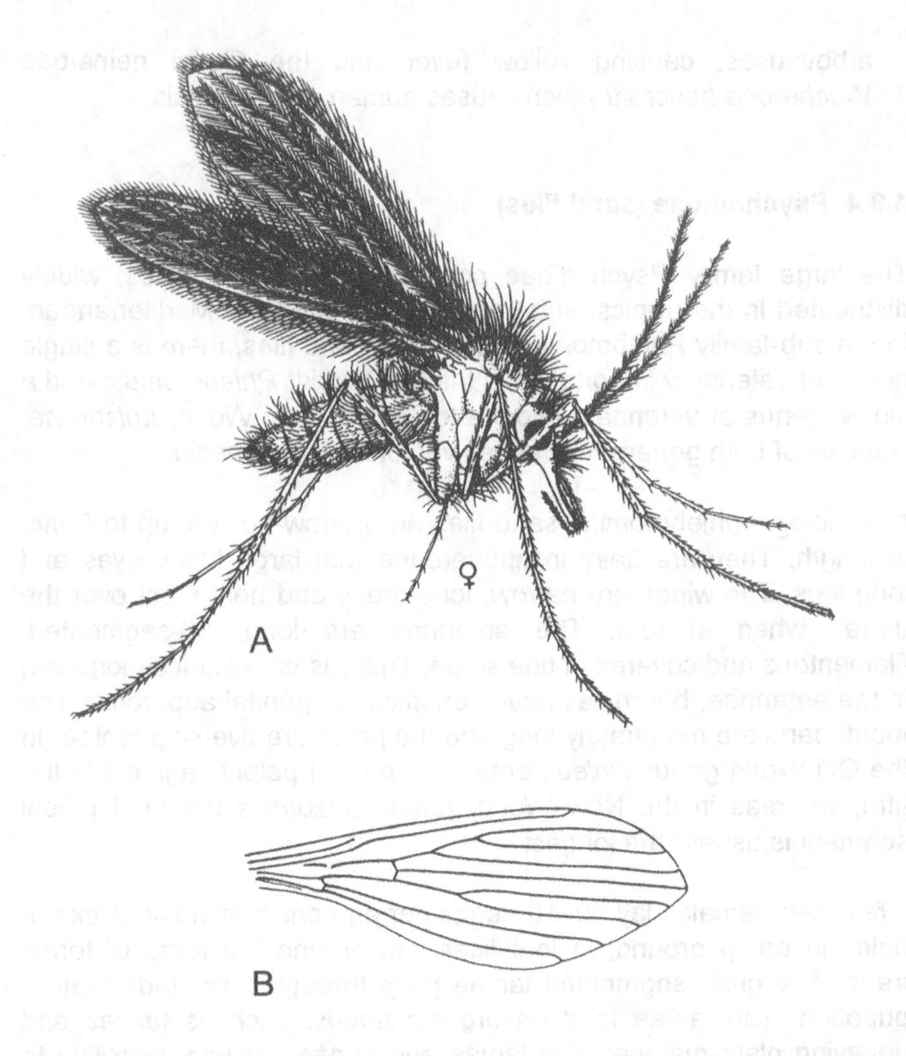

Fig. 4.25 (A) Adult female sand fly, *Phlebotomus papatasi*. (B) Wing venation typical of species of *Phlebotomus* (Psychodidae). (Reproduced from Lane, 1993b.)

fly when wind speeds are low.

The rate of life-cycle development is usually slow, taking at least 7–10 weeks, with many Palaearctic species having only two generations per year.

Pathology: sand flies are important as vectors of various pathogens. Of particular importance is leishmaniasis in humans and dogs, caused by the protozoa, *Leishmania* spp. The diseases caused in humans are commonly classified as either visceral (kala-azar) or cutaneous infections. Dogs, rodents and other wild animals act a reservoirs of infection. Dogs affected with cutaneous leishmaniasis have a nonpruritic, exfoliative dermatitis with alopecia and peripheral lymphadenopathy. Systemic leishmaniasis leads to splenomegaly, hepatomegaly, generalized lymphadenopathy, lameness, anorexia, weight loss and death. The disease has been reported in cats. In North America, sand flies may also act as vectors of vesicular stomatitis of cattle and horses, which is caused by a rhabdovirus.

4.10 OTHER DIPTERA OF VETERINARY INTEREST

4.10.1 Eye gnats

Eye gnats of the genus *Hippelates*, are acalypterate Diptera of the family Cloropidae. They are small (1.5–2.5 mm in length), shiny, dark flies. They are extremely persistent, non-biting flies and feed on body fluids such as tears, oil secretions, pus and blood. They land some distance from the intended feeding site and then crawl over the skin to reach their feeding site, usually around the eyes, genitals and any open sores. They have been associated with bovine keratoconjunctivitis and mastitis in cattle and, for this reason, are of some importance.

4.11 FURTHER READING

Allan, S.A., Day, J.F. and Edman, J.D. (1987) Visual ecology of biting flies. *Annual Review of Entomology*, **32**, 297–316.

Axtell, R.C. and Arends, J.J. (1990) Ecology and management of arthropod pests of poultry. *Annual Review of Entomology*, **35**, 101–26.

Ball, S.G. (1984) Seasonal abundance during summer months of some cattle-visiting Muscidae (Diptera) in North-East England. *Ecological Entomology*, **9**, 1–10.

Berlyn, A.D. (1978) The field biology of the adult sheep headfly, *Hydrotaea irritans* (Fallén) (Diptera: Muscidae), in south-western Scotland. *Bulletin of Entomological Research*, **68**, 431–6.

Boorman, J. (1993) Biting midges (Ceratopogonidae), in *Medical Insects and Arachnids* (eds R.P. Lane and R.W. Crosskey), Chapman & Hall, London, pp. 288–309.

Bowen, M.F. (1991) The sensory physiology of host-seeking behaviour in mosquitoes. *Annual Review of Entomology*, **36**, 139–58.

Bram, R.A. (1978) *Surveillance and Collection of Arthropods of Veterinary Importance.* USDA, Department of Agriculture Handbook, No. 518.

Burger, J.F. and Pechuman, L.L. (1986) A review of the genus *Haematopota* (Diptera: Tabanidae) in North America. *Journal of Medical Entomology*, **23**, 345–52.

Chainey, J.E. (1993) Horse-flies, deer-flies and clegs (Tabanidae), in *Medical Insects and Arachnids* (eds R.P. Lane and R.W. Crosskey), Chapman & Hall, London, pp. 310–32.

Cole, F.R. (1969) *The flies of Western North America*, University of California, Berkley and Los Angeles.

Colyer, C.N. and Hammond, C.O. (1968) *Flies of the British Isles*, 2nd edn, Fredrick Warne, London.

Crosskey, R.W. (1993a) Introduction to the Diptera, in *Medical Insects and Arachnids* (eds R.P. Lane and R.W. Crosskey), Chapman & Hall, London, pp. 51–77.

Crosskey, R.W. (1993b) Blackflies (Simuliidae), in *Medical Insects and Arachnids* (eds R.P. Lane and R.W. Crosskey), Chapman & Hall, London, pp. 241–87.

Crosskey, R.W. and Lane, R.P. (1993) House-flies, blow-flies and their allies (calypterate Diptera), in *Medical Insects and Arachnids* (eds R.P. Lane and R.W. Crosskey), Chapman & Hall, London, pp. 403–28.

DeFoliart, G.R., Grimstad, P.R. and Watts, D.M. (1987) Advances in mosquito-borne arbovirus/vector research. *Annual Review of Entomology*, **32**, 479–505.

Downes, J.A. (1969) The swarming and mating flight of Diptera. *Annual Review of Entomology*, **14**, 271–98.

Edwards, F.W., Oldroyd, H. and Smart, J. (1939) *British Blood-Sucking Flies*, British Museum, London.

Graham-Smith, G.S. (1930) *Parasitology*, Cambridge University Press, Cambridge.

Gullan, P.J. and Cranston, P.S. (1994) *The Insects. An Outline of Entomology*, Chapman & Hall, London.

Harwood, R.F. and James, M.T. (1979) *Entomology in Human and Animal Health*, Macmillan, New York.

Hillerton, J.E., West, J.G.H. and Shearn, M.F.H. (1992) The cost of summer mastitis. *Veterinary Record*, **131**, 315–17.

Jordan, A.M. (1986) *Trypanosomiasis Control and African Rural Development*, Longman, New York.

Jordan, A.H. (1993) Tsetse-flies (Glossinidae), in *Medical Insects and Arachnids* (eds R.P. Lane and R.W. Crosskey), Chapman & Hall, London, pp. 333–88.

Ketttle, D.S. (1977) Biology and bionomics of blood sucking ceratopogonids. *Annual Review of Entomology*, **22**, 33–51.

Lane, R.P. (1993a) Introduction to the arthropods, in *Medical Insects and Arachnids* (eds R.P. Lane and R.W. Crosskey), Chapman & Hall, London, pp. 30–48.

Lane, R.P. (1993b) Sandflies (Phlebotominae), in *Medical Insects and Arachnids* (eds R.P. Lane and R.W. Crosskey), Chapman & Hall, London, pp. 78–119.

McAlpine, J.F. (1961) Variation, distribution and evolution of the *Tabanus (Hybomitra) frontalis* complex of horse flies (Diptera: Tabanidae). *Canadian Entomologist*, **93**, 894–924.

Morgan, C.E. and Thomas, G.D. (1974) Annotated bibliography of the horn fly, *Haematobia irritans* (L.), including references to the buffalo fly, *H. exigua* (de Meijere), and other species belonging to the genus *Haematobia*. *United States Department of Agriculture Miscellaneous Publication*, **1278**, 1–134.

Newstead, R., Evans, A.M. and Potts, W.H. (1924) Guide to the study of tsetse flies, in *Liverpool School of Tropical Medicine Memoirs*, University of Liverpool Press, Liverpool, and Hodder and Stoughton, London.

Oldroyd, H. (1970) Diptera I. Introduction and key to families, 3rd edn. *Handbooks for the Identification of British Insects*, **9**(1), 1–104.

Ribeiro, J.M.C. (1987) Role of saliva in blood-feeding by arthropods. *Annual Review of Entomology*, **32**, 463–78.

Service, M.W. (1993) Mosquitoes (Culicidae), in *Medical Insects and Arachnids* (eds R.P. Lane and R.W. Crosskey), Chapman & Hall, London, pp. 120–240.

Skidmore, P. (1985) *The biology of the Muscidae of the world,* W. Junk Publishers, Dordrecht.

Smith, K.G.V. (1989) An introduction to the immature stages of British Flies; Diptera larvae, with notes on eggs, puparia and pupae. *Handbooks for the Identification of British Insects,* **10**, (12), 1–280.

Smith, G.V. (1993) Insects of minor medical importance, in *Medical Insects and Arachnids* (eds R.P. Lane and R.W. Crosskey), Chapman & Hall, London. pp. 576–93.

Soulsby, E.J.L. (1982) *Helminths, Arthropods and Protozoa of Domesticated Animals*, Baillière Tindall, London.

Steelman, C.D. (1976) Effects of external and internal arthropod parasites on domestic livestock production. *Annual Review of Entomology,* **21**, 155–78.

Tabachnick, W.J. (1996) *Culicoides variipennis* and bluetongue-virus epidemiology in the United States. *Annual Review of Entomology*, **41**, 23–44.

Tarry, D. (1985) The control of sheep headfly disease. *Proceedings of the Sheep Veterinary Society*, **2**, 51–6.

Tarry, D.W., Bernal, L. and Edwards, S. (1991) Transmission of bovine virus diarrhoea virus by blood feeding flies. *The Veterinary Record*, **128**, 82–4.

Tarry, D.W., Sinclair, I.J. and Wassall, D.A. (1992) Progress in the British hypodermosis eradication programme: the role of serological surveillance. *Veterinary Record*, **131**, 310–12.

Titchener, R.N., Newbold, J.W. and Wright C.L. (1981) Flies associated with cattle in South-West Scotland during the summer months. *Research in Veterinary Science*, **30**, 109–13.

Urquhart, G.M., Armour, J., Duncan, J.L. *et al.* (1987) *Veterinary Parasitology*, Longman, Harlow.

Zumpt, F. (1965) *Myiasis in Man and Animals in the Old World,* Butterworths, London, pp. 267.

Zumpt, F. (1973) *The Stomoyniae biting flies of the world. Diptera. Muscidae. Taxonomy, biology, economic importance and control measures,* Gustav Fischer, Stuttgart.

5

Myiasis

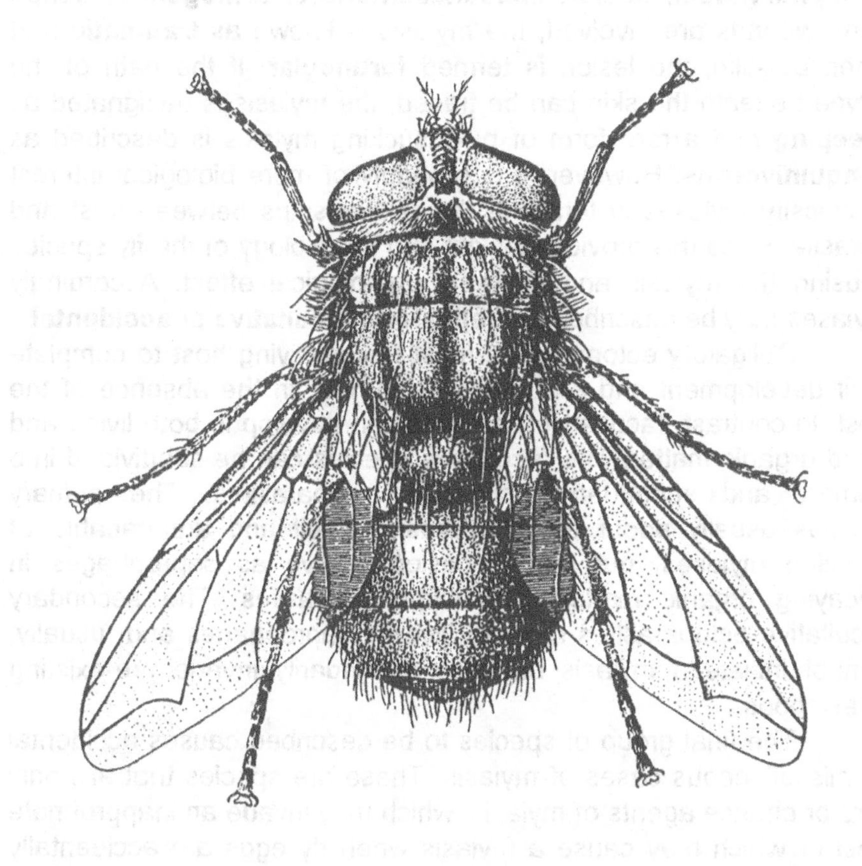

Adult female screwworm fly, *Cochliomyia hominivorax* (reproduced from Herms and James, 1961).

5.1 INTRODUCTION

Myiasis is the infestation of the organs or tissues of host animals by the larval stages of dipterous flies, usually known as **maggots** or **grubs**. The fly larvae feed directly on the host's necrotic or living tissue. The hosts are usually mammals, occasionally birds and, less commonly, amphibians or reptiles.

Myiases are often classified according to the anatomical position in, or on, the animal that the larvae infest. Broadly speaking, they may be described as **dermal**, **sub-dermal** or **cutaneous**, **nasopharyngeal**, **ocular**, **intestinal/enteric** or **urinogenital**. When open wounds are involved, the myiasis is known as **traumatic** and when boil-like, the lesion is termed **furuncular**. If the path of the larvae beneath the skin can be traced, the myiasis is designated as **creeping** and a rare form of bloodsucking myiasis is described as **sanguinivorous**. However, it is probably of more biological interest to classify myiases in terms of the relationships between host and parasite, since this provides insight into the biology of the fly species causing the myiasis and its likely pathological effect. Accordingly myiases may be described as **obligatory**, **facultative** or **accidental**.

Obligatory ectoparasites must have a living host to complete their development and are unable to survive in the absence of the host. In contrast, facultative parasites can develop in both living and dead organic matter. The facultative species can be subdivided into primary and secondary facultative ectoparasites. The primary species usually adopt an ectoparasitic habit and are capable of initiating myiases, but may occasionally live as saprophages in decaying organic matter and animal carcasses. The secondary facultative ectoparasites normally live as saprophages and, usually, cannot initiate a myiasis but may secondarily invade pre-existing infestations.

The final group of species to be described causes accidental or miscellaneous cases of myiasis. These are species that are only rare or chance agents of myiasis, which may invade an inappropriate host or which may cause a myiasis when fly eggs are accidentally ingested. These species are primarily of interest from a medical point of view and will not be discussed further in this book.

5.2 MORPHOLOGY

The body of the immature cyclorrhaphous dipteran larva is usually conical, pointed anteriorly and truncated posteriorly (Fig. 5.1). However, this shape may be modified, with the larvae of some species being barrel-like or, occasionally, flattened. The body is clearly divided, usually into 12 segments. There is a head, three thoracic segments and eight abdominal segments. However, no distinction between the thoracic and abdominal segments is usually apparent. The cuticle is typically soft and unsclerotized but is often covered by spines or scales arranged in circular bands. Although legless, in some species, the body may have a number of fleshy protuberances which aid in locomotion.

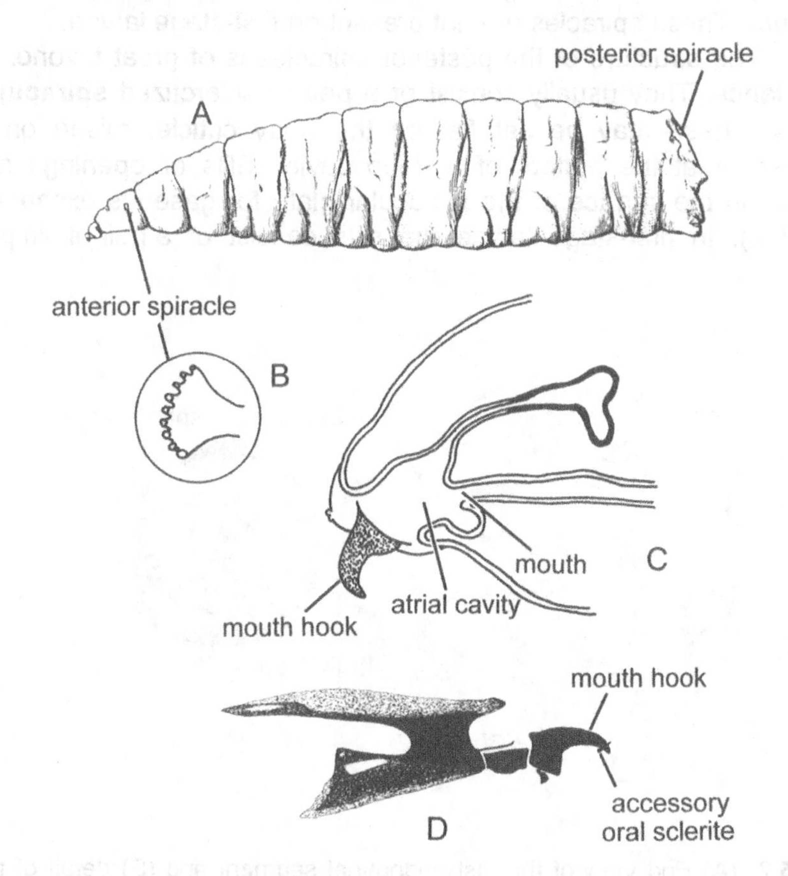

Fig. 5.1 Structure of a cyclorrhaphous fly larva, (A) lateral view, (B) detail of anterior spiracle (reproduced from Hall and Smith, 1993). (C) Transverse section through the head and mouthparts. (d) Cephalopharyngeal skeleton.

The head segment is divided by a ventral furrow into left and right cephalic lobes, with the mouth opening at the base of the furrow. The head also bears two peg-like sensory organs. The true head is completely invaginated into the thorax. The functional mouth is at the inner end of the pre-oral cavity, the atrium, from which a pair of sclerotized **mouth-hooks** protrude. The mouth-hooks are part of a complex sclerotized structure, known as the **cephalopharyngeal skeleton**, to which muscles are attached (Fig. 5.1).

There is a pair of **anterior spiracles** on the prothoracic segment, immediately behind the head (Fig. 5.1) and a pair of **posterior spiracles** on the twelfth segment (Fig. 5.2). The anterior spiracles of calliphorids and sarcophagids protrude externally as a fan-shaped series of finger-like lobes, each ending in a small aperture. These spiracles are not present on first-stage larvae.

The structure of the posterior spiracles is of great taxonomic importance. They usually consist of a pair of sclerotized **spiracular plates**. These may be set flat on the body cuticle, raised on a process or at the bottom of a deep cavity. Slits or openings are present in the surface of the spiracular plate for gaseous exchange (Fig. 5.2). In first-stage larvae the slits consist of a pair of simple

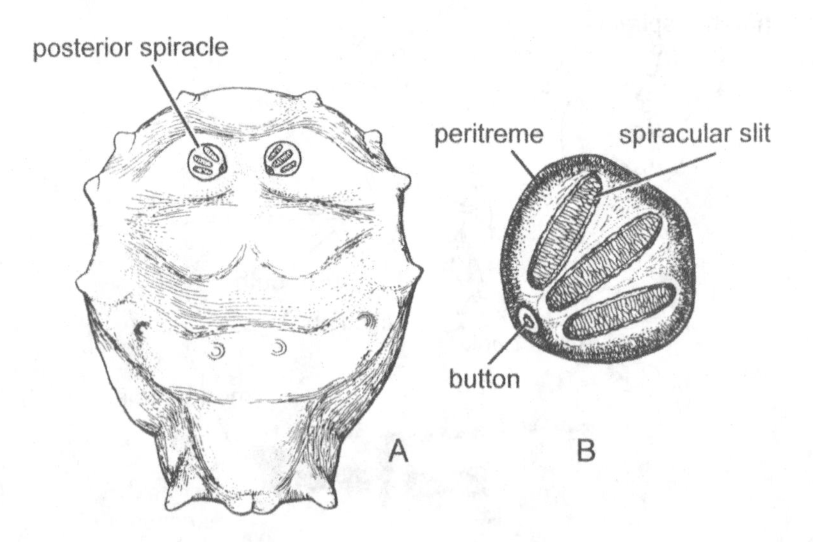

Fig. 5.2 (A) End view of the last abdominal segment and (B) detail of the posterior spiracles of a third-stage larva of *Calliphora vicina* (reproduced from Hall and Smith, 1993).

holes. In second- and third-stage larvae, each of the sclerotized plates contain two or three slits, respectively. The exception to this pattern is the Oestridae, where instead of slits the posterior spiracles occur as a large number of small pores. The outer rim of the spiracular plate may be heavily sclerotized and is known as the **peritreme** (Fig. 5.2). The peritreme may form a complete ring or may be incomplete. In second- and third-stage larvae a rounded structure, known as the **button**, may be visible on the sclerotized plate in a ventral and lateral position. This structure is the scar left from the spiracle of the previous stage after moulting. The last segment also bears a number of tubicles, typically three above each spiracle and three below.

5.3 LIFE HISTORY

Most of the adult agents of myiasis are oviparous, laying large numbers of eggs either directly on to the host or on to vegetation at a site where they are likely to be picked up by the passing host. A number, notably species of *Oestrus* and *Sarcophaga,* are ovoviviparous, depositing live first-stage larvae directly on to their host.

The egg stage is usually brief, generally lasting about 24 hours, and is followed by three larval stadia, during which feeding occurs. When feeding is complete the, third-stage larva enters a wandering phase when it leaves the host and locates a suitable site for pupariation, usually burrowing into the ground. After pupation, the newly emerged adult breaks out of the puparium and works its way to the surface using its ptilinum (see section 4.2). The adults of most species of Calliphoridae (blow flies) and Sarcophagidae (flesh flies) are anautogenous and require a protein meal to initiate egg production. However, in many of the Oestridae (warble and bot flies) the mouthparts are functionless and adults do not feed.

Two broad life-history groups can be detected within the dipteran agents of myiasis. One group is composed of highly specialized, obligate and relatively host-specific ectoparasites, typified by the oestrids. The second group is composed of flies that are generally facultative ectoparasites with a relatively broad host range, typifed by the calliphorid blowflies. These two groups probably reflect quite different evolutionary origins of the myiasis habit.

The specialist oestrids may have arisen from less specialized

Diptera that had ectoparasitic, blood-sucking larvae. Examples of existing blood-feeding species are the Congo floor maggot (*Auchmeromyia* spp.) which feeds on mammals and *Protocalliphora* spp. which feed on birds. From initially puncturing the skin, the larvae may have moved to invading and ultimately burrowing under the skin. The morphological and physiological adaptations which were required to allow larvae to survive in these hostile environments necessitated the development of a number of highly specialized features and resulted in the evolution of a high degree of host specificity and an obligate ectoparasitic habit.

In contrast, among the calliphorids and sarcophagids, generalized free-living saprophagous carrion feeders probably formed the ancestral origins of the parasitic habit. These may have given rise to species which were attracted to warm-blooded animals with suppurating wounds or where the skin was soiled by faeces. From initially simply feeding as adults at these sites, flies may have been able to move to ovipositing and allowing their larvae to infest dying or diseased animals. From this intermediate stage, the primarily facultative ectoparasites which were capable of attacking healthy animals may have developed, followed ultimately by the obligate ectoparasites. Hence, the evolution of the myiasis habit from saprophagous feeders probably involved a change in the timing of attack from dead animals, to moribund or clinically diseased animals and, finally, to healthy animals.

5.4 PATHOLOGY

The direct pathological effects of myiasis may vary considerably and depend on the species of ectoparasite, the number of larvae and the site of the infestation. In many cases infestation by small numbers of fly larvae may have little or no discernible clinical effect on the host. However, a heavier burden of parasites may produce effects ranging from irritation, discomfort and pruritus, resulting in reduced feeding, weight loss, reduced fertility and loss of general condition. Ultimately, heavy infestation may lead rapidly to host death from direct tissue damage, haemorrhage, bacterial infection, dehydration, anaphylaxis and toxaemia. Myiasis from a range of species also has been shown to produce a marked immunological response in the host.

5.5 CLASSIFICATION

All the flies that act as economically important agents of veterinary myiasis are calypterate Diptera in the superfamily **Oestroidea**. Within this super-family there are three major families of myiasis-producing flies: **Oestridae**, **Calliphoridae** and **Sarcophagidae**.

The Oestridae contains about 150 species, known as the bots and warbles. There are four sub-families of importance: **Oestrinae**, **Gasterophilinae**, **Hypodermatinae** and **Cuterebrinae**. The sub-family Oestrinae contains one genus of major importance, ***Oestrus***, and three genera of lesser importance, ***Rhinoestrus***, ***Cephenemyia*** and ***Cephalopina***. The sub-family Gasterophilinae contains a single genus of importance, ***Gasterophilus***. The sub-family Hypodermatinae contains one genus of major importance, ***Hypoderma***, and a second, less widespread genus, ***Przhevalskiana***. The sub-family Cuterebrinae contains two genera of interest, ***Cuterebra*** and ***Dermatobia***. All are obligate parasites, showing a high degree of host specificity. Their larvae spend their entire period of larval growth and development feeding within their vertebrate hosts, causing nasopharyngeal, digestive tract, or dermal–furuncular myiases.

The Calliphoridae, known as blow flies, are a large family, composed of over 1000 species divided between 150 genera. At least 80 species have been recorded as causing traumatic, cutaneous myiasis. These species are found largely in four important genera: ***Cochliomyia***, ***Chrysomya***, ***Lucilia*** and ***Calliphora***. The genera ***Protophormia*** and ***Phormia*** also each contain a single species of importance. Most of these species are either primary or secondary facultative invaders. Only two species, *Chrysomya bezziana* and *Cochliomyia hominivorax*, are obligate agents of myiasis.

The family Sarcophagidae, known as flesh flies, contains over 2000 species in 400 genera. Most species of Sarcophagidae are of no veterinary importance, breeding in excrement, carrion and other decomposing organic matter. The only genus containing species which act as important agents of veterinary myiasis is ***Wohlfahrtia***.

5.6 RECOGNITION OF DIPTEROUS AGENTS OF MYIASIS

The larvae of most species of Diptera are extremely difficult to identify, especially as first- or second-stage larvae. Indeed, the species of a number of genera are, to all intents and purposes, indistinguishable given our current knowledge of their morphology. The host, geographical location and type of myiasis, therefore, may be important clues to identification. It is particularly helpful, where possible, to rear specimens through to emergence so that the adult fly can be used to help confirm the identification.

Guides to the recognition of the key genera of larvae and adult Oestridae, Calliphoridae and Sarcophagidae likely to be found in myiases of domestic animals are presented below (modified from Zumpt, 1965; Crosskey and Lane, 1993; and Hall and Smith, 1993; the first of which is especially recommended if a more detailed key is required). The guide to larvae presented below applies specifically to recognition of the third stage. This stage is usually of the longest duration and, since the larvae are approaching their maximum size or are beginning to wander, is usually the stage when they are most commonly observed. It should be noted, that because the external structure of larvae may change over the course of their growth and development, first- and second-stage larvae may not key out appropriately.

5.6.1 Guide to third-stage larvae causing myiasis in domestic animals

1 Body more or less cylindrical; no obvious head capsule **2**

Fly larvae with an obvious head capsule; rarely found associated with livestock myiasis

Diptera Nematocera or Brachycera

2 Body with obvious fleshy processes **3**

Body without fleshy processes **4**

3 Third-stage larvae large, up to 18 mm long; large, pointed fleshy processes laterally and dorsally (Fig. 5.15); posterior spiracular plate without button; peritremes with a narrow

opening; in carrion or secondarily in cutaneous myiasis of sheep; distribution, Afrotropical, Australasian and Oriental

Chrysomyia albiceps and C. rufifaces (Calliphoridae)

Third-stage larvae 7–8 mm in length; body flattened, with long processes (Fig. 5.19C); posterior spiracles on short stalks on terminal segment; uncommon in livestock myiasis **Fanniidae**

4 Posterior spiracles with a large number of small pores or many short intertwining slits arranged in three groups on each spiracular plate (Fig. 5.3D) **5**

Posterior spiracles with up to three straight or curved slits (Fig. 5.2B) **7**

5 Mouth-hooks well developed, strongly hooked **6**

Mouth-hooks poorly developed; third-stage larvae 20–30 mm in length (Fig. 5.9); in subcutaneous swellings, or warbles (Fig. 5.10); on cattle or deer ***Hypoderma* spp. (Oestridae)**

6 Body with weak spines in distinct regions; posterior spiracles with many small pores (Fig. 5.3); in nasal myiasis of sheep; distribution, worldwide ***Oestrus ovis* (Oestridae)**

Body spines stronger and more evenly distributed; posterior spiracles with many small slits; found in dermal myiases; in rodents and rabbits; distribution, New World

***Cuterebra* spp. (Oestridae)**

7 Posterior spiracles with straight or arced slits **8**

Posterior spiracles with serpentine slits; anterior spiracles in the form of membranous stalks bearing finger-like processes; body with obvious spines; furuncular myiases of dogs, rats and humans; distribution, sub-Saharan Africa

***Cordylobia* spp. (Calliphoridae)**

Posterior spiracles with serpentine slits (Fig. 5.19A); anterior spiracles not as above; uncommon in livestock myiasis

Muscidae

8 Posterior spiracles sunk in a deep cavity which may conceal them (Fig. 5.21B); slits more or less parallel **9**

Posterior spiracles visible, either exposed on surface or set in a ring of tubercles **11**

9 Body with strong spines **10**

Body with short spines; obligate agent of cutaneous myiasis; primarily in sheep and goats; distribution, worldwide **_Wohlfahrtia_ spp. (Sarcophagidae)**

10 Posterior spiracles with slits bowed outwards at the middle; body oval (Fig. 5.6G); found in the pharynx or digestive tract of equids **_Gasterophilus_ spp. (Oestridae)**

Posterior spiracles with slits relatively straight; body enlarged anteriorly and tapering posteriorly (Fig. 5.11A); distribution, New World **_Dermatobia hominis_ (Oestridae)**

11 Posterior spiracles with straight slits (Fig. 5.2) **12**

Posterior spiracles with arced slits; uncommon in livestock myiasis **Muscidae**

12 Posterior spiracles with a fully closed peritremal ring (Fig. 5.17B) **13**

Posterior spiracles with an open peritremal ring (Fig. 5.12B) **14**

13 Cephalopharyngeal skeleton with pigmented accessory oral sclerite (Fig. 5.19B); distribution, worldwide **_Calliphora_ spp. (Calliphoridae)**

Cephalopharyngeal skeleton without pigmented accessory oral sclerite (Fig. 5.17C); distribution, worldwide **_Lucilia_ spp. (Calliphoridae)**

14 Tracheal trunks leading from posterior spiracles without dark pigmentation **15**

Tracheal trunks leading from posterior spiracles with conspicuous dark pigmentation extending forwards as far as the 9th or 10th segment (Fig. 5.12A); obligate, primary agent of traumatic livestock myiasis; distribution, Neotropical and Nearctic ***Cochliomyia hominivorax*** **(Calliphoridae)**

15 Posterior margin of segment 11 with dorsal spines **16**

Posterior margin of segment 11 without dorsal spines; a secondary facultative agent of cutaneous livestock myiasis; distribution, Neotropical and Nearctic

Cochliomyia macellaria **(Calliphoridae)**

16 Posterior spiracles with distinct button **18**

Posterior spiracles without distinct button **17**

17 Body without fleshy processes; segments with belts of strongly developed spines (Fig. 5.13A); anterior spiracle with 4–6 branches; an obligate, primary agent of cutaneous livestock myiasis; distribution, Afrotropical and Oriental

Chrysomya bezziana **(Calliphoridae)**

Anterior spiracle with 11–13 branches; largely saprophagous; an occasional facultative ectoparasite causing cutaneous myiasis; distribution, Oriental and Australasian

Chrysomya megacephala **(Calliphoridae)**

18 Posterior margins of segment 10 with dorsal spines; length of the larger tubercles on upper margin of posterior face of terminal segment greater than half the width of a posterior spiracle; causes facultative, cutaneous myiasis of cattle, sheep and reindeer; distribution, northern Holarctic

Protophormia terraenovae **(Calliphoridae)**

Posterior margins of segment 10 without dorsal spines; length of the larger tubercles on upper margin of posterior face of terminal segment less than half the width of a posterior spiracle; distribution, Holarctic

Phormia regina **(Calliphoridae)**

5.6.2 Guide to genera of adult Diptera causing myiasis in domestic animals

1 Insects with one pair of wings on the mesothorax and a pair of club-like halteres on the metathorax; antennae composed of three segments, third segment usually with an arista; foot with two pads; frons with ptilinal suture; second antennal segment usually with a groove; thoracic transverse suture strong; thoracic squamae (see Chapter 4) usually well developed **Calypterate Diptera 2**

2 Mouthparts small, usually functionless; head bulbous; antennae small; flies more or less covered with soft hair **3**

Mouthparts usually well developed; antennae not small; flies with strong bristles; hypopleural bristles present; post-scutellum weak or absent **7**

3 Vein M bent towards vein R_{4+5} **4**

Vein M not bent towards vein R_{4+5}; squamae small; cross-vein dm-cu absent; ovipositor strongly developed in female (Fig. 5.5) **Gasterophilinae spp. (Oestridae)**

4 Sharp bend of vein M towards vein R_{4+5} but the two do not meet before the margin **5**

Vein M joins vein R_{4+5} before the margin; vein dm-cu in line with deflection of vein M; vein A_1+CuA_2 does not reach the margin (Fig. 5.4B); frons enlarged; frons, scutellum and dorsal thorax bear small wart-like protuberances; eyes small (Fig. 5.4A); abdomen brownish-black **Oestrinae spp. (Oestridae)**

5 Blue-black colour **Cuterebrinae spp.**

Not blue-black **6**

6 Vein A_1+CuA_2 reaches the margin; vein dm-cu in line with deflection of vein M (Fig. 5.8B); hairy bee-like flies with a

light–dark colour pattern; fan of yellow hypopleural hairs; palps absent **Hypodermatinae spp (Oestridae)**

7 Metallic, iridescent appearance (blue-black, violet-blue, green) **8**

Dull grey appearance; three black stripes on the scutum; abdomen usually with chequered or spotted pattern **13**

Flies of predominantly reddish-yellow or reddish-brown colour, not metallic; distribution, tropical Africa

***Cordylobia* spp. (Calliphoridae)**

8 Wing with stem vein (base of R) entirely bare **9**

Wing with stem vein with fine hairs along margin **10**

9 Flies with metallic green or coppery green thorax and abdomen; thoracic squamae bare; found in cutaneous myiasis, particularly of sheep; distribution, worldwide

***Lucilia* spp. (Calliphoridae)**

Flies with black-blue thorax and blue or brown abdomen; thoracic squamae with long dark hair on upper surface; may be secondary invaders of cutaneous myiasis; distribution, worldwide ***Calliphora* spp. (Calliphoridae)**

10 Head with almost entirely black ground colour and black hair; thoracic squamae bare; alar squamae hairy on outer half or dorsal surface **12**

Head with ground colour of at least lower half entirely or mainly orange or orange-red and with white, yellow or orange hair; thoracic squamae bare on dorsal surface **11**

11 Thoracic squamae hairy on whole dorsal surface; scutum of thorax without bold black stripes; distribution, Afrotropical, Oriental, Australasian, southern Palaearctic

***Chrysomya* spp. (Calliphoridae)**

Thoracic squamae hairy only at the base, usually concealed by the alar squamae; scutum of thorax with three bold, black stripes; distribution: Nearctic and Neotropical
Cochliomyia spp. (Calliphoridae)

12 Thorax with anterior spiracle black or reddish-brown; alar squamae with obvious dark hair dorsally; distribution, Palaearctic and Nearctic only
Protophormia spp. (Calliphoridae)

Thorax with anterior spiracle yellow or orange; thoracic squamae with white-yellow hair dorsally; distribution, Palaearctic and Nearctic only
Phormia spp. (Calliphoridae)

13 Arista almost bare; abdomen with pattern of black spots (Fig. 5.21C) **_Wohlfahrtia_ spp. (Sarcophagidae)**

Arista with long and conspicuous hairs, at least on the basal half; abdomen with dark and light chequered pattern (Fig. 5.21A) **_Sarcophaga_ spp. (Sarcophagidae)**

5.7 OESTRIDAE

The family Oestridae contains flies commonly known as bots and warbles. All are obligate parasites and most show a high degree of host specificity. The larvae are characterized by posterior spiracular plates containing numerous small pores. The larvae all develop exclusively in the nasopharyngeal cavities or skin-boils (warbles) of mammals. The adults have primitive, usually non-functional mouthparts and are short lived.

5.7.1 Oestrinae

The sub-family Oestrinae contains 34 species in nine genera. One genus, *Oestrus*, is of major economic veterinary importance and three other genera, *Rhinoestrus*, *Cephenemyia* and *Cephalopina*, are of lesser or local significance as parasites of domestic animals.

Oestrus

Oestrus is a small genus, containing only five species, the most well known and economically important of which is *Oestrus ovis*, the sheep nasal bot fly. The other species are parasites of antelope and goats, primarily in Africa. The larvae of *O. ovis* develop in the head sinuses and nasal passages of sheep and goats. Although originally Palaearctic, it is now found in all sheep-farming areas of the world, having been spread with sheep as they were transported worldwide.

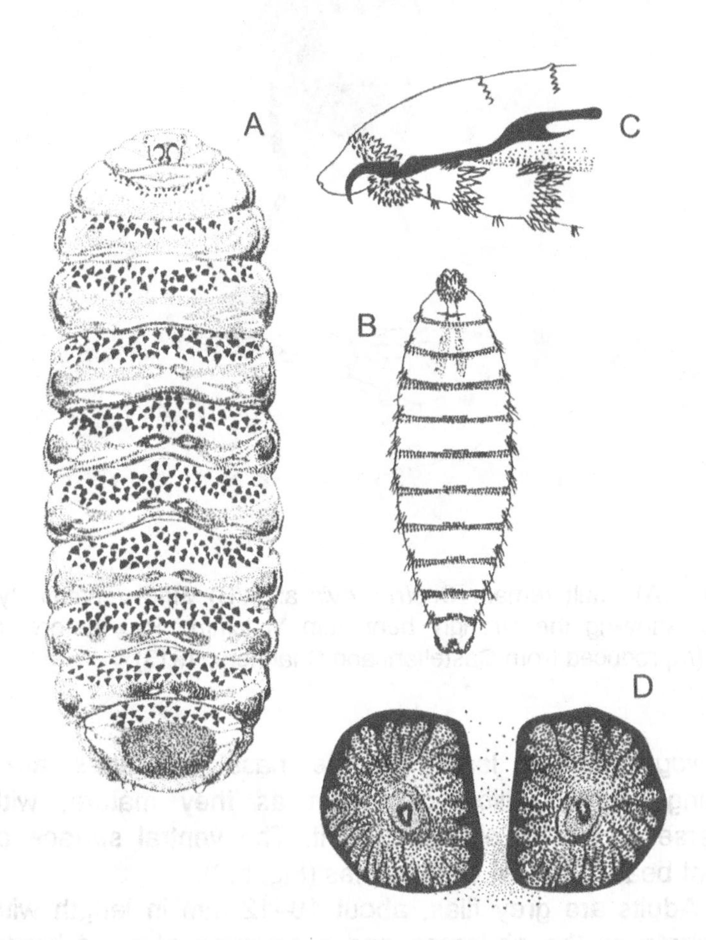

Fig. 5.3 *Oestrus ovis*, ventral view of (A) third-stage larva and (B) first-stage larva. (C) Mouthparts of first-stage larva in lateral view and (D) posterior spiracles of third-stage larva. (Reproduced from Hall and Smith, 1993.)

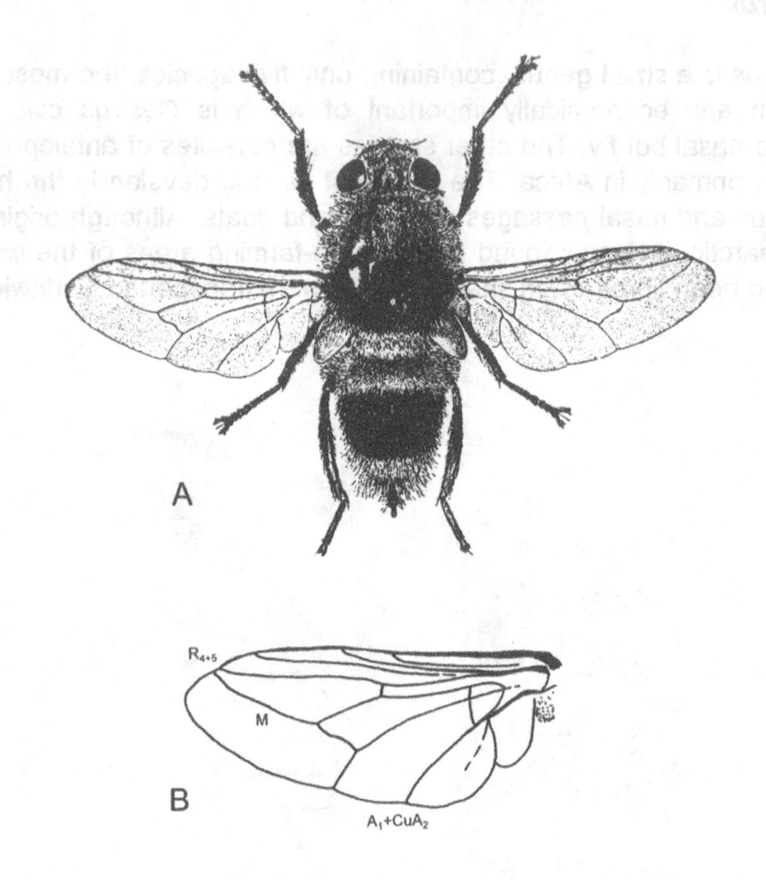

Fig. 5.4 (A) Adult female *Oestrus ovis* and (B) wing venation typical of *Oestrus*, showing the strongly bent vein M joining R_{4+5} before the wing margin (reproduced from Castellani and Chalmers, 1913).

Morphology: mature larvae in the nasal passages are white, becoming slightly yellow or brown as they mature, with dark transverse bands on each segment. The ventral surface of each segment bears a row of small spines (Fig. 5.3).

Adults are grey flies, about 10–12 mm in length with small black spots on the abdomen and a covering of short brown hairs (Fig. 5.4). The head is broad, with small eyes and the frons, scutellum and dorsal thorax bear small wart-like protuberances. The segments of the antennae are small and the arista bare. The mouthparts are reduced to small knobs.

Life cycle: females are oviviparous, depositing up to 25 live first-stage larvae at a time, in or on the nostrils of the host, during flight. The adults are particularly active during hot, dry weather and this activity can lead to considerable disturbance and panic in a flock.

First-stage larvae are about 1 mm long and crawl up the nasal cavity and attach to the mucous membranes. Here they feed on mucus and desquamated cells. The first-stage larvae enter the frontal sinuses via the ethmoid process before moulting. The second-stage larvae are 4–12 mm in length. Second-stage larvae then pass into the frontal sinuses where they moult for the final time. When mature, at up to 20 mm in length, the third-stage larvae re-enter the nasal cavities where they are sneezed out. On reaching the ground the larvae pupariate.

Development can take as little as 25–35 days and, in the warmer parts of its range, up to three generations per year may be recorded. In southern Europe, generation peaks in *O. ovis* populations have been recorded in March–April, June–July and September–October. More commonly, however, there are only two generations per year, with adults present in late spring and late summer. Larval development ceases in autumn and the first-stage larvae overwinter within the head of the host and will not migrate to the frontal sinuses until the following spring.

Pathology: as they deposit larvae, the activity of adult *O. ovis* may annoy sheep, leading to a loss of grazing time, reduced weight gain in lambs and loss of condition. However, in general, infestations are relatively light, only an average of 2–20 larvae being present in the frontal sinus of infested animals at any one time.

The parasitic rhinitis caused by the larvae of *O. ovis* is characterized by a sticky, mucoid nasal discharge which at times may be haemorrhagic. Histopathological changes in the nasal tissues of infected sheep include catarrh, infiltration of inflammatory cells and squamous metaplasia, characterized by conversion of secretory epithelium to stratified squamous type. Immune responses by the host to infestation by *O. ovis* have been recorded.

Clinical symptoms of infestation may range from mild discomfort, nasal discharge, sneezing, nose rubbing or head shaking. Dead larvae in the sinuses can cause allergic and inflammatory responses, followed by bacterial infection and sometimes death. Larvae may occasionally penetrate the olfactory mucosa and enter the brain – causing ataxia, circling and head pressing.

Infestation prevalence tends to be highly localized. In individual sheep flocks infestation rates of up to 44–88% have been recorded in France and as low as 0.75% in Britain. Infestation rates of 6–52% have been recorded in Zimbabwe, 69% in India and 100% in Morocco, South Africa and Brazil. Infestation by *O. ovis* has been associated with losses in weight gain of between 1 and 4.5 kg, losses in wool production of up to 200–500 g and a reduction in milk production of up to 10%.

Other Oestrinae of veterinary interest

The genus *Cephenemyia* is restricted to the Holarctic, and the larvae develop exclusively in deer. Parasitism may affect in excess of 70% of a herd. Of particular interest are *Cephenemyia trompe* and *Cephenemyia auribarbis* which are found in reindeer and caribou. *Cephenemyia phobifer* and *Cephenemyia stimulator* are found in red and roe deer, respectively; *Cephenemyia pratti* may occur in the mule deer; *Cephenemyia jellisoni* in the whitetail and Pacific blacktail deer; and *Cephenemyia apicata* in Californian deer. Females deposit live first-stage larvae in the nostrils of the host and the larvae subsequently move to the pharyngeal and nasal cavities. Mature larvae migrate to the anterior pharynx before leaving the host. The activity of the larvae causes nasal discharge, sneezing, coughing and restlessness.

The genus *Rhinoestrus* is highly host specific and infests equines and large African mammals. Eleven species have been described, four of which are restricted to equines. Adult *Rhinoestrus* resemble *Oestrus*, but have more conspicuous tubercles on the head and thorax. In a study in the Caspian region of the former USSR, 97–100% of horses were found to be infested with *Rhinoestrus*, the major species being *Rhinoestrus latifrons* and *Rhinoestrus purpureus*. In this study, a mean of 154 larvae were found in the head cavities of each infested horse, with a maximum infestation recorded of 899 larvae. *Rhinoestrus usbekistanicus* parasitizes horses and donkeys in the Palaearctic.

The camel nasal bot fly, *Cephalopina titillator,* is a parasite of camels in dry parts of the Palaearctic and Oriental regions, and Australia. Infestation of up to 90% of camel herds has been recorded. The presence of larvae may cause considerable irritation and breathing difficulty.

5.7.2 Gasterophilinae

The sub-family Gasterophilinae contains 18 species in five genera, although only one genus, *Gasterophilus,* contains species of veterinary significance.

Gasterophilus

Species of *Gasterophilus* are obligate parasites of horses, donkeys, mules, zebras, elephants and rhinoceroses. Nine species are recognized in total, six of which are of interest as veterinary parasites of equids in temperate habitats. All were originally restricted to the Palaearctic and Afrotropical regions, but three species, *Gasterophilus nasalis*, *G. haemorrhoidalis* and *G. intestinalis,* have been inadvertently introduced into the New World.

Morphology: the adult flies are large, 11–15 mm in length, and the

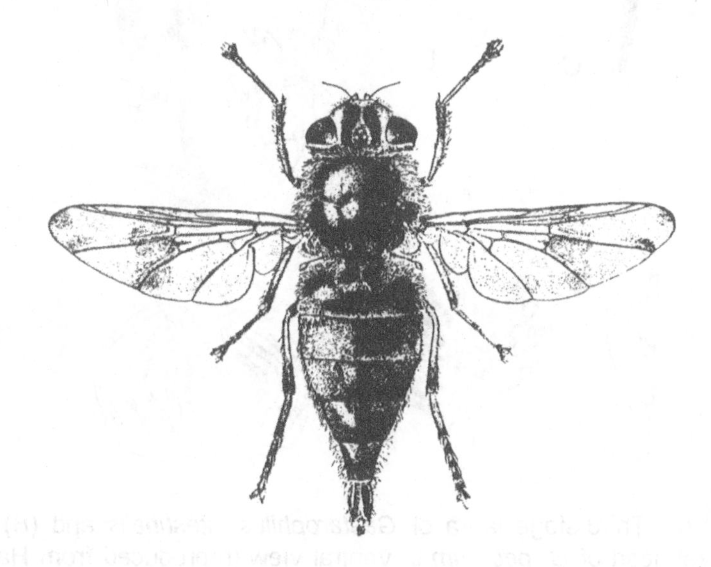

Fig. 5.5 Adult female *Gasterophilus intestinalis* (reproduced from Castellani and Chalmers, 1913).

body is densely covered with yellowish hairs (Fig. 5.5). In most species, the hairs of the mesonotum are dark, producing the appearance of a transverse dark band or dark patches behind the suture. In the female the ovipositor is strong and protuberant. The wings of adult *Gasterophilus* characteristically have no cross-vein dm-cu.

The third-stage larvae of the most important species may be

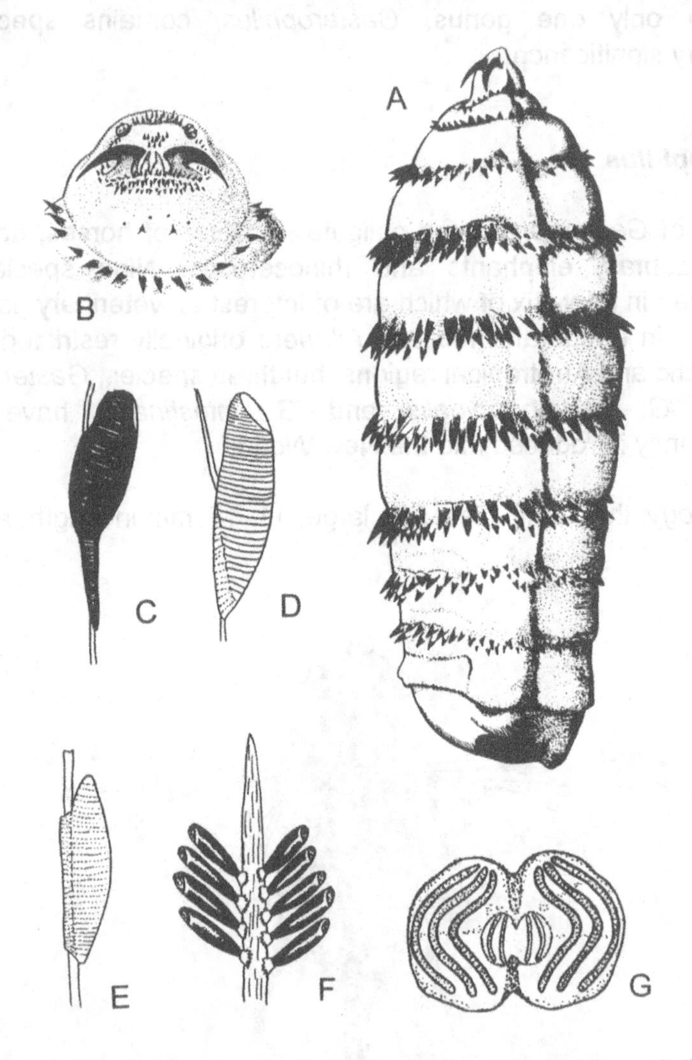

Fig. 5.6 (A) Third-stage larva of *Gasterophilus intestinalis* and (B) third-stage larval head of *G. pecorum* in ventral view (reproduced from Hall and Smith, 1993). Eggs of (C) *G. haemorrhoidalis,* (D) *G. intestinalis,* (E) *G. nasalis* and (F) *G. pecorum* (reproduced from Smith, 1989). (G) Posterior spiracles of a third-stage larva of *G. intestinalis.*

distinguished using the following guide (from Zumpt, 1965):

1 Spines on the ventral surface of segments arranged in a single row **2**

Spines on the ventral surface of segments arranged in two rows **3**

2 First three body segments more or less conical; third segment always with a dorsal row of spines and sometimes with ventral spines (Fig. 5.7C); distribution, worldwide, particularly in the Holarctic ***Gasterophilus nasalis* (Oestridae)**

First three body segments more or less cylindrical, showing sharp constrictions posteriorly; third segment without spines dorsally or ventrally; distribution, eastern Palaearctic ***Gasterophilus nigricornis* (Oestridae)**

3 Head segment with only lateral groups of denticles; dorsal row of spines on the 8th segment not broadly interrupted medially; at least 10th segment with dorsal spines **4**

Head segment with two lateral groups of denticles and one central group, the latter situated between the antennal lobes and mouth-hooks (Fig. 5.6B); dorsal rows of spines broadly interrupted medially on the 7th and 8th segments; 10th and 11th segments without spines; distribution, Palaearctic and Afrotropical ***Gasterophilus pecorum* (Oestridae)**

4 Mouth-hooks uniformly curved dorsally; body spines sharply pointed **5**

Mouth-hooks not uniformly curved dorsally, but with a shallow depression (Fig. 5.7A); body spines with blunt tips (Fig. 5.7A); distribution, worldwide ***Gasterophilus intestinalis* (Oestridae)**

5 Mouth-hooks strongly curved, their tips directed backwards and approaching the base (Fig. 5.7B); body segment 3 ventrally with three complete rows of spines; body segment 11 with one row of spines, interrupted by a broad median gap; distribution, Palaearctic ***Gasterophilus inermis* (Oestridae)**

Mouth-hooks directed more laterally (Fig. 5.7D); body segment 3 ventrally with one medially interrupted row of spines; body segment 11 with one row of a variable number of spines which are not interrupted medially; distribution, worldwide ***Gasterophilus haemorrhoidalis*** **(Oestridae)**

Life cycle: all the species of *Gasterophilus* have a similar general life cycle. Adults are short lived, with an effective adult life span of only a few days. Adults have non-functional mouthparts and do not feed. Females, therefore, are autogenous and may mate and begin oviposition very rapidly after emergence. The eggs are usually laid directly on the host, attached to the hairs in particular body regions, most commonly on the head. The precise site of oviposition, however, varies between the species of *Gasterophilus*. A combination of the presence of an attachment organ and an adhesive ensures effective long-term egg attachment (Fig. 5.6).

After hatching the larvae burrow into the tissues of the host, eventually moving, by a variety of routes, into the alimentary tract of the host. The larvae feed on tissue exudates and do not imbibe blood. When mature, the larvae detach and are voided from the host in the faeces. The larvae then burrow into the ground and pupariate. Adult bot flies then emerge and egg laying begins in early summer.

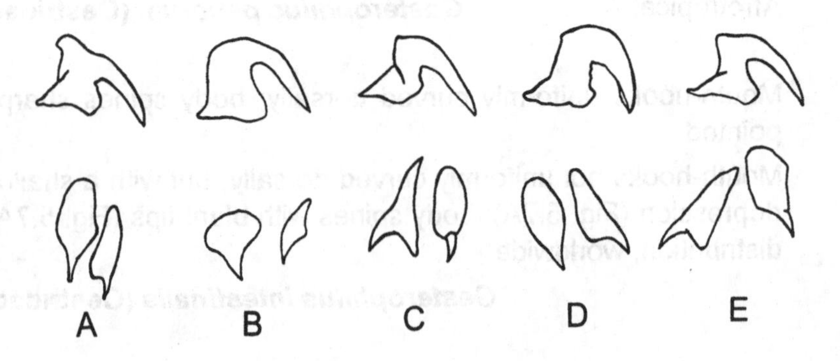

Fig. 5.7 Mouth-hooks and (below) ventral spines of the fifth segment of (A) *Gasterophilus intestinalis*, (B) *G. inermis*, (C) *G. nasalis*, (D) *G. haemorrhoidalis* and (E) *G. pecorum* (reproduced from Zumpt, 1965).

There is usually only a single generation each year in temperate habitats. Species-specific variations in these life cycles are given below.

Gasterophilus intestinalis: each female attaches its light yellow-coloured eggs to the hairs, usually on the inner forelegs, of horses and donkeys. Several eggs may be glued to each hair and up to 1000 eggs may be deposited by a female *G. intestinalis* during its lifetime of only a few days. During grooming, the rise in temperature, moisture and friction brought about by contact with the tongue or lips of the horse stimulate the eggs to hatch. After hatching, the larvae burrow into the dorsal mucosa at the anterior end of the tongue where they excavate galleries in the subepithelial layer of the mucus membrane. They migrate through the tongue for 24–28 days. After exiting the tongue and moulting, second-stage larvae attach for a few days to the sides of the pharynx and then move to the oesophageal portion of the stomach where they cluster at the boundary of glandular and non-glandular epithelium. Here they moult and remain until the following spring or early summer when they detach and pass out in the dung. Pupariation takes place shortly after in the soil or dry dung.

Gasterophilus haemorrhoidalis: in mid to late summer, adult females lay batches of about 150–200 eggs, attached to the hairs fringing the lips of horses, donkeys or zebra. Moisture from licking stimulates the hatching of the larvae. The larvae burrow into the epidermis of the lips and migrate into the mouth. After moulting, the second-stage larvae move to the stomach and duodenum where they moult again. In the early spring, the third-stage larvae move to the rectum and reattach very close to the anus. Some time later they detach and are passed out in the faeces.

Gasterophilus nasalis: oviposition takes place in late spring and early summer and the eggs are laid in small batches underneath the jaw of horses, donkeys and zebra. The eggs hatch in 5–10 days and the first-stage larvae travel along the jaw, entering the mouth between the lips. The larvae burrow into the spaces around the teeth and between the teeth and gums. This may result in the development of pus-sockets and necrosis in the gums. The first larval stage lasts 18–24 days, following which larvae moult and second-stage larvae move to the pyloric portion of the stomach or anterior portion of the duodenum where they attach. Third-stage larvae are eventually passed out with the faeces about 11 months after hatching.

Gasterophilus pecorum: up to 2000 eggs are laid in batches of 10–115 and distributed on pasture vegetation. The eggs are highly resistant and the developed larva may remain viable for months within its egg until ingested by horses. In the mouth, the eggs hatch within 3–5 minutes. First-stage larvae immediately penetrate the mucus membrane of the lips, gums, cheeks, tongue and hard palate and burrow towards the root of the tongue and soft palate where they may remain for 9–10 months until fully developed. They may also be swallowed and settle in the walls of the pharynx, oesophagus or stomach.

Gasterophilus nigricornis: eggs are laid on the cheeks of horses and donkeys. The larvae hatch in 3–9 days and burrow directly into the skin. They then burrow to the corner of the mouth and penetrate the mucous membranes inside the cheek. Once they have reached the central part of the cheek, about 20–30 days after hatching, they moult and leave the mucous membranes. The second-stage larvae are then swallowed and attach themselves to the wall of the duodenum. About 60–90 days later they moult again and the third-stage larvae are eventually passed out in the faeces in early spring the following year.

Gasterophilus inermis: the adult female lays up to 300 eggs, each attached individually to the base of a hair on the cheeks of horses and zebra. After hatching, the larvae burrow into the epidermis, and migrate towards the mouth. The migration route of the larvae can be detected by the presence of a track, along which the hair has fallen out. The larvae enter at the corner of the mouth and penetrate the mucous membranes of the cheek. The second- and third-stage larvae are found in the rectum.

Pathology: light infestations of bots are believed to have little pathogenic effect and are tolerated well. However, bots may cause obstruction to the food passing from the stomach to the intestine, particularly when the larvae are in or near the pylorus. The penetration of the mouth-hooks at the site of attachment may result in erosions, ulcers, nodular mucosal proliferation, stomach perforation, gastric abscesses, peritonitis and, in heavy infections, general debilitation and even rectal prolapse.

The oviposition behaviour of gasterophilids may cause considerable disturbance and panic to horses. Ovipositing females may be remarkably tenacious, laying eggs on mobile as well as stationary animals. Females will pursue galloping horses and immediately resume ovipositing when the horse stops. The high value of many horses, the recurrent treatments and possible self-injury by horses incurred when trying to avoid ovipositing females make *Gasterophilus intestinalis* a major economic pest of equines in North America. In the Old World, *G. pecorum* is reported to be the most important pathogenic species. Mortality of horses associated with infestations by *G. pecorum*, and swallowing difficulties associated with oesophageal constriction and hypertrophy of the musculature have been recorded.

5.7.3 Hypodermatinae

The sub-family Hypodermatinae contains 32 species in 11 genera, one genus of which, *Hypoderma,* is of major veterinary importance.

Hypoderma

The genus *Hypoderma* contains the species known variously as heel flies, warble flies or cattle grubs. The larvae live as subcutaneous parasites and are relatively host specific. The word 'warble' is Anglo-Saxon for boil. The primary hosts of species of *Hypoderma* are cattle, deer and reindeer. There are some reports of *Hypoderma* infestations of horses and sheep, but infestation is not sufficiently frequent to be important in these hosts. The two most important species affecting cattle are *Hypoderma bovis* and *Hypoderma lineatum.* They are widely distributed throughout the USA, Europe and Asia.

Morphology: adult female *H. bovis* are about 15 mm in length whereas *H. lineatum* are about 13 mm in length. Both are bee-like in appearance, covered with dense hair in a characteristic light–dark colour pattern (Fig. 5.8). This colour pattern is particularly marked in *H. bovis* and the thorax of *H. bovis* is more profusely hairy than that

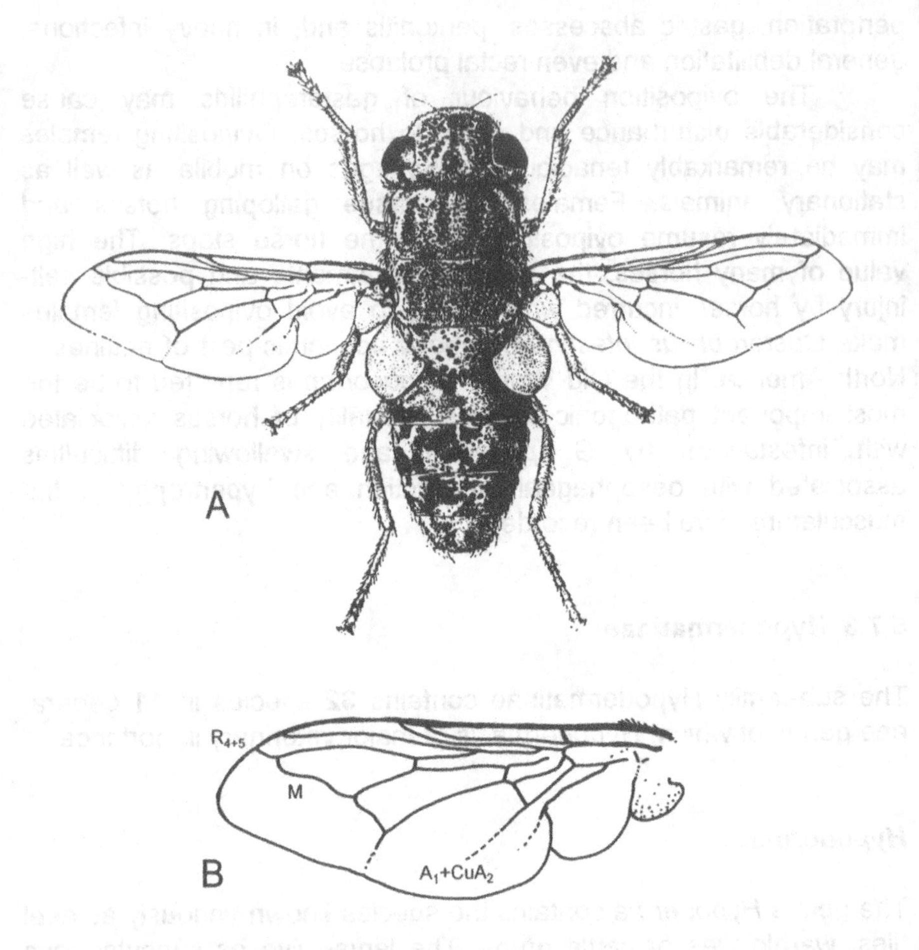

Fig. 5.8 (A) Adult female of *Hypoderma bovis* (reproduced from Castellani and Chalmers, 1913) and (B) wing venation typical of *Hypoderma*, showing the strongly bent vein M not joining R_{4+5} before the wing margin, and vein A_1+CuA_2 reaching the wing margin.

of *H. lineatum*. The mouthparts of both species are are small and lack palps.

The third-stage larvae may be distinguished from other species of *Hypoderma* by examination of the posterior spiracular plate which, in these species, is completely surrounded by small spines. The two species may be distinguished from each other by the observation that in *H. bovis* the posterior spiracular plate surrounding the button has a narrow funnel-like channel whereas, in *H. lineatum,* it has a broad channel (Fig. 5.9).

Life cycle: adults are active only on warm, sunny days, in the northern hemisphere, from April to June in the case of *H. lineatum* and from mid-June to early September in the case of *H. bovis*. Mating takes place off the host at aggregation points where females are intercepted in flight. Females are reproductively well adapted to their short, non-feeding life; they emerge with all their eggs fully developed and the capacity to mate immediately and oviposit soon after.

A single female may lay as many as 300–600 eggs during its life of only a few days. Eggs are deposited most commonly on the lower regions of the legs and lower body, where they are glued to the hairs of the host animal, either singly, by *H. bovis*, or in batches of up to 15, by *H. lineatum*. Eggs are attached by an outgrowth of the egg case, known as the attachment organ (Fig. 5.9B,C).

Eggs hatch within a week and the first-instar larvae, which are less than 1 mm in length, crawl down the hairs and either burrow

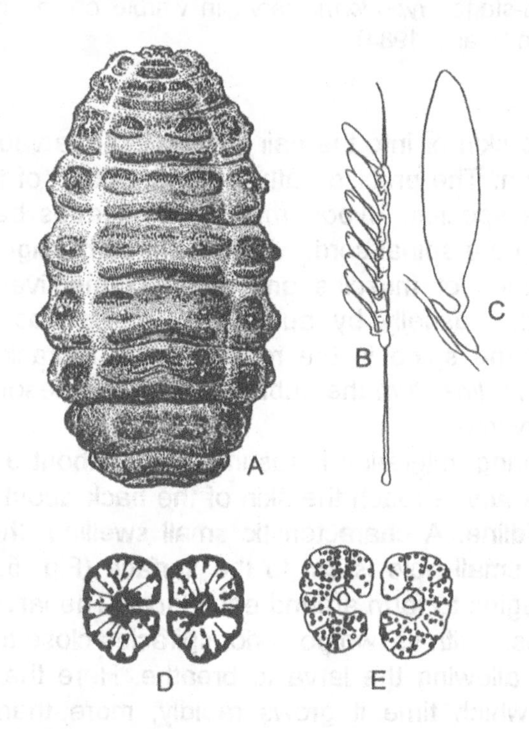

Fig. 5.9 (A) Third-stage larva of *Hypoderma bovis* (reproduced from Hall and Smith, 1993). Eggs of (B) *H. lineatum* and (C) *H. bovis* (reproduced from Smith, 1989). Posterior spiracles of third-stage larvae of (D) *H. bovis* and (E) *H. lineatum*.

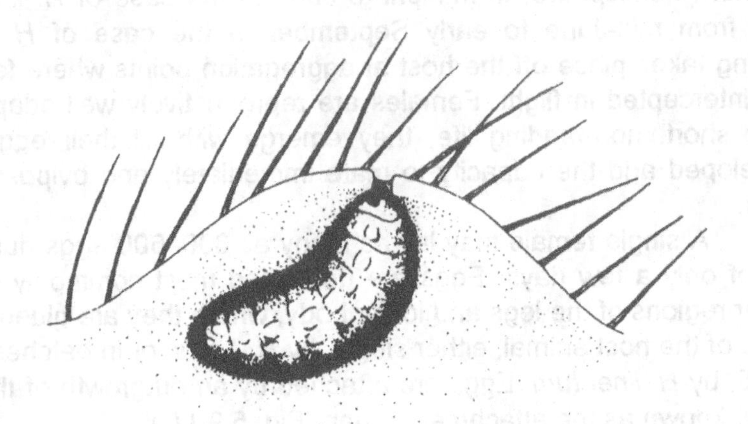

Fig. 5.10 Third-stage *Hypoderma* larva in warble on the back of a cow (reproduced from Evans, 1984).

directly into the skin or into the hair follicles. The larvae then burrow beneath the skin. The precise pathway and pattern of this migration depend on the species. *Hypoderma bovis* migrates below the skin along nerves to the spinal cord, while *H. lineatum* migrates between the fascial planes of muscles and along connective tissue. After about 4 months, usually by autumn, *H. bovis* has reached the epidural fat of the spine in the region of the thoracic and lumbar vertebrae and *H. lineatum* the submucosa of the oesophagus. Here the larvae overwinter.

Next spring, migration is resumed until, about 9 months after oviposition, the larvae reach the skin of the back about 25 cm either side of the midline. A characteristic small swelling, the 'warble', is formed and a small hole is cut to the surface (Fig. 5.10). A cystic nodule then begins to form around each larva. The larva reverses its position and rests with its two posterior spiracles close to the opening in the warble, allowing the larva to breathe. Here the larva moults twice, during which time it grows rapidly, more than doubling in length.

In the northern hemisphere the warbles of *H. lineatum* appear between January and February up to April, while those of *H. bovis* appear between March and June. After a period of 30–60 days the third-stage larva works its way out of the skin, drops to the ground and pupariates. A fully grown larvae of *H. bovis* may be about 27–28 mm in length and *H. lineatum* about 25 mm in length. The entire life cycle requires about a year.

Mortality of larvae is high and only about 3–5% of eggs are believed to give rise to adults in the following generation. The numbers of larvae in individual infested cattle is usually relatively low, with averages ranging between 5 and 20 larvae per infested animal. However, maximum infestations of up to 250 larvae per animal have been recorded.

Pathology: the greatest economic problem caused by *Hypoderma* results from the breathing holes cut by the larvae. These greatly reduce the value of the hide. Occasionally oedema and inflammation occur at the site of entry of first-stage larvae. The exit hole of third-stage larvae may also be prone to infection and it may be attractive to other insects.

First-stage larvae of *H. lineatum* migrating through the connective tissues produce yellow or greenish, gelatinous, oedematous areas with an overwhelming eosinophil infiltration, sometimes known as 'butchers jelly'. Hypersensitivity may also cause anaphalactic shock, particularly if larvae die; though this is rare. However, the direct effects of infestation by *Hypoderma* on weight gain of beef cattle are believed to be negligible. Younger animals appear to be more prone to infestation than older animals.

The characteristic noise and persistent ovipositing behaviour by *H. bovis* can be recognized by the cattle and results in dramatic avoidance behaviour known as 'gadding'. Gadding behaviour may result in self-inflicted wounding, spontaneous abortion, retarded growth, loss of condition and reduced milk yields. In contrast, the approach of *H. lineatum* to cattle is made in a series of silent hops along the ground, hence the common name of 'heel fly'.

The incidence and prevalence of *Hypoderma* can be locally variable, but may be high. In continental Europe, for example, 40–90% of cattle may be infested by warbles in some areas. Each infested animal may be parasitized by an average of 15–20 (range 0–200) third-stage larvae. In the 1980s in the USA, *Hypoderma* were estimated to cause annual losses of between $60 million and $600 million, primarily from loss of hide value and meat trim at slaughter. However, after intensive control campaigns, populations of *Hypoderma* spp. have been eliminated from the UK and Ireland and suppressed in parts of the USA and Canada, with only occasional cases, originating from imported cattle or fly immigration, reported in these areas.

Other Hypodermatinae of veterinary interest

Hypoderma diana is an important parasite of deer in the Palaearctic. In the northern hemisphere the fly is most active in May and June. The larvae move to the back of deer via the spine, like *H. bovis,* and the hide damage is similar to that of cattle. *Hypoderma diana* will not infect cattle; it may infest sheep but larvae do not mature properly. Infestations of up to 300 larvae per animal have been recorded.

Hypoderma tarandi is a common parasite of reindeer in northern Eurasia and North America. In Canada this species may have a significant effect on caribou activity, increasing restlessness and decreasing the amount of time the caribou spend feeding and resting. *Hypoderma tarandi* is also considered to be one of the most important reindeer parasites in Alaska and in Norway. Eggs are attached to the hairs of the lower legs and flanks and larvae migrate beneath the skin directly to the site of warble formation, usually on the rump. Infestations of up to 432 larvae per animal have been recorded.

In the genus *Przhevalskiana,* the goat warble, *Przhevalskiana silenus,* is common around the Mediterranean basin. The larvae burrow into the skin of goats in the same manner as the larvae of *Hypoderma*. Infested animals lose condition. In southern Italy, infestation rates ranging between 30 and 81% have been recorded in goat herds, with a mean of 5 larvae per animal. Younger animals appear to be more prone to infestation than older animals.

5.7.4 Cuterebrinae

The sub-family Cuterebrinae contains over 70 species in six genera. Two genera are of veterinary interest *Cuterebra* and *Dermatobia*. They are found exclusively in the Nearctic and Neotropical regions and do not occur in the Old World.

Cuterebra

Species of the genus *Cuterebra* are largely dermal parasites of rodents and rabbits but may occasionally infest dogs and cats. The larvae cause sub-dermal nodules.

Morphology: the larvae have strongly curved mouth-hooks and numerous strong body spines. The adults are large flies, up to 20 mm in length, covered by dense, short hair and with a blue-black-coloured abdomen. They have small, non-functional mouthparts and do not feed as adults.

Life cycle: females lay eggs on the ground near or within the entrance of host nests or on grass near trails used by hosts. These are picked up by the passing host. The larvae enter the body, directly through the skin or through one of the orifices, and then migrate sub-dermally. At their final, species-specific resting site the larvae eventually form a warble-like swelling. In rodents the warble is often formed near the anus, scrotum or tail. Larval development may require between 3 and 7 weeks. When mature, the larvae leave the host and drop to the ground where they pupariate. In most regions there is only a single generation per year.

Pathology: in the warble formed around each larva, a thin layer of necrotic tissue develops and the larva feeds off the tissue debris and exudate. In general, the cuterebrid species are of little economic veterinary importance. However, occasional fatal cases of infestation have been recorded in cats and dogs.

Dermatobia

The genus *Dermatobia* contains a single species, *Dermatobia hominis,* which infests domestic animals and humans. *Dermatobia hominis*, also known as the torsalo, berne or human bot fly, is a Central and South American species, the larvae of which create boil-like swellings where they enter the skin.

Morphology: the larvae of *D. hominis* have a distinctive shape, being narrowed at the posterior end, particularly in the second-stage larva. The third-stage larva is more oval in shape (Fig. 5.11) with prominent

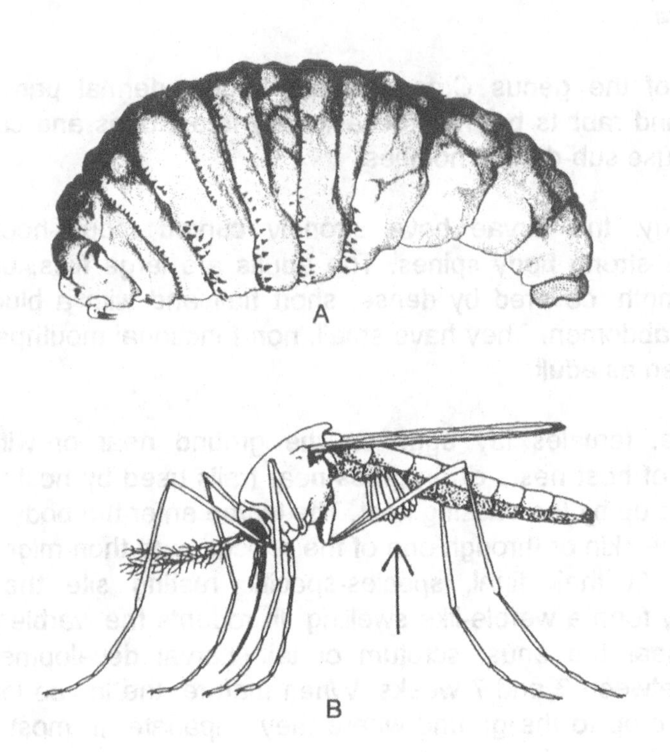

Fig. 5.11 (A) Third-stage larva of *Dermatobia hominis* (reproduced from Hall and Smith, 1993). (B) Adult mosquito carrying *D. hominis* eggs (reproduced from Evans, 1984).

flower-like anterior spiracles. Adults are large, blue-black flies, with a yellow-orange head and legs. The thorax of the adults possesses a sparse covering of short setae. The arista of the antennae has setae on the outer side only.

Life cycle: the life cycle of *D. hominis* is highly unusual; the adult female captures another carrier insect, usually a mosquito or muscid, on or near cattle and lays its eggs on to the carrier. The eggs are glued side by side along their axis, usually on one side of the carrier insect (Fig. 5.11). The eggs incubate and, when the carrier next visits a host, first-stage larvae hatch in response to the sudden temperature rise near the host's body. The larvae invade the host's subcutaneous tissues where they remain for 35–42 days. The larvae do not wander and at the site of penetration a small nodule of host tissue develops around each larva. Each female *D. hominis* may lay up to 1000 eggs during its lifetime, with an average of about 28 eggs per carrier insect. When fully developed, the larvae break throughout the host's skin, and drop to the ground where they pupariate.

Pathology: *D. hominis* is a serious pest of cattle in Central America. The cutaneous swellings can be pruritic and the exit holes may attract other myiasis flies. Infestation may result in damage to the hide and reduction in meat and milk production. Sheep, and occasionally dogs and cats, may also be attacked.

5.8 CALLIPHORIDAE

The Calliphoridae are medium to large flies, almost all of which have a metallic-blue or green sheen. There are over 1000 species divided between 150 genera. The majority of species in the family are saprophages, living in decaying organic material. About 80 species, belonging to the genera, *Cochliomyia, Calliphora*, *Chrysomya, Cordylobia, Lucilia*, *Protophormia* and *Phormia,* may be found in myiases of domestic animals.

Most of the calliphorid agents of myiasis have broadly similar life cycles. All are oviparous and, with the exception of species of *Cordylobia*, eggs are laid in wounded, infected or faecally soiled skin of warm-blooded vertebrate hosts. The larvae pass through three instars while feeding on the host tissues, causing cutaneous or traumatic myiasis. When mature, the larvae enter a wandering stage in which they migrate away from the strike focus and drop to the ground. After a period of dispersal they pupate in the substrate, to emerge eventually as adults.

5.8.1 *Cochliomyia*

Species of the genus *Cochliomyia* are green to violet-green blowflies with three prominent black, longitudinal stripes on the thorax and short palps. The genus contains four species, two of which, *Cochliomyia hominivorax* and *Cochliomyia macellaria,* are of particular importance, infesting cattle, horses, sheep, goats, pigs, dogs and humans.

Cochliomyia hominivorax

Known as the New World screwworm fly, *C. hominivorax* is an obligate ectoparasite and will infest almost all warm-blooded livestock, wildlife and humans. The natural range of the fly extends from the southern states of the USA through Central America and the Caribbean Islands to northern Chile, Argentina and Uruguay.

Morphology: the adult fly has a deep greenish-blue metallic colour with a yellow, orange or reddish face and three dark stripes on the dorsal surface of its thorax. In the larvae, the tracheal trunks leading from the posterior spiracles have a dark pigmentation extending forwards as far as the ninth or tenth segment (Fig. 5.12). This pigmentation is most conspicuous in fresh specimens. Mature third-stage larvae are about 15 mm in length.

Life cycle: females are autogenous and mate during early vitellogenesis. They lay batches of about 200–500 eggs every 2–3 days during adult life, which is on average 7–10 days in length. Eggs are deposited at the edges of open wounds or in body orifices, such as the nostrils, eyes, mouth, ears, anus and vagina. Shearing, castration or dehorning wounds are common oviposition sites, as are the navels of newly born calves. Even wounds the size of a tick bite are reported to be sufficient to attract oviposition. The eggs hatch within 24 hours and the larvae start to feed gregariously, burrowing head-down into the living tissue. The resulting wound may rapidly become extensive, attracting other female *C. hominivorax* and secondary agents of myiasis. The larvae are mature in 5–7 days following which they leave the wound, fall to the ground and pupariate. The entire egg to adult life cycle may be completed in about 24 days, but this may be extended under cooler conditions. There is no true diapausing stage and *C. hominivorax* cannot survive overwinter in cool temperate habitats.

Pathology: if untreated, repeated infestation by *C. hominivorax* may quickly lead to the death of the host. In the USA, *C. hominivorax* used to spread north and west each summer into more temperate zones from its overwintering areas near the Mexican border. The infestation of livestock by *C. hominivorax* was a major economic problem and, in Texas alone in the epidemic year of 1935, there were approximately 230,000 infestations in livestock and 55 in humans. The annual cost of *C. hominivorax* control in the United States in 1958 was estimated to be about $US140 million per year.

As a result of the economic cost of this pest, large-scale screwworm fly control was initiated in the south-eastern states of the USA in 1957–59. This was achieved by the release of large numbers of male *C. hominivorax* which had been sterilized by radiation. Sterilized males mate with wild females which are in turn rendered infertile. Subsequent control operations spread the area of sterile male release and in 1966 effective control of *C. hominivorax* in the US was declared. Despite a number of sporadic, but significant, outbreaks, effective control has been maintained. The eradication programme has subsequently been directed against the fly in Mexico, Puerto Rico, Vieques and the Virgin Islands.

In 1988, *C. hominivorax* were discovered in an area 10 km south of Tripoli in Libya. This was the first known established population of this species outside the Americas. The fly quickly spread to infest about 25,000 km^2. In 1989 there were about 150 cases of myiasis by *C. hominivorax* but by 1990, a total of 12,068 confirmed cases of screwworm fly myiasis were recorded and, at its peak, almost 3000 cases were seen in the single month of

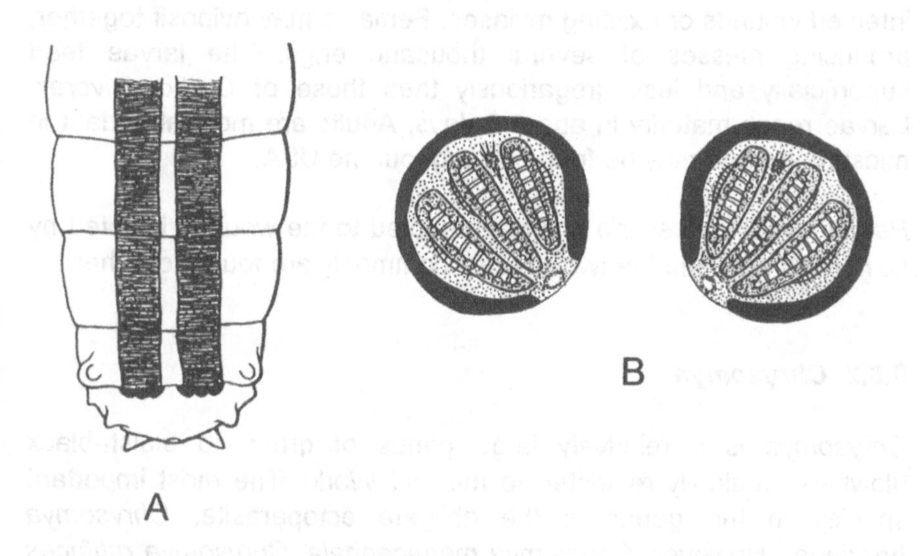

Fig. 5.12 (A) Pigmented dorsal tracheal trunks of *Cochliomyia hominivorax*, (B) posterior spiracles of *Cochliomyia macellaria* (reproduced from Hall and Smith, 1993).

September 1990. It was estimated that if unchecked the infestation could cost the Libyan livestock industry about US$30 million per year and the North African region approximately US$280 million per year. This led to the implementation of a major international control programme which has successfully eradicated the fly from this area, again using the release of sterile males.

Cochliomyia macellaria

Cochliomyia macellaria is a ubiquitous carrion breeder and occurs from Canada to Argentina. However, it can act as a secondary invader of strikes, and is known as the secondary screwworm fly.

Morphology: adults are extremely similar in appearance to *C. hominivorax*, but possess a number of white spots on the last segment of the abdomen. The larvae may be distinguished from those of *C. hominivorax* by the absence of pigmented tracheal trunks leading from the posterior spiracles.

Life cycle: adult females lay batches of about 250 eggs in carrion, infected wounds or existing myiases. Females may oviposit together, producing masses of several thousand eggs. The larvae feed superficially and less gregariously then those of *C. hominivorax*. Larvae reach maturity in about 6 days. Adults are most abundant in midsummer and may be found throughout the USA.

Pathology: *C. macellaria* is often attracted to the wounds initiated by *C. hominivorax* and the two species commonly are found together.

5.8.2 *Chrysomya*

Chrysomya is a relatively large genus of green to bluish-black blowflies, originally restricted to the Old World. The most important species in the genus is the obligate ectoparasite, *Chrysomya bezziana*. However, *Chrysomya megacephala, Chrysomya rufifaces* and *Chrysomya albiceps* are also of interest as secondary invaders of domestic animals in various parts of the world.

Chrysomya bezziana

The Old World screwworm fly, *Chrysomya bezziana,* is an obligate ectoparasite which occurs throughout much of Africa, India, the Arabian peninsula, South-East Asia, the Indonesian and Philippine islands and New Guinea. In the Old World it occupies much the same niche as *C. hominivorax* does in the New World.

Morphology: in the adults, the body is green or blue with narrow dark bands along the posterior margins of the abdominal segments and two faint dark stripes on the thorax. The legs are dark, the thoracic squamae are waxy white and the anterior spiracle is black-brown or dark orange. The adults have a pale-coloured face.

The first-stage larvae are creamy white and about 1.5 mm in length. The second- and third-stage larvae are 4–9 mm and 18 mm in length, respectively, and are similar in appearance; each segment carrying a broad, encircling belt of strongly developed spines (Fig. 5.13). The anterior spiracle has 4–6 branches.

Life cycle: *C. bezziana* will infest almost all warm blooded-livestock, wildlife and humans, and not carrion. Eggs are laid at the edges of a pre-existing wound or in body orifices, such as eyes, ears or nostrils. Females are normally autogenous and lay batches of, on average, 175–200 eggs every 2 or 3 days during their adult life of about 9 days. The eggs hatch in 12–18 hours at 37°C and the first-instar larvae migrate to the wound or moist tissue where they aggregate and begin feeding on the wound fluids. During the 6–7 days during which the larvae feed, they burrow deeply into the host's tissues so that only the posterior segment and spiracles of the larvae are exposed. When mature, the third-stage larvae leave the wound, drop to the ground and pupariate.

Pathology: in cattle, infestation by *C. bezziana* causes intermittent irritation and pyrexia, followed by the production of a cavernous lesion. The tissue shows progressive liquefaction, necrosis and haemorrhage, before the larvae leave the wound. Histologically, two distinct phases are observed; the first being intense neutrophil infiltration and haemorrhage associated with the growth of the larvae. The second is a fibroplastic healing phase in which mast cells and eosinophils are prominent.

The precise status of *C. bezziana* as a clinical and economic

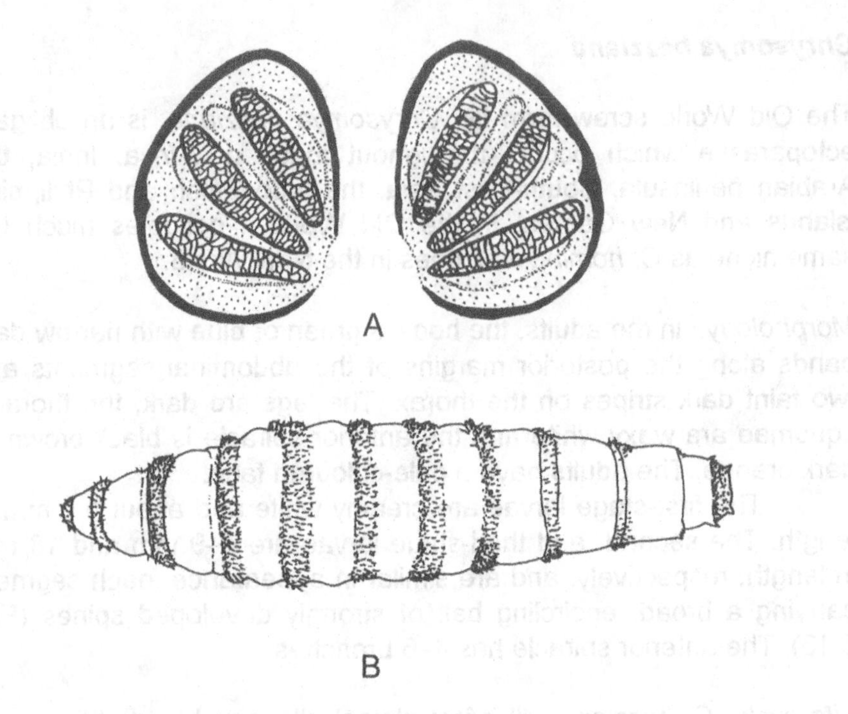

Fig. 5.13 (A) Posterior spiracles and (B) third-stage larva of *Chrysomya bezziana* (reproduced from Zumpt, 1965).

pest is uncertain, particularly in sub-Saharan Africa, and few studies have been able to obtain quantitative estimates of myiasis incidence, its clinical or economic importance. The absence of livestock throughout much of its range in sub-Saharan Africa, due to the presence of trypanosomiasis and its vector the tsetse fly, may substantially limit its economic impact. However, *C. bezziana* has been inadvertently introduced into several countries in the Middle East, and such an introduction is believed to pose a major economic threat to the pastoral industry of Australia where it has been estimated that its cost to the livestock industry would be up to A$430 million in 1990 values.

There is no evidence to support the assertion that *C. bezziana* is a less aggressive species than *C. hominivorax* or that the effects of its myiasis are less pathogenic.

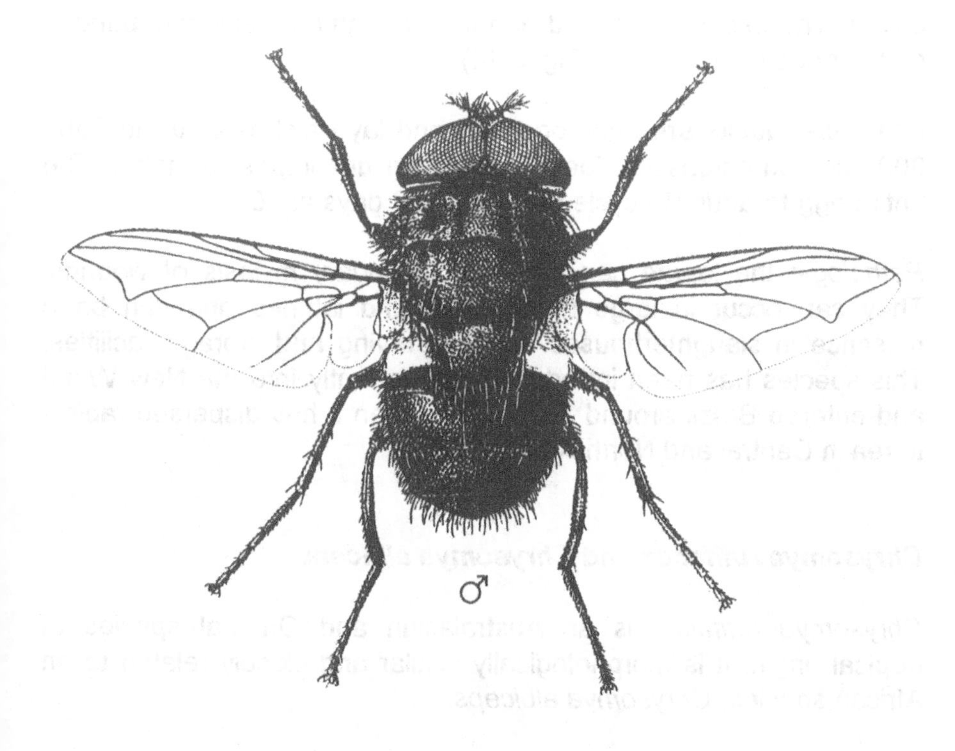

Fig. 5.14 Adult male of *Chrysomya megacephala* (reproduced from Crosskey and Lane, 1993).

Chrysomya megacephala

Chrysomya megacephala is a native of Australasian and Oriental regions. Although largely saprophagous, this species is occasionally found in myiases of domestic animals and humans. It is commonly known as the Oriental latrine fly because it can breed in faeces as well as on carrion.

Morphology: third-stage larvae are about 16 mm in length and may be easily distinguished from those of *C. bezziana* by the anterior spiracle, which in *C. megacephala* has 11–13 branches. In the adults, the body is green-blue, the legs are dark and the thoracic squamae are brown or grey. The anterior spiracle is dark-brown in colour. The second and third abdominal segments are dark-banded on their posterior margins (Fig. 5.14).

Life cycle: adults are anautogenous and lay batches of up to 250–300 eggs on carcasses, faeces and other decomposing matter. The entire egg-to-adult life cycle takes about 8 days at 30°C.

Pathology: the larvae can cause a secondary myiasis of wounds. They can occur in large numbers around latrines and can be a nuisance in slaughterhouses and fish drying and storage facilities. This species has been introduced inadvertently into the New World and entered Brazil around 1975. Since then it has dispersed rapidly to reach Central and North America.

Chrysomya rufifaces and *Chrysomya albiceps*

Chrysomya rufifaces is an Australasian and Oriental species of tropical origin. It is morphologically similar and closely related to an African species, *Chrysomya albiceps*.

Morphology: the larvae bear a number of fleshy projections on each segment, which give these species their common name of 'hairy maggot blowflies'. The projections become longer on the dorsal and lateral parts of the body. Small spines are present on the stalks of at least some of the projections of *C. rufifaces* but not *C. albiceps*. Third-stage larvae are about 18 mm in length. In the adult the body is metallic green or bluish in colour and the hind margin of the abdominal segments have blackish bands. The anterior spiracle is white or light yellow.

Life cycle: both species are saprophagous, normally laying batches of about 200 eggs on carcasses. However, they also may act as facultative ectoparasites. First-stage larvae are entirely necrophagous but second and third instars may also be facultatively predaceous on other dipteran larvae. The entire egg-to-adult life cycle may take as little as 9–12 days, the precise duration depending on temperature.

Pathology: in Australia and New Zealand, *C. rufifaces* will act as a

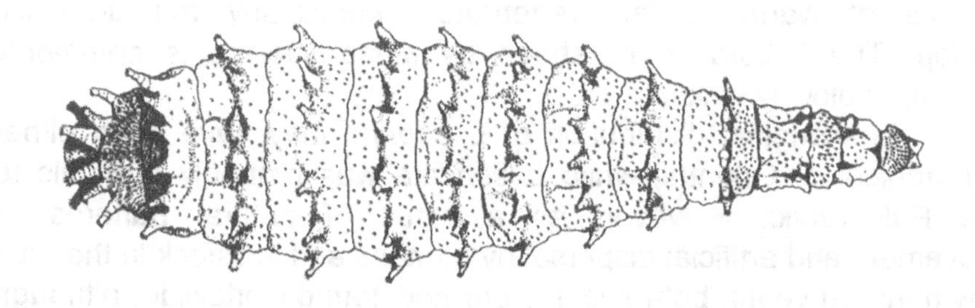

Fig. 5.15 Third-stage larva of *Chrysomya albiceps* (reproduced from Hall and Smith, 1993).

secondary invader of sheep myiasis following initial infestation, usually by *Lucilia cuprina*. Similarly, in southern Africa, *C. albiceps* will also act as a secondary invader of sheep myiasis. In 1978, *C. rufifaces* was introduced into Central America and has since dispersed through Central America and into the southern states of the USA.

5.8.3 *Lucilia*

There are at least 27 species in the genus *Lucilia*. However, only two species, *Lucilia sericata* and *Lucilia cuprina*, are of major clinical and economic significance worldwide as primary agents of cutaneous myiasis, particularly affecting sheep, although they may also strike a range of other wild and domestic animals and humans. *Lucilia sericata* and *L. cuprina* may be described, incorrectly, as members of a sub-genus *Phaenicia* by some authors. A number of other species, particularly *Lucilia caesar*, *Lucilia illustris* and *Lucilia ampullacea*, may also be found occasionally as facultative agents of myiasis.

Lucilia sericata and *Lucilia cuprina*

The sheep blowflies *Lucilia cuprina* and *Lucilia sericata* are facultative ectoparasites. Their larvae infest and feed on the living tissues of warm-blooded vertebrates, particularly the domestic sheep. The infestation of sheep by these species is commonly known as blowfly strike.

The original distribution of *L. cuprina* may have been either Afrotropical or Oriental, while *L. sericata* was probably endemic to the Palaearctic. However, as a result of natural patterns of movement and artificial dispersal by humans and livestock in the past few hundred years, both species are now found worldwide, although in general, *L. sericata* is more common in cool-temperate and *L. cuprina* in warm-temperate and sub-tropical habitats.

Lucilia sericata is the most important agent of sheep myiasis throughout northern Europe and was first recorded as an ectoparasite in England in the fifteenth century. *Lucilia cuprina* is absent from most of Europe, although it has been recorded from southern Spain and North Africa. *Lucilia cuprina* was probably introduced into Australia towards the middle or end of the nineteenth century and it is now the dominant sheep myiasis species for mainland Australia and Tasmania, being present in 90–99% of flystrike cases. Although *Lucilia sericata* is present in Australia, it is generally confined to more urban habitats.

Lucilia sericata arrived in New Zealand, over 100 years ago and soon established itself as the primary myiasis fly in this country, occurring in 75% of all cases of sheep strike. However, in the early 1980s *L. cuprina* was discovered, probably introduced from Australia, and now, despite its low abundance, in northern areas of New Zealand it appears to be displacing *L. sericata* to become the most important primary cause of flystrike in sheep.

In southern Africa the primary myiasis fly of sheep is *L. cuprina*. Although this species had been known in South Africa since 1830, little sheep strike was recorded until the early decades of the twentieth century, possibly as a result of the introduction of more susceptible Merino breeds or changes in husbandry practices. In North America, *L. sericata* (syn. *Phaenicia sericata*) is an important agent of sheep myiasis. Interestingly, although *L. cuprina* (syn. *Phaenicia cuprina* = *Phaenicia pallescens*) is known to be present in the USA, it does not appear to be important in sheep myiasis.

It would seem likely that the myiasis habit in *L. sericata* and

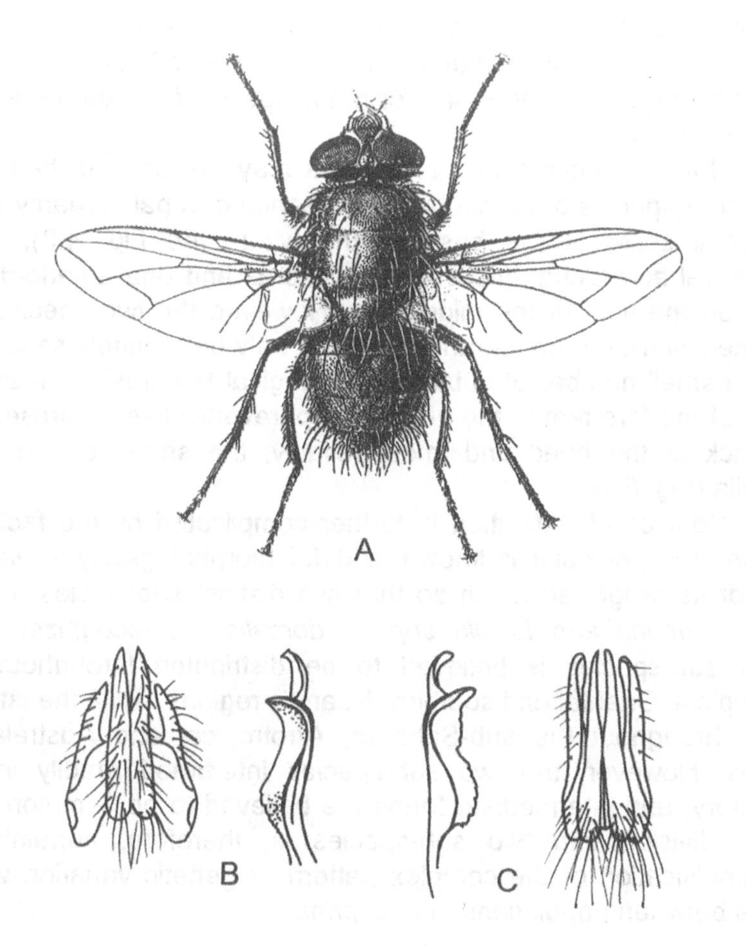

Fig. 5.16 (A) Adult *Lucilia sericata* (reproduced from Shtakelbergh, 1956). Male genitalia (aedeagus in lateral view and forceps in dorsal view) of (B) *Lucilia sericata* and (C) *Lucilia cuprina* (reproduced from Aubertin, 1933).

L. cuprina has arisen relatively recently and independently in geographically isolated populations, perhaps in response to changes in sheep husbandry, with *L. cuprina* becoming the predominant pathogenic species in sub-tropical and warm-temperate habitats (e.g. Australia) and *L. sericata* in cool-temperate habitats (e.g. Europe and New Zealand).

Morphology: *Lucilia* spp. are metallic green-coloured flies, characterized by the presence a bare stem-vein, bare squamae and the presence of three pairs of postsutural, dorso-central bristles (see Fig. 4.2) on the thorax. All the species in this genus bear a very close resemblance to each other and for many species females are almost indistinguishable.

Both *L. sericata* and *L. cuprina* may be distinguished from most other species of *Lucilia* by the presence of a pale creamy-white basicostal scale at the base of the wing (see Fig. 4.3), three postsutural acrostichal bristles on the thorax and one anterio-dorsal bristle on the tibia of the middle leg. However, the two species are extremely similar in appearance and can only be routinely separated using a small number of subtle morphological features, such as the colour of the fore femur, the number of paravertical setae present on the back of the head and, most reliably, the shape of the male genitalia (Fig. 5.16).

Species identification is further complicated by the fact that *L. cuprina* in particular is known to differ morphologically in various parts of its range; so much so that two distinct subspecies, *Lucilia cuprina cuprina* and *Lucilia cuprina dorsalis* are recognized. The former sub-species is believed to be distributed throughout the Neotropical, Oriental and southern Nearctic regions, while the latter is found throughout the sub-Saharan, Afrotropical and Australasian regions. However, the two sub-species interbreed readily in the laboratory, and intermediate forms are believed to be common. The simple division into two sub-species is, therefore, certainly an oversimplification of the complex pattern of genetic variation which occurs between populations of *L. cuprina*.

Life cycle: both *L. sericata* and *L. cuprina* are anautogenous and must obtain a protein meal before maturing their eggs. When protein is freely available, females deposit batches of 225–250 eggs at three-day intervals throughout their life. Adults are diurnal and are most active in open sunny areas. The average longevity of an adult female is about 7 days.

In contrast to the feeding behaviour of screwworm larvae, the larvae of *L. cuprina* and *L. sericata* usually feed superficially on the epidermis and lymphatic exudate or necrotic tissue. Only when crowded will they begin to feed on healthy tissue. The mouth-hooks are used to macerate the tissues and digestion occurs extraorally by means of amylase in the saliva and proteolytic enzymes in the larval excreta. The larvae pass through three stages before wandering from the lesion and dropping to the ground where they pupariate.

The time required to complete the life cycle from egg to adult is highly dependent on ambient temperature, but is commonly in the order of 4–6 weeks.

Throughout most of its range, *L. sericata* is seasonally abundant, in the northern hemisphere passing through three or four generations between May and September and overwintering as third-stage larvae. In Australia, however, there may be up to eight generations per year and in some areas *L. cuprina* is able to breed continuously.

Pathology: blowfly strike by *L. sericata* occurs most commonly in the perineal and inner thigh region and is strongly associated with faecal soiling. Faecal soiling is also an important predisposing factor for strike by *L. cuprina*. However, in Australia, body strike is frequently the main form of myiasis. Body strike occurs most commonly around the shoulders and back region and is frequently associated with the incidence of bacterial dermatophilosis. Dermatophilosis is a chronic dermatitis, caused by the bacterium *Dermatophilus congolensis* which invades the epidermis. Body strike in Australia is also often associated with bacterial fleece rot, a superficial dermatitis induced by moisture and proliferation of the bacteria *Pseudomonas*

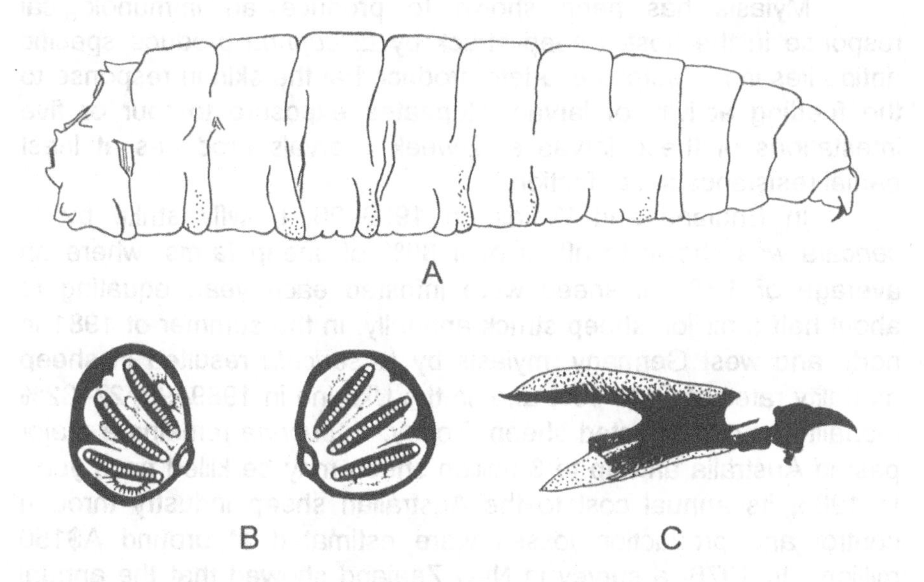

Fig. 5.17 Third-stage larva of *Lucilia sericata*, (A) body in lateral view (reproduced from Smith, 1989). (B) Posterior peritremes and (C) cephalopharyngeal skeleton (reproduced from Hall and Smith, 1993).

aeruginosa on the skin, resulting in a matted band of discoloured fleece. It is possible that dermatophilosis and fleece rot act synergistically in attracting blowflies and their subsequent oviposition. There is little recorded involvement of either form of dermatitis in predisposing sheep to strike by *L. sericata* in northern Europe.

The risk of myiasis by *L. sericata* has been shown to increase with increasing flock size and stocking density, and to decrease with increasing farm altitude. In Australia, flystrike has been shown to depend on the number and activity of gravid *L. cuprina* and to be higher in warm, humid conditions with low wind speeds; crutch strike may become more important than body strike under dryer conditions and when fly densities are low.

Sheep struck by *Lucilia* have a rapid increase in body temperature and respiratory rate, accompanied by loss of weight and anorexia. The animals become anaemic and suffer severe toxaemia, with both kidney and heart tissues affected. The feeding activity of the larvae may cause extensive tissue damage which, in combination with the larval proteases produced, results in the development of inflamed, abraided or undermined areas of skin. This may result in considerable distress to the struck animal, a loss of fertility and, if untreated, rapidly leads to death from toxaemia.

Myiasis has been shown to produce an immunological response in the host. Sheep struck by *L. cuprina* produce specific antibodies in the serous exudate produced at the skin in response to the feeding activity of larvae. Repeated exposure to four or five infestations of these larvae at 2-week intervals produces at least partial resistance to reinfection.

In England and Wales, in 1989–90, blowfly strike by *L. sericata* was shown to affect over 80% of sheep farms, where an average of 1.5% of sheep were infested each year, equating to about half a million sheep struck annually. In the summer of 1981 in north and west Germany, myiasis by *L. sericata* resulted in sheep mortality rates of up to 10% and, in the Ukraine in 1989–90, 27–32% mortality among affected sheep. Today, *L. cuprina* remains a major pest in Australia and up to 3 million sheep may be killed each year. In 1985, its annual cost to the Australian sheep industry through control and production losses were estimated at around A$150 million. In 1976, a survey in New Zealand showed that the annual prevalence of flystrike in sheep was 1.7% in the North Island and 0.7% in the South Island and the annual cost to farmers was at least NZ $1.7 million.

Blowfly strike of domestic rabbits and occasionally other domestic mammals and birds may be common, particularly if dirty, debilitated by clinical disease or wounded.

Other species of *Lucilia* of veterinary interest

In Europe, *Lucilia caesar* may become more common in sheep in more northerly latitudes. In Norway, from 27 cases of sheep myiasis the primary species was found to be *L. caesar*, occurring alone or in combination with *Lucilia illustris*. *Lucilia ampullacea* also may occasionally be found in myiases of sheep and other mammals. *Lucilia bufonivora* is an obligate parasite of toads.

5.8.4 *Phormia* and *Protophormia*

These two genera are closely related and each contains a single species of veterinary interest: *Phormia regina* and *Protophormia terraenovae*. Both are found in the northern Holarctic.

Morphology: both *P. regina* and *P. terraenovae* are similar in appearance. The adults are dark metallic blue-black in colour. In *P. terraenovae* the anterior thoracic spiracle is black or black-brown and is difficult to distinguish from the general body colour. In contrast, in *Phormia regina* the anterior spiracle is yellow or orange and stands out clearly against the dark background colour of the thorax.

The third-stage larvae of both *P. terraenovae* and *P. regina* are characterized by strongly developed, fairly pointed tubercles on the posterior face of the last segment (Fig. 5.18). In third-stage larvae of *P. terraenovae* the tubercles on the upper margin of the last segment are longer than those of *P. regina,* being longer than half the width of a posterior spiracle, whereas in *P. regina* they are less than half the width of a posterior spiracle in length. The larvae of *P. terraenovae* also possess dorsal spines on the posterior margins of segment 10 which are absent in larvae of *P. regina*.

Life cycle: these species prefer cool temperatures and more northerly habitats. They usually breed in carrion, but both can be found in myiases of livestock. *Protophormia terraenovae* is abundant in early

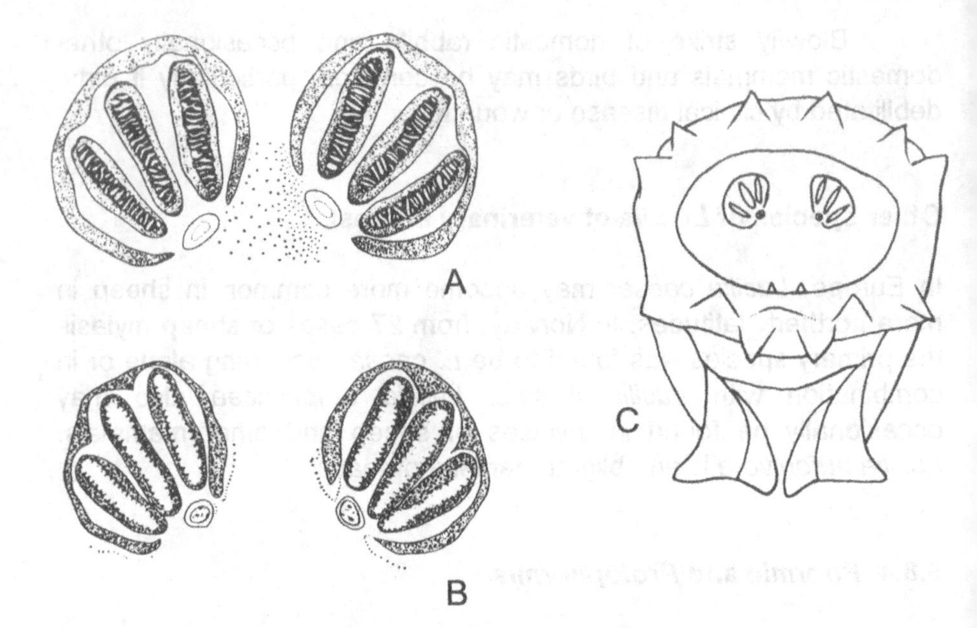

Fig. 5.18 Posterior spiracles of third-stage larvae of (A) *Protophormia terraenovae* and (B) *Phormia regina.* (C) Tubercles on the posterior face of the last segment of third-stage *Protophormia terraenovae* (reproduced from Hall and Smith, 1993).

spring in Finland and is the dominant blowfly in the Arctic and subarctic.

Pathology: *P. terraenovae* may be common in sheep strike in northerly areas of Britain and Canada and can be a serious parasite of cattle, sheep and reindeer. Although present in the Palaearctic, *P. regina* is more common in livestock myiases, particularly of sheep, in northern USA and Canada.

5.8.5 *Calliphora*

There are a great many species in this widely distributed genus, particularly in the Holarctic and Australasian regions.

Morphology: this is a large and diverse genus. Adults are usually about 8–14 mm in length. A number of species, such as *Calliphora vicina* and *Calliphora vomitoria*, are commonly referred to as bluebottles, because of the predominant metallic, blue or blue-black colouration of the adult. In other Australasian species of interest, such as *Calliphora albifrontalis, Calliphora nociva* and *Calliphora stygia,* the thorax is non-metallic blue-black in colour but the abdomen is predominantly brown or brown-yellow. *Calliphora vicina* and *C. vomitoria* may be distinguished from each other by the presence of yellow-orange jowls with black hairs in the former and black jowls with predominantly reddish hairs in the latter.

The third-stage larvae of all species of *Calliphora* of veterinary interest possess a cephalopharyngeal skeleton with a pigmented accessory oral sclerite (Fig. 5.19).

Life cycle: all species of *Calliphora* develop in decomposing organic matter, but a number may act as secondary or tertiary agents of myiasis. They are commonly found around houses and livestock facilities since adults are attracted to faeces and will enter buildings. Eggs are laid in batches in carrion or other waste material. The life cycle is typical, with three larval stages, followed by wandering of the third-stage larva and pupariation in the ground.

Pathology: the Western Australian brown blowfly, *C. albifrontalis* and the lesser brown blowfly, *C. nociva*, are important native Australasian species found as secondary or tertiary invaders of sheep strikes in the Australasian region. In western Australia *C. albifrontalis* may be responsible for up to 10% of single-species strikes. In New Zealand, *Calliphora stygia* may be a common secondary invader of ovine myiasis, being present in strikes from October to May. Clearly, these species have been able to adapt relatively rapidly to breed on sheep soon after their introduction into Australia and New Zealand.

The two most abundant species of *Calliphora* in the Holarctic region are *C. vicina* and *C. vomitoria*. They are now also present in many parts of the Oriental and Australasian regions. *Calliphora vicina* has been recorded as laying eggs on living small mammals and may occur as a secondary invader of sheep myiases. Attempts to induce primary sheep strike by *C. vicina* have proved unsuccessful and it has been suggested that this species may be physiologically unable to infest sound sheep, either because the sheep body temperature is fatally high or because larvae are unable to feed on the animal tissues without the prior activity of *Lucilia* larvae. *Calliphora vomitoria*

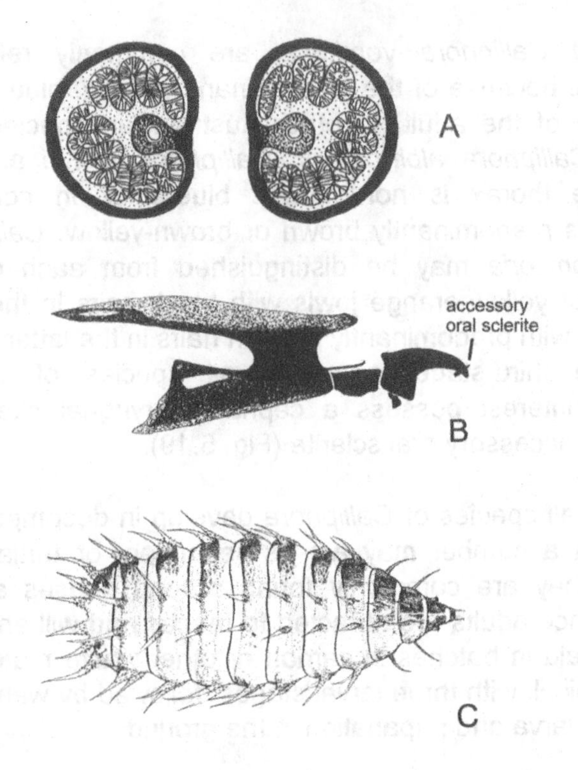

Fig. 5.19 (A) Posterior spiracles of third-stage larva of the house fly *Musca domestica*. (B) Cephalopharyngeal skeleton of *Calliphora vicina*. (C) Third-stage larva of the lesser house fly *Fannia canicularis* (reproduced from Hall and Smith, 1993).

is the most abundant species of fly larva found in carrion in more northerly and upland sites and is only rarely found in myiases of living animals.

5.8.6 *Cordylobia*

There are three species in this genus, the most well known of which is *Cordylobia anthropophaga*, the Tumbu fly of sub-Saharan Africa. Also found in sub-Saharan Africa is *Cordylobia rodhaini*, which infests rodents.

Cordylobia anthropophaga

Morphology: the adults of both sexes are large, stout flies up to 12 mm in length, with large, fully developed mouthparts. The body is not metallic-coloured like blowflies, instead they are generally dull yellow-brown or red-brown in colour. The arista of the antenna has setae on both sides. The thoracic squamae are without setae and the stem vein of the wing is without bristles. The larvae are stout, up to 15 mm in length in the third stage and covered with small, dark spines (Fig. 5.20).

Life cycle: eggs are laid in batches of 200–300 on dry, shaded ground, particularly sandy ground contaminated with urine or faeces. Eggs are laid in early morning or late evening. The larvae hatch in 1–3 days and can remain alive, without feeding, for 9–15 days hidden just beneath the soil surface. A sudden rise in temperature, vibration or carbon dioxide, which might signify the presence of a host, activates the larvae. They attach to the host and immediately burrow into the skin. The larvae do not wander sub-dermally. A swelling develops around the larva at the point of entry. The swelling has a hole through which the larva breathes. The three larval stages are completed in the host in 8–15 days. The larva then exits the lesion, falls to the ground and pupariates.

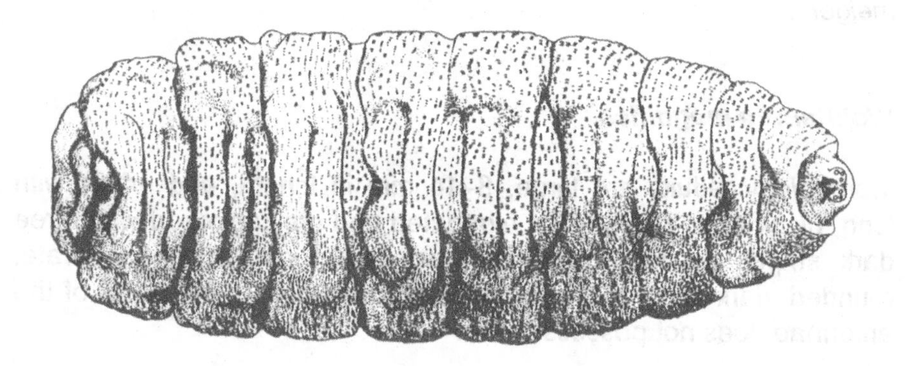

Fig. 5.20 Third-stage larva of *Cordylobia anthropophaga* (reproduced from Hall and Smith, 1993).

Pathology: this species causes a warble-like myiasis, which is initially pruritic, becoming more painful as the larva grows. Serous fluid may exude from the lesion. Dogs are frequently attacked and are important reservoirs of infection. Many other wild and domestic animals and humans, particularly rats, may also be infested.

5.9 SARCOPHAGIDAE

The Sarcophagidae, known as flesh flies, are grey-black, non-metallic, medium to large flies with prominent stripes on the thorax (Fig. 5.21). There are over 2000 species in the family Sarcophagidae, divided into 400 genera. Most species of *Sarcophaga* are of no veterinary importance, breeding in excrement, carrion and other decomposing organic matter. The only important genus of the family Sarcophagidae containing species which act as agents of veterinary myiasis is *Wohlfahrtia*.

5.9.1 *Wohlfahrtia*

The most economically important species in the genus *Wohlfahrtia* is *Wohlfahrtia magnifica*. This is an obligate larval parasite causing traumatic cutaneous myiasis of warm-blooded vertebrates throughout the Mediterranean basin, eastern and central Europe and Asia Minor. Also of note are *Wohlfahrtia nubia, Wohlfahrtia vigil* and *Wohlfahrtia meigeni*.

Wohlfahrtia magnifica

Morphology: adults are large, 8–14 mm in length, and bristly with long, black legs. The body is elongated and grey coloured with three dark stripes on the thorax and a number of distinct, separate, rounded, dark patches on the abdomen (Fig. 5.21). The arista of the antennae does not possess setae.

Life cycle: *W. magnifica* is an obligate agent of myiasis. Female flies deposit 120–170 first-stage larvae on the host, in wounds or next to body orifices. The larvae feed and mature in 5–7 days, moulting twice, before leaving the wound and dropping to the ground where

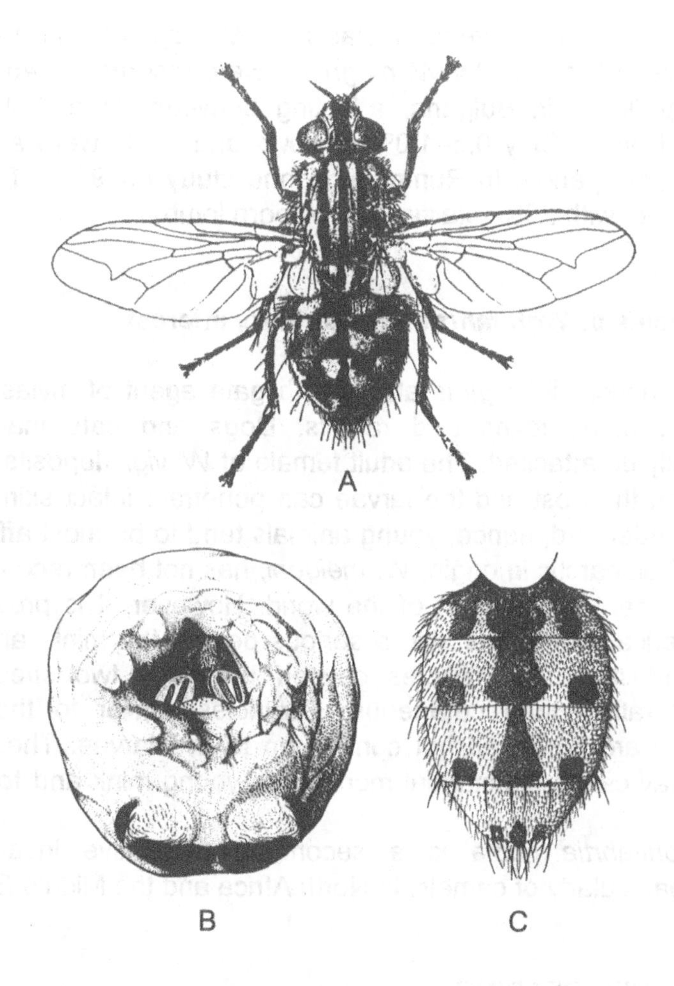

Fig. 5.21 (A) Adult of the flesh fly *Sarcophaga carnaria*, (B) rear view of the last segment of *Sarcophaga* showing the posterior spiracles deeply sunk in a cavity and (C) abdomen of *Wohlfahrtia magnifica* (reproduced from Smith, 1973).

they pupariate.

Pathology: W. magnifica can cause rapid and severe myiasis in most livestock, particularly sheep and camels and also poultry, although cattle, horses, pigs and dogs may also be infested. Levels of infestation appear to be high, particularly in sheep in eastern Europe. Faecal soiling in sheep has been recorded as an important predisposing factor for breech myiasis by *W. magnifica.* In a 4-year period, cases of myiasis by *W. magnifica* were recorded in 45 out of 195 sheep flocks in Bulgaria, affecting between 23 and 41% of sheep each year. Only 0.5–1.0% of cows and goats were affected over the same period. In Rumania, in one study 80–95% of sheep were infested, with 20% fatalities of newborn lambs.

Other species of *Wohlfahrtia* of veterinary interest

In North America, *W. vigil* is also an obligate agent of myiasis and may infest mink, foxes and rabbits. Dogs and cats may also occasionally be attacked. The adult female of *W. vigil* deposits active maggots on the host and the larvae can penetrate intact skin if it is thin and tender and, hence, young animals tend to be most affected. Although Palaearctic in origin, *W. meigeni,* has not been recorded as a myiasis agent in that part of the world. However, it is present in North America, where it is also a serious pest to the mink- and fox-farming industry. The myiases caused by these two species is furuncular rather than cutaneous. Furuncles similar to those of *Dermatobia* are produced but contain up to five larvae. These two species may cause substantial mortality to young mink and foxes in fur farms.

Wohlfahrtia nubia is a secondary facultative invader of wounds, particularly of camels, in North Africa and the Middle East.

5.10 FURTHER READING

Ashworth, J.R. and Wall, R. (1994) Responses of the sheep blowflies *Lucilia sericata* and *L. cuprina* (Diptera: Calliphoridae) to odour and the development of semiochemical baits. *Medical and Veterinary Entomology*, **8**, 303–9.

Aubertin, D. (1933) Revision of the genus *Lucilia* R.-D. (Diptera, Calliphoridae). *Journal of the Linnean Society*, **33**, 389–436.

Barnard, D.R. (1977) Skeletal-muscular mechanisms of the larva of *Lucilia sericata* (Meigen) in relation to feeding habit. *The Pan-Pacific Entomologist*, **53**, 223–9.

Baron, R.W. and Colwell, D.D. (1991) Mammalian immune responses to myiasis. *Parasitology Today*, **7**, 353–5.

Baumgartner, D.L. (1993) Review of *Chrysomya rufifaces* (Diptera: Calliphoridae). *Journal of Medical Entomology*, **30**, 338–52.

Baumgartner, D.L. and Greenberg, G. (1984) The genus *Chrysomya* (Diptera: Calliphoridae) in the New World. *Journal of Medical Entomology*, **21**, 105–13.

Beck, T., Moir, B. and Meppen, T. (1985) The cost of parasites to the Australian sheep industry. *Quarterly Review of Rural Economics*, **7**, 336–43.

Brinkmann, A. (1976) Blowfly myiasis of sheep in Norway. *Norwegian Journal of Zoology*, **24**, 325–30.

Castellani, A. and Chalmers, A.J. (1913) *Manual of Tropical Medicine*, London.

Crosskey, R.W. and Lane, R.P. (1993) House-flies, blow-flies and their allies (calypterate Diptera), in *Medical Insects and Arachnids* (eds. R.P. Lane and R.W. Crosskey), Chapman & Hall, London, pp. 403–28.

Erzinclioglu, Y.Z. (1987) The larvae of some blowflies of medical and veterinary importance. *Medical and Veterinary Entomology*, **1**, 121–5.

Erzinclioglu, Y.Z. (1989) The origin of parasitism in blowflies. *British Journal of Entomology and Natural History*, **2**, 125–8.

Evans, H.E. (1984) *Insect Biology. A Textbook of Entomology*, Adison Wesley, Massachusetts.

French, N.P., Wall, R., Cripps, P.J. and Morgan, K.L. (1992) The prevalence, regional distribution and control of blowfly strike in England and Wales. *The Veterinary Record*, **131**, 337–42.

French, N.P., Wall, R., Cripps, P.J. and Morgan, K.L. (1994) Blowfly strike in England and Wales: the relationship between prevalence and farm management factors. *Medical and Veterinary Entomology*, **8**, 51–6.

French, N.P., Wall, R. and Morgan, K.L. (1995) The seasonal pattern of blowfly strike in England and Wales. *Medical and Veterinary Entomology*, **9**, 1–8.

Hall, M.J.R. and Smith, K.G.V. (1993) Diptera causing myiasis in man, in *Medical Insects and Arachnids* (eds R.P. Lane and R.W. Crosskey), Chapman & Hall, London, pp. 429–69,

Hall, M.J.R. and Wall, R. (1994) Myiasis of humans and domestic animals, in *Advances in Parasitology* (eds J.R. Baker, R. Muller and D. Rollinson), Academic Press, London, Vol. 35, pp. 258–334.

Herms, W.B. and James, M.T. (1961) *Medical Entomology*, Macmillan, New York.

Humphrey, J.D., Spradbery, J.P. and Tozer, R.S. (1980) *Chrysomya bezziana*: pathology of old world screw-worm fly infestations in cattle. *Experimental Parasitology*, **49**, 381–97.

James, M.T. (1947) The flies that cause myiasis in man. *United States Department of Agriculture Miscellaneous Publications*, **631**, 1–175.

Krafsur, E.S., Whitten, C.J. and Novy, J.E. (1987) Screwworm eradication in north and central America. *Parasitology Today*, **3**, 131–7.

Shtakelbergh, A.A. (1956) *Diptera associated with man from the Russian fauna*. Moscow.

Smith, K.G.V. (1973) *Insects and Other Arthropods of Medical Importance*, British Museum (Natural History), London.

Smith, K.G.V. (1989) An introduction to the immature stages of British flies: Diptera larvae, with notes on eggs, puparia and pupae. *Handbooks for the Identification of British Insects*, **10**, (14), 1–280.

Spradbery, J.P. (1994) Screw-worm fly: a tale of two species. *Agricultural Zoology Reviews*, **6**, 1–62.

Stevens, J. and Wall, R. (1996) Classification of the genus *Lucilia*: a taxonomic study using parsimony analysis. *Journal of Natural History*, **30**, 1087–94.

Stevens, J.R. and Wall, R. (1996) Species, sub-species and hybrid populations of the blowflies *Lucilia cuprina* and *Lucilia sericata* (Diptera: Calliphoridae). *Proceedings of the Royal Society (B)*, **263**, 1335–41.

Sutherst, R.W., Spradbery, J.P. and Maywald, G.F. (1989) The potential geographical distribution of the Old World screw-worm fly, *Chrysomya bezziana*. *Medical and Veterinary Entomology*, **4**, 273–80.

Tarry, D.W. (1986) Progress in warbly fly eradication. *Parasitology Today*, **2**, 111–16.

Tenquist, J.D. and Wright, D.F. (1976) The distribution, prevalence and economic importance of blowfly strike in sheep. *New Zealand Journal of Experimental Agriculture*, **4**, 291–5.

Thomas, D.B. and Mangan, R.L. (1989) Oviposition and wound visiting behaviour of the screwworm fly, *Cochliomyia hominivorax* (Diptera: Calliphoridae). *Annals of the Entomological Society of America*, **82**, 526–34.

Van Emden, F.I. (1954) Diptera Cyclorrapha, Calyptrata (I) Section (a) Tachinidae and Calliphoridae. *Handbooks for the Identification of British Insects,* **10**, (4), 1–133.

Wall, R. (1993) The reproductive output of the blowfly *Lucilia sericata*. *Journal of Insect Physiology*, **39**, 743–50.

Wall, R., French, N.P. and Morgan, K.L. (1993) Predicting the abundance of the sheep blowfly *Lucilia sericata* (Diptera: Calliphoridae). *Bulletin of Entomological Research*, **83**, 431–6.

Zumpt, F. (1965) *Myiasis in Man and Animals in the Old World*, London, Butterworths.

6

Fleas (Siphonaptera)

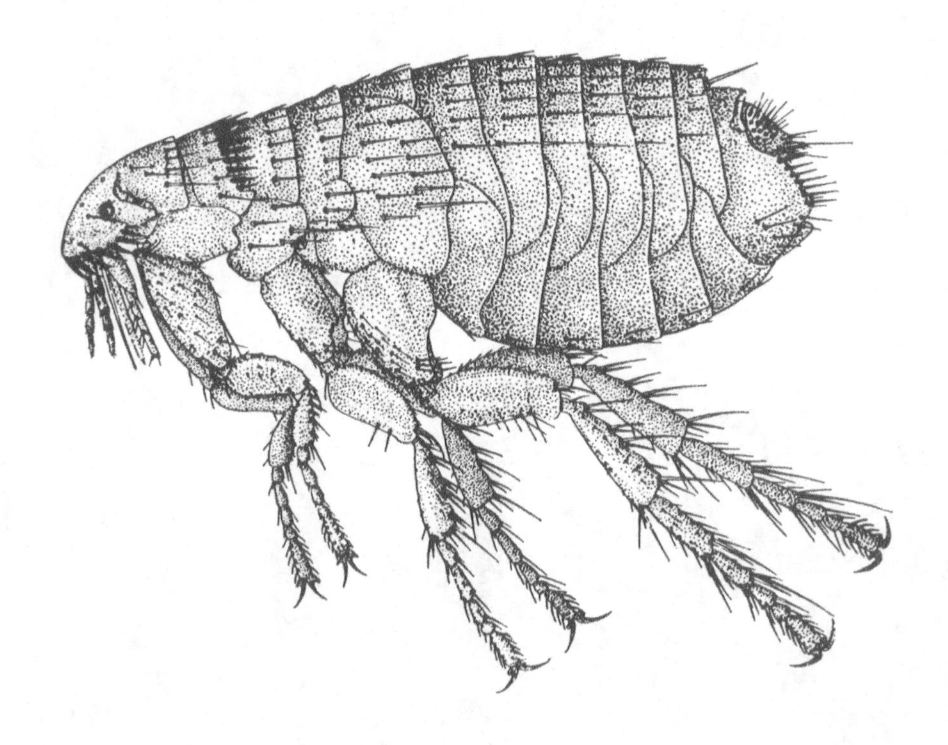

Adult cat flea, *Ctenocephalides felis felis* (reproduced from Lewis, 1993).

6.1 INTRODUCTION

The fleas (order Siphonaptera) are small, wingless, obligate blood-feeding insects. Over 95% of flea species are ectoparasites of mammals, while the others are ectoparasites of birds. The order is relatively small with about 2500 described species, almost all of which are morphologically extremely similar.

Adult fleas usually live in temporary association with their host; flea morphology, physiology and behaviour are intimately associated with hosts that are only intermittently available, temporary episodes of feeding and the microhabitat of the nest, burrow or dwelling of the host animal. Mammals and birds which do not build nests, or return regularly to specific bedding, lairs or burrows generally do not have fleas. Hence, fleas are common on rodents, bats, carnivores and rabbits and virtually absent on ungulates. Through their close association with the habitations of humans and their companion and domestic animals, a number of species of flea have become distributed worldwide and now proliferate in previously inhospitable habitats.

Many species of flea are able to parasite a range of hosts. This, combined with their mobility which allows them to move easily between hosts, makes them parasites of considerable medical and veterinary importance and makes them difficult to control. Blood-feeding may have a range of damaging effects on the host animal, causing inflammation, pruritus or anaemia. Fleas may also act as vectors of bacteria, protozoa, viruses and tapeworms. However, in veterinary entomology fleas are probably of most importance as a cause of cutaneous hypersensitivity reactions.

The distinctiveness of fleas and the existence of small isolated families suggest that they are a very ancient order, and fleas probably became parasites of mammals relatively early in the history of their hosts. The first flea fossils are known from the Cretaceous (135–65 million years ago) and fleas almost identical to living species have been found embedded in amber, more than 50 million years old.

Fleas have no close relatives and their evolution is the subject of some debate. It is thought that they may have evolved from a common origin of the scorpion flies (Mecoptera). The scorpion flies are medium-sized, usually winged insects, with long, narrow legs adapted for walking and with the head elongated into a broad

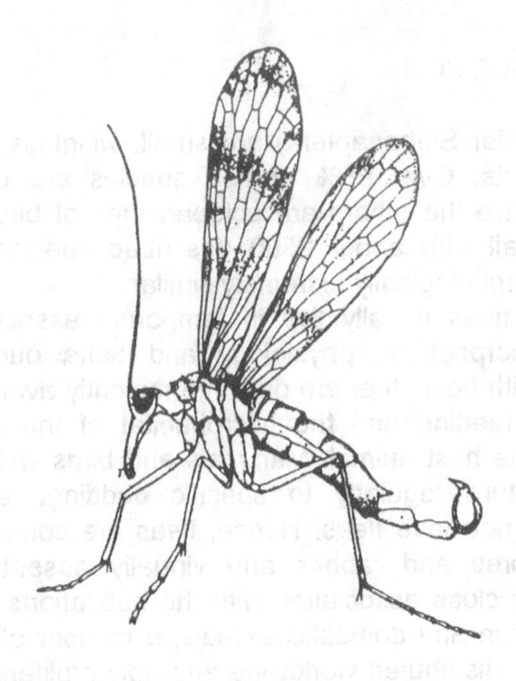

Fig. 6.1 Adult scorpion fly (*Mecoptera*) (reproduced from Gullan and Cranston, 1994).

rostrum incorporating the mouthparts (Fig. 6.1). The ancestors of fleas may have resembled the mecopteran family Boreidae which, like fleas, are wingless and capable of jumping. They feed on detritus and other arthropods. We can speculate that the mecopteroid ancestors of fleas may have accumulated in the burrows of mammals, where excrement and other arthropods were abundant and subsequently switched to using their piercing mouthparts, adapted for preying on other insects, for feeding on the blood of the mammalian owner of the burrow. After evolving initially as parasites of mammals, a small number of species appear to have become secondarily adapted to parasitize birds.

6.2 MORPHOLOGY

The adults are highly modified for an ectoparasitic life and are structurally very different from most other insects. In contrast to lice or ticks, the flea body is laterally compressed (Fig. 6.2). Adults are wingless and usually between 1 and 6 mm in length, with females being larger than males. The body colour may vary from light brown to black. The body is armed with spines, which are directed backward. The function of the spines is of some debate. They may serve simply to protect the articular membranes at joints between tergites. Alternatively, as is more commonly suggested, they may allow movement through the hair or feathers of the host while preventing dislodgement during grooming. This idea is supported by the relationships that exist between number and size of spines and the nature of the fur, feathers and habits of the host. Indeed, it has even been suggested that the distance between pronotal spines may be directly related to the diameter of individual hairs and the density of the host's hair.

As with all insects, the body is divided into head, thorax and abdomen. The head is immobile and its shape is highly variable and may be useful in species identification. Some caution is needed however, since head shape may also vary between the sexes and individuals of a species. In some species, approximately midway along the anterior margin of the frons is a small bump known as the

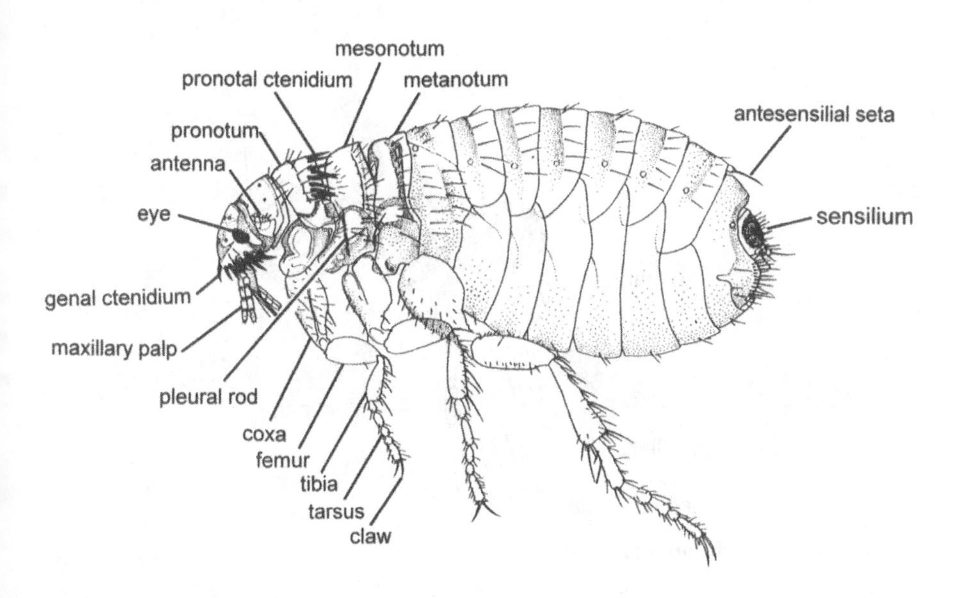

Fig. 6.2 Morphological features an an adult flea (reproduced from Lewis, 1993).

frontal tubercle, which is variable in shape and structure. The ventral portion of the anterior part of the head is known as the **gena**. It may extend backwards forming a distinct genal lobe, below which is found the small, three-segmented, backwardly directed antenna, in a groove known as the **antennal fossa**. A conspicuous comb of spines is often present on the gena, known as the **genal ctenidium**. The spines of the ctenidium are heavily sclerotized outgrowths of cuticle rather than setae. Eyes are absent in some species of nest flea. If present, however, the eyes are usually simple and found on the head in front of the antennae. These eyes probably are reduced compound eyes not displaced ocelli.

Ventrally there is a pair of broad maxillary lobes, known as stipes, bearing long **maxillary palps** (Fig. 6.3). Below these structures are the mouthparts, known as the **fascicle**, consisting of a pair of fine grooved **laciniae**, sometimes bearing course teeth. The grooved laciniae are closely opposed, forming a groove, in which lies the **labrum-epipharynx**. The laciniae are supported on each side by a pair of five-segmented **labial palps**. The laciniae are used to puncture the host's skin and the tip of the labrum-epipharynx

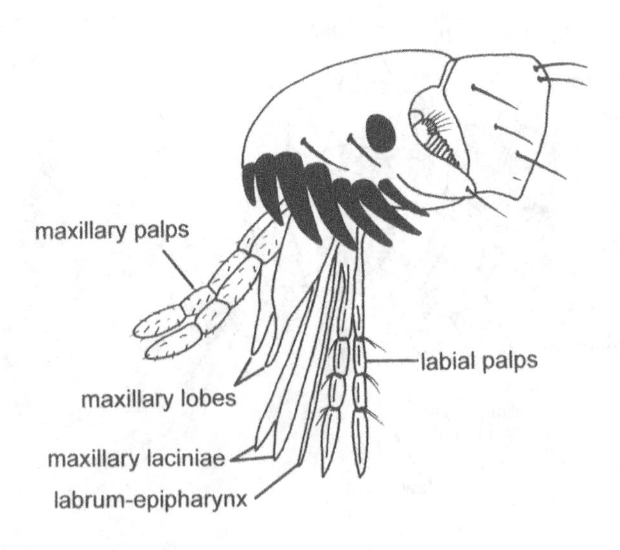

Fig. 6.3 Head and mouthparts on an adult flea.

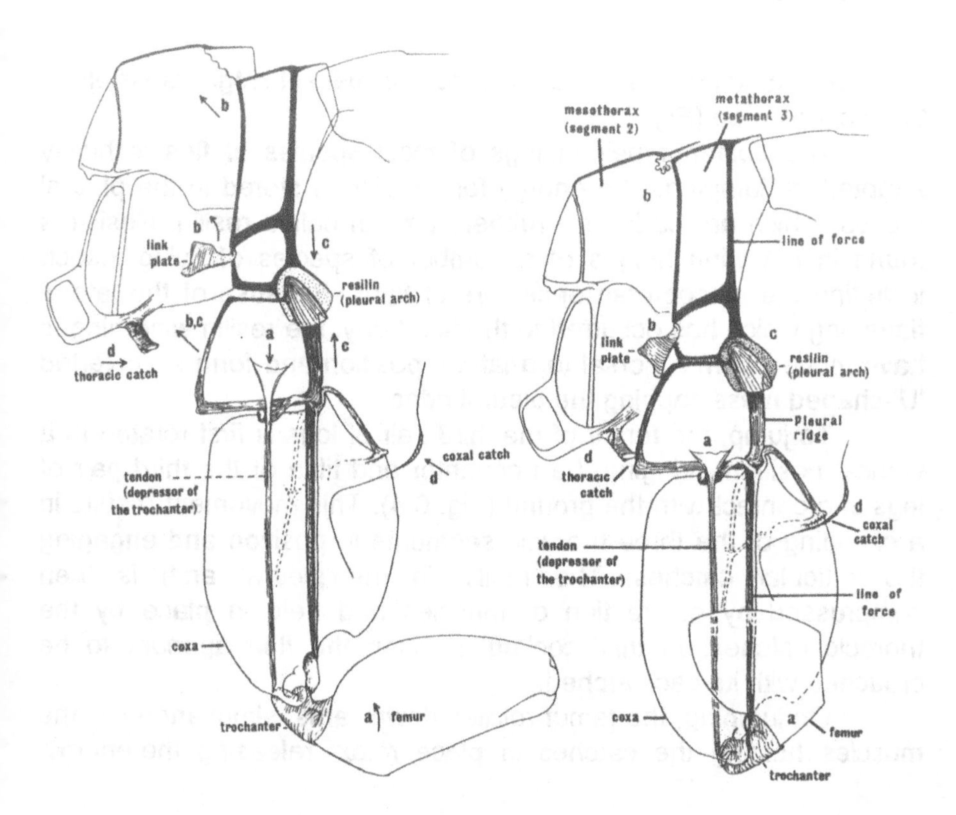

Fig. 6.4 (Left) Metathorax and third leg in the relaxed position. The letters indicate the sequence of movements in preparation for a jump. (a) The raising of the femur pulling the tendon of the trochanteral depressor downwards. (b) Rotation of link plate raising the third segment which brings it into line with the second segment. (c) Compression of the resilin. (d) Engagement of thoracic catch in the metasternum and coxal/abdominal catch. (Right) Metathorax and third leg showing the different sclerites in position when the flea is ready to jump. (a) Femur raised and the tendon of the trochanteral depressor held in the funnel-shaped socket. (b) Link plate fully elevated. (c) Resilin compressed. (d) Thoracic and coxal/abdominal catches engaged. (Reproduced from Rothschild *et al.*, 1972.)

enters the capillary allowing blood to flow up the food canal. Blood-feeding may take 2–10 minutes to complete, with females taking almost twice as much blood as males.

The dorsal sclerites of the three thoracic segments are distinct (pronotum, mesonotum and metanotum). A **pronotal ctenidium** is present in some genera. In some genera, a characteristic, vertical, cuticular rod, known as the **pleural rod** or **pleural ridge**, is visible in the mesopleuron (Fig. 6.2).

The posterior pair of legs of most species of flea is highly adapted for jumping. The energy for jumping is stored in the pleural arches, which are pads of a rubbery protein called resilin. Resilin is found in the wing hinges of a number of species of flying insect, including the mecopteran ancestors of fleas. Because of the lateral flattening which has occurred to the flea body, the resilin wing hinges have moved from a dorsal to a lateral position and form an inverted 'U'-shaped mass capping the pleural ridge.

To jump, the femur of the third pair of legs is first rotated to a vertical position, bringing the trochanter and tibia of the third pair of legs into contact with the ground (Fig. 6.4). This movement results in a clamping of the three thoracic segments in position and engaging the cuticular catches. The resilin in the pleural arch is then compressed by contraction of muscles and held in place by the thoracic catches. In this 'cocked' position the flea appears to be crouched with its back arched.

On jumping, the femur rotates downwards, simultaneously the muscles holding the catches in place relax, releasing the energy

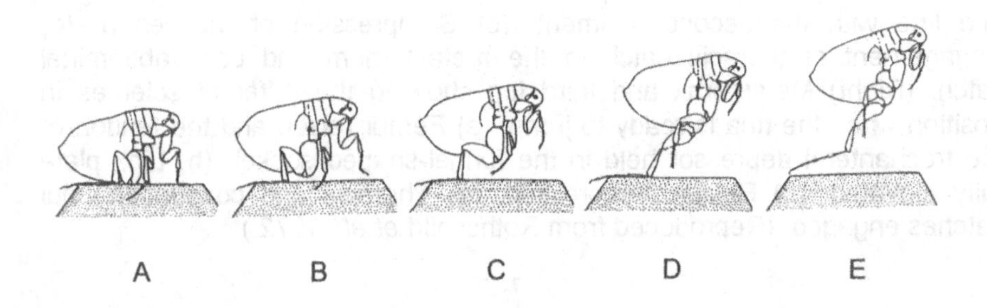

Fig. 6.5 Sequence of events in the jump of a rat flea (A to E) (from Rothschild *et al.*, 1972).

stored in the resilin. Because the tendon of the trochanteral depressor becomes wedged in its socket and forms an inextensible anchor, force is transmitted down the vertical ridges of the notum, pleuron and coxa to the trochanter. The leverage for this movement is supplied by the point of attachment of the tendon of the trochanteral depressor acting against the coxa-trochanteral hinge located at the base of the line of force. After the trochanter leaves the ground the thrust is continued as the leg straightens and the descending femur exerts force through the tibia (Fig. 6.5).

The peak acceleration of the rat flea, *X. cheopis* is 1350 m/s^2, producing forces acting on the flea of approximately 140 *G*. Astronauts in a space rocket experience forces of up to 8 *G*. Resilin overcomes two disadvantages of muscle: first is the slow energy release of muscle and second is the fact that muscle acts slowly at low temperatures. Resilin is much less affected by temperature, which is why flea jumping ability is not strongly impaired by temperature. The amount of resilin is linked to jumping ability and this in turn is linked to the size of the host and the host-finding behaviour of the flea. The fleas of burrow-dwelling animals or nest-dwelling birds tend to have limited jumping ability. For example, the semi-sessile rabbit flea *S. cuniculi* can only jump about 4 cm. In contrast, the fleas of deer, which have no well-defined nest, have relatively large amounts of resilin in their pleural arches and can make correspondingly large jumps. Similarly, the rat flea, *X. cheopis*, which weighs about 0.2–0.4 mg has an average jump of about 18 cm, with a maximum jump of about 30 cm.

The flea abdomen is clearly divided into segments (Fig. 6.2). There are 10 segments but the last three are highly modified in the terminal portion of the abdomen. Each of the eight visible segments bears a pair of spiracles. The shape of the abdomen may be used to distinguish the sexes. In female fleas both the ventral and dorsal surfaces are rounded. In the male flea the dorsal surface is relatively flatter and the ventral surface greatly curved. In both sexes, there is an organ called the **sensilium** found on the dorsal surface of the terminal portion of the abdomen. The sensilium consists of a flat plate bearing a number of dome-shaped structures from which bristles project. This organ probably serves some sensory function, but its role has yet to be established. Behind the sensilium is a pair of anal stylets, bearing apical setae and a number of shorter bristles.

In the male the penis, known as the aedeagus, is a complex, coiled structure consisting of one or more penis rods.

The larvae are white and maggot-like (Fig. 6.6). They have a well-developed, though eyeless, head, with well-developed chewing mouthparts. There are 13 body segments, each with a circle of backwardly directed bristles. There are no appendages. Fully grown larvae may be 4–10 mm in length.

6.3 LIFE HISTORY

Two broad trends in the life cycles of fleas are evident. A simple association with the nest habitat is preserved in many groups of the family Ceratophyllidae, characterized by infrequent and brief associations with the host and often considerable adult movement between hosts and nests. In contrast, many groups of the family Pulicidae show prolonged adult associations with the host. However, within these broad categories a high degree of co-evolution between individual flea species and their hosts may exist and the variation in flea life cycles may be considerable. The rate of life-cycle completion

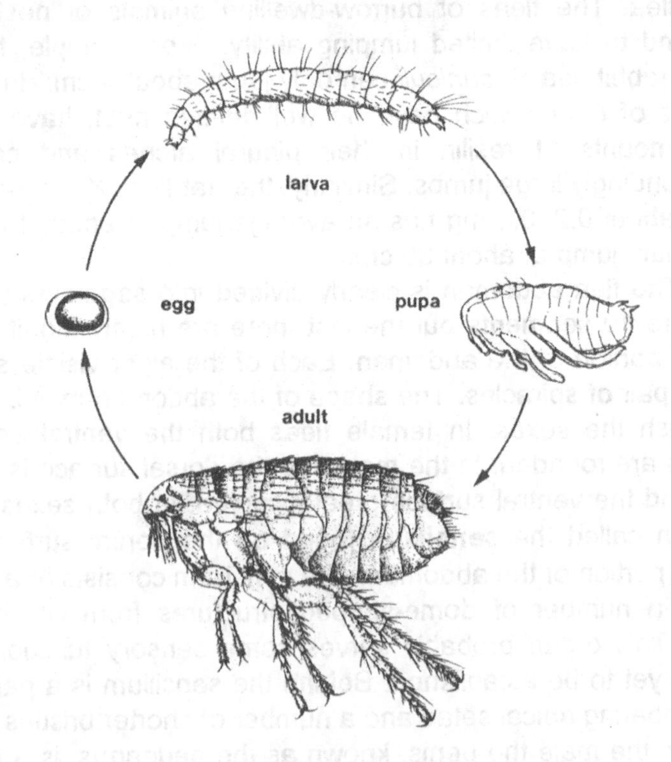

Fig. 6.6 Life cycle of a typical flea (reproduced from Lewis, 1993).

is dependent on temperature, relatively high humidity and host availability.

In general, a female flea can produce several hundred eggs in its lifetime, with batches of between 2 and 25 oviposited at intervals of 1–2 days. The eggs are usually large and may supply many of the vitamins and sterols required for subsequent larval development. Eggs may be laid on the ground, in the host's nest or bedding or on the host itself. In some fleas, such as the Oriental rat flea, the eggs are sticky; in others, such as the sticktight flea, they are dry. Eggs generally hatch within a few days of oviposition, provided the humidity is high, usually above 70%.

The flea embryo is equipped with a sharp spine on the head to help it to burst through the eggshell. Flea larvae are active and feed on proteinaceous organic debris, such as hair, feathers or adult flea faeces. Flea larvae usually live in the host nest or bedding, but may occasionally be found on the host itself. The larvae of *Hoplopsyllus* live as ectoparasites in the fur of the arctic hare, for example. Larvae are particularly susceptible to desiccation. The larvae usually moult twice until, when fully grown, the third-stage larva spins a thin cocoon of silk. Particles of debris stick to the freshly spun silk, serving to camouflage it. After a number of days of quiescence, the cocooned larva transforms into a pupa. The duration of the pupal stage is highly dependent on ambient temperature though less dependent on high humidity than previous stages. After emerging from the pupal cuticle, the adult flea remains within the cocoon until stimulated by a rise in temperature or other stimuli caused by a host returning to the nest.

Both males and females are obligate blood-feeders and females require a blood meal before they can begin to mature their eggs. Most species of flea are not host specific and will try to feed on any available animal. It is probably adaptation to local microclimate within a nest or habitat which makes fleas more or less common on particular hosts. However, while many species will attempt to feed on any available animal, and such a meal may be advantageous in helping to keep the adult alive, in many cases full fertility is only achieved after feeding on specific hosts.

Generally, adult fleas do not actively search for hosts, rather awaiting the approach of a host. They may remain motionless untilvibrations or a sudden rise in temperature or humidity signal the proximity of a host animal and trigger jumping. Adults of the

European chicken flea, *Ceratophyllus gallinae*, will leave the nest and ascend vegetation, jumping in response to the passage of shadows.

During feeding adults excrete faeces which are rich in partially digested blood and which form an important food source for the larvae. Adult fleas can also survive for long periods (up to 6 months) between feeds, allowing them to locate new hosts or to await the return of a host to its nest site.

Most fleas feed before mating. However, others, particularly bird fleas, copulate before taking an initial blood meal. Egg production may begin within a few days of adult emergence, provided a host has been located. In most flea species oviposition takes place as soon as appropriate conditions of temperature and humidity occur. In others, such as *Spilopsyllus cuniculi* in Europe and *Cediopsylla simplex* in North America, oviposition is intimately synchronized with the reproductive cycle of wild and domestic rabbits, as will be described later. Adult fleas are often described as long lived, and may survive for several months in the laboratory. However, in the wild they probably only continue to reproduce actively for a few days.

After contact has been made between the male and female, in mobile species such as the chicken flea, *C. gallinae,* the male grasps the female abdomen from beneath using adhesive organs on the inner surfaces of the erected antennae. Copulation may last for an average of 3 hours. In many species of flea a single penis rod is inserted into the spermathecal duct. However, in the rabbit flea, *S. cuniculi,* the two penis rods operate together. The two penis rods enter the female and the shorter, thicker rod lodges in the bursa copulatrix of the female and serves to guide the longer rod, which has sperm wound around its notched tip, into the spemathecal duct. In sessile species such as *S. cuniculi* the male antennae lack the adhesive discs and mating takes place as the female is feeding.

6.4 PATHOLOGY

The feeding behaviour of fleas causes significant veterinary problems worldwide. In 1995 the market for flea control products was worth over US$1 billion in the USA.

- Although blood-meal size is small, repeated feedings and high infestation can cause significant blood loss, and heavy infestations may cause fatal iron-deficiency anaemia in very young animals.

- Inflammation and pruritus may occur at the site of a flea bite, leading to self-wounding from scratching or biting by the host animal. Fleas are of primary veterinary importance as the agents responsible for provoking flea-bite allergic dermatitis, particularly in dogs and cats. When a flea feeds, saliva is injected into the dermis. The saliva may contain substances which soften and spread the host's dermal tissue and which prevent coagulation of the blood. The first flea bite may cause no observable skin reaction, but the host may develop a hypersensitivity to antigens in the flea saliva. Subsequent bites over an extended period may trigger an allergic dermatitis. Clinical signs differ between individuals. In dogs, primary lesions are discrete crusted papules which cause intense pruritus and scratching, which produce areas of alopecia or pyotraumatic dermatitis. In cats, two distinct clinical manifestations are associated with flea allergy: milary dermatitis and feline symmetrical alopecia. So-called feline hormonal alopecia is now known to be due to self-trauma, usually as the result of fleas, and is termed feline symmetrical alopecia. In feline symmetrical alopecia the skin appears normal but microscopic examination of the distal hair tips shows that they have been fractured by excessive grooming. However, in milary dermatitis, the skin, especially across the dorsum and trunk, is covered in crusted papules with variable degrees of hair loss.
- Cat fleas (*Ctenocephalides felis felis*), dog fleas (*Ctenocephalides canis*) and human fleas (*Pulex irritans*) can act as intermediate hosts of *Dipylidium caninum,* the double-pored dog tapeworm. Tapeworm eggs are passed out in faeces of the vertebrate host, following which they are eaten by flea larvae along with general organic debris. The tapeworm eggs hatch in the midgut of the flea larva and the worm larvae penetrate the gut wall, passing into the haemocoel. The tapeworm larvae develop within the flea body cavity throughout larval, pupal and adult flea development, eventually encapsulating as an infective larva, known as a cysticercoid. During grooming, adult fleas are often eaten by the vertebrate host. The cysticercoids are then liberated and develop into tapeworms in the digestive tract.
- Fleas are also vectors of viral and bacterial infections, particularly of diseases such as plague and tularaemia. In animals, the rabbit flea, *S. cuniculi,* is the primary vector of myxomatosis in some parts of the world.

6.5 CLASSIFICATION

The Siphonaptera is a small and very distinct order. More than 2500 species have been described and the order may eventually be shown to contain as many as 3000 species. There are generally considered to be 15 or 16 families and 239 genera. Only two families contain species of veterinary importance: the **Ceratophyllidae** and the **Pulicidae**. The Ceratophyllidae is a large family, at present thought to contain over 500 species, of which about 80 species are parasites of birds and the remainder of which are parasites of small rodents. Most species of the family are Holarctic in distribution. The Pulicidae are parasites of a range of mammals. They are distributed worldwide.

6.6 RECOGNITION OF FLEAS OF VETERINARY IMPORTANCE

The physical differences between species and even between families tend to be small and there may be considerable variation between individuals within a species. Identification, therefore, is often difficult. The following is a general diagnostic guide to the adults of the most common species of veterinary importance found as parasites on domestic and companion animals in temperate habitats (modified from Lewis, 1993).

6.6.1 Guide to the flea species of veterinary importance

1	Ctenidia absent	2
	Ctenidium present, at least on the pronotum	4
2	Pleural ridge absent	3
	Pleural ridge present	***Xenopsylla cheopis***
3	Frons sharply angled (Fig. 6.10); head behind the antenna with two setae and, in the female, usually with a well-developed occipital lobe; the maxillary laciniae are broad and coarsely serrated; adult females embedded in the skin in aggregations on bare areas; found on birds, especially	

poultry, also on cats, dogs, rabbits and humans **_Echidnophaga gallinacea_**

Frons smoothly rounded; head behind antennae with only one strong seta; conspicuous ocular seta below the eye; a single, much reduced spine on the genal margin (Fig. 6.11); on pigs, badgers, humans **_Pulex irritans_**

4 Genal ctenidium present **5**

Genal ctenidium absent; pronotal ctenidium with 18–20 spines; head with a row of three strong setae below the eye (Fig. 6.14); frontal tubercle on head of both sexes conspicuous; 3–4 conspicuous bristles on the inner surface of the hind femur; on rodents **_Nosopsyllus fasciatus_**

Genal ctenidium absent; pronotal ctenidium with more than 24 spines; head with a row of three strong setae below the eye (Fig. 6.13); on poultry **_Ceratophyllus_ spp.**

5 Genal ctenidium formed of eight or nine spines oriented vertically **6**

Genal ctenidium with 4–6 oblique spines; frontal tubercle conspicuous on head of both sexes (Fig. 6.9); on rabbits **_Spilopsyllus cuniculi_**

6 Head strongly convex anteriorly in both sexes and not noticeably elongate; hind tibia with eight seta-bearing notches along the dorsal margin (Fig. 6.8); on cats and dogs **_Ctenocephalides canis_**

Head not strongly convex anteriorly and distinctly elongate, especially in the female; hind tibia with six seta-bearing notches along the dorsal margin (Fig. 6.7); on cats and dogs **_Ctenocephalides felis_**

6.7 PULICIDAE

6.7.1 *Ctenocephalides*

The eleven species of this genus are found largely in Africa and Eurasia. Their hosts are primarily carnivores, though they may also be found on some lagomorphs and goats in the Mediterranean region.

Ctenocephalides felis felis

The cat flea, *C. felis,* is widely distributed worldwide There are four recognized sub-species, all of which are primarily parasites of carnivores: *Ctenocephalides felis felis* is now found worldwide, *Ctenocephalides felis strongylus* and *Ctenocephalides felis damarensis* are found in Africa, and *Ctenocephalides felis orientis* is found from India to Australia.

Morphology: the cat flea has both genal and pronotal ctenidia. The genal ctenidium consists of 7–8 spines and the pronotal ctenidium about 16 spines (Fig. 6.7). The teeth of the genal ctenidium are all about the same length. In the female *C. f. felis* the head is twice as long as high and pointed anteriorly. In the male *C. f. felis* the head is as long as wide but is also slightly elongate anteriorly. On the dorsal

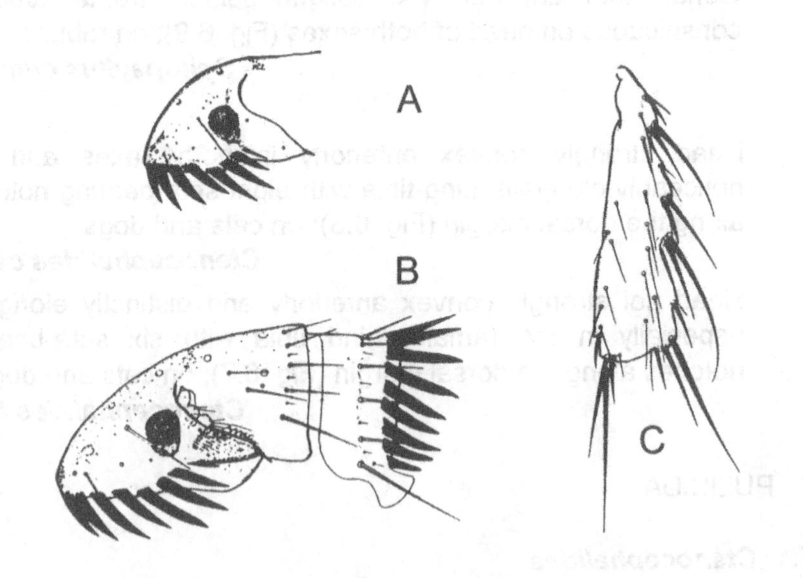

Fig. 6.7 The cat flea, *Ctenocephalides felis felis*: (A) front of male head; (B) female head and pronotum; (C) hind tibia (reproduced from Lewis, 1993).

border of the hind (metathoracic) tibia in both sexes of *C. f. felis* there are only six notches bearing setae (Fig. 6.7).

Life cycle: adult cat fleas are positively phototactic and negatively geotactic, which causes them to move to the top of carpets, grass or other substrate. Here they await a passing host. Visual and thermal cues trigger jumping, and the presence of carbon dioxide increases the orientation of fleas towards the host animal. The jumping response is greatly enhanced when the light source to which they are orienting is suddenly and intermittently interrupted, as would be caused by a passing host. After locating a host, adult females mate and feed. Newly emerged female *C. f. felis* increase in weight by up to 75% after feeding on a cat for 12 hours. Within 10 minutes of feeding adults begin to produce faeces. The excreted blood quickly dries into reddish-black faecal pellets, known as 'flea dirt'.

Once on its host *C. f. felis* tends to become a permanent resident, but may only live for a week or more. Within 24–48 hours of the first blood meal females begin to oviposit. The eggs are pearly white and oval and about 0.5 mm in length. In the laboratory, an adult female *C. f. felis* can produce an average of about 30 eggs per day and a maximum of 50 eggs per day, over a life of about 50–100 days. However, on a cat the average life span is probably substantially lower than this, and is probably less than 1 week. Eggs are usually deposited on the host, but fall to the ground within a few hours. Only those eggs that fall into an appropriate environment will ultimately develop into adults and viable flea eggs tend to accumulate in areas where the hosts sleep or rest. At 70% relative humidity and 35 °C, 50% of eggs hatch within 1.5 days. At 70% relative humidity and 15 °C it takes 6 days for 50% of eggs to hatch. Eggs cannot survive below 50% relative humidity.

Within the host lair or bedding the larvae of *C. f. felis* exist in a protected environment, with relatively high humidity, buffered from the extreme fluctuations of ambient temperatures and provided with detritus and a source of adult flea faecal blood. The larvae have limited powers of movement (probably less than 20 cm before pupation) and crawl about their environment largely at random, but are negatively phototactic and positively geotactic. In the domestic environment this behaviour often takes them to the base of carpets where they can encounter food and are sheltered from light and mechanical damage. At 24 °C and 75% relative humidity the duration of the three larval instars is about 1 week and at 13 °C and 75%

relative humidity larval development takes about 5 weeks. The larvae are extremely susceptible to desiccation and mortality is high below 50% relative humidity. The areas within a building with the necessary humidity for egg and larval survival are limited. Sites outdoors are even less common and flea larvae cannot develop in areas exposed to the hot sun.

When fully developed, the mature third-stage larva empties its gut and spins a thin, silk cocoon. The cocoon is oval and about 5 mm in length. Within the cocoon the larva pupates. At 24°C and 78% relative humidity the duration of the pupal stage is about 8–9 days. When fully developed, adults emerge but may remain within the cocoon. Adults may remain in this state for up to 140 days at 11°C and 75% relative humidity. Emergence of the adult from the cocoon is triggered by stimuli such as mechanical pressure, vibrations or heat. Adult emergence may be extremely rapid, when provided with appropriate conditions. The ability to remain within the cocoon for extended periods is essential for a species such as *C. f. felis* since its mobile hosts may only return to the lair or bedding at infrequent intervals. Newly emerged adults can survive for several days without feeding, provided the relative humidity is above about 60%.

At 24°C and 78% relative humidity, with a plentiful food supply, the total developmental cycle of *C. f. felis* may be completed in as little as 17–22 days for females and 20–26 days for males. Under most household conditions *C. f. felis* will complete the developmental cycle in 3–5 weeks. However, under adverse conditions this can be extended to as long as 190 days.

Pathology: the cat flea, *C. f. felis,* is the most common species of flea found on domestic cats and dogs throughout North America and northern Europe. More than 90% of fleas found on dogs or cats are likely to be *C. f. felis*. It may cause a significant dermatitis and occasionally iron-deficiency anaemia in severe infestations. Anaemia caused by *C. f. felis* is particularly prevalent in young animals and has been reported in cats and dogs and, very rarely, goats, cattle and sheep.

During feeding *C. f. felis* injects saliva which contains non-protein haptenic materials (incomplete antigens). These combine with the host's skin collagen to form the complete allergens which are responsible for the production of hypersensitivity. Flea allergy dermatitis is one of the most common cause of dermatological disease of dogs and is a major cause of miliary dermatitis of cats.

Dermatitis associated with allergy to flea bites is characterized by intense pruritus, resulting in licking, chewing and scratching. Dogs chronically infested with *C. f. felis* rarely develop a state of natural tolerance resulting in loss of clinical signs.

In normal grooming behaviour, a cat may ingest almost 50% of its resident flea population within a few days. As a result, *C. f. felis* is an important intermediate host of the tapeworm *Dipylidium caninum.* Tapeworm eggs are passed out in faeces of the vertebrate host, following which they are eaten by flea larvae along with general organic debris. The tapeworm eggs hatch in the midgut of the flea larva and the worm larvae penetrate the gut wall, passing into the haemocoel. The tapeworm larvae develop within the flea body cavity throughout larval, pupal and adult flea development, eventually encapsulating as an infective larva, known as a cysticercoid. After ingestion of the adult flea by the host, cysticercoids are liberated and develop into tapeworms in the digestive tract. The removal of fleas during grooming also reduces the chance of finding fleas during a skin and coat examination. This is a particular diagnostic problem in cats with a low flea burden but marked flea-bite hypersensitivity. In such cases examination of the mouth may reveal fleas caught in the spines of the cats tongue.

Ctenocephalides felis felis also acts as an intermediate host of the non-pathogenic, subcutaneous filaroid nematode of dogs *Dipetalonema reconditum,* which adults may ingest during blood-feeding.

Ctenocephalides canis

Morphology: the dog flea, *C. canis,* is closely related and is morphologically very similar to the cat flea, *C. f. felis*, although they cannot interbreed and, therefore, are truly distinct species. Like *C. f. felis,* the dog flea has both genal and pronotal ctenidia. The genal ctenidium consists of 7–8 spines and the pronotal ctenidium about 16 spines (Fig. 6.8). However, in both female and male *C. canis* the head is more rounded than in *C. f. felis* and the first spine of the genal ctenidium is shorter than the rest. On the dorsal border of the hind (metathoracic) tibia in both sexes of *C. canis* there are eight notches bearing stout setae (Fig. 6.8).

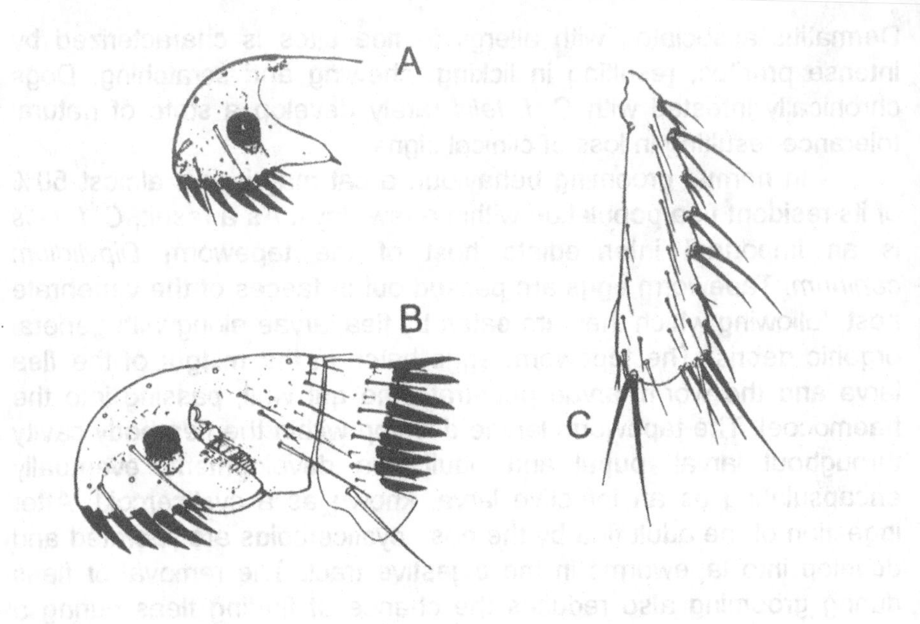

Fig. 6.8 The dog flea, *Ctenocephalides canis*; (A) front of male head; (B) female head and pronotum; (C) hind tibia (reproduced from Lewis, 1993).

Life cycle: the life cycle of *C. canis* is similar to that of *C. f. felis*. The behavioural difference between the two species seems largely to involve the range of environmental conditions which their larvae are capable of tolerating. While household dogs in northern Europe and North America are more likely to be infested by the cat flea, working dogs in kennels and dogs in rural areas are more likely to be infested by *C. canis*. Further south in Europe, pets become more likely to have the dog flea. Dogs, cats, rats, rabbits, foxes and humans have been recorded as hosts.

Pathology: as in the case of *C. f. felis*, the dog flea *C. canis* is responsible for the production of hypersensitivity and flea allergy dermatitis. Although less common than *C. f. felis, C. canis* is still found on a significant number of dogs in the UK. *Ctenocephalides canis* may also act as an important intermediate host of the tapeworm *D. caninum*. *Ctenocephalides canis* is able to mature eggs after feeding on human blood, but extended periods of blood-feeding are required to achieve maturation of the ovaries.

6.7.2 *Spilopsyllus*

Spilopsyllus cuniculi

Morphology: the rabbit flea, *S. cuniculi,* has both pronotal and genal ctenidia, the latter being composed of 4–6 oblique spines (Fig. 6.9). Eyes are present and the frons at the front of the head is rounded with the frontal tubercle conspicuous. There are two stout spines beneath the eye.

Life cycle: the rabbit flea, *S. cuniculi,* occurs on rabbit ears. It is more sedentary than most other species of flea and remains for long periods with its mouthparts embedded in the host.

Reproduction is under the control of hormones in the blood of the mammalian host. The presence of progesterones inhibits or delays flea maturation. Following mating, the adult female rabbit ovulates and, about 10 days before parturition, the levels of oestrogens and corticosteroids in the blood increase. These hormones stimulate development of the eggs of the female flea. When the young rabbits are born, the fleas move down the face and on to the young rabbits on which they feed, mate and lay their eggs. Copulation of *S. cuniculi* only takes place in the presence of young rabbits (1–10 days old). An airborne kairomone emanating from the newborn rabbits and their urine boosts copulation and reproduction. The hormones of the host also cause adult fleas to increase the rate of defecation by about five times. This provides a greater source of food for the newly hatched larvae.

Adult female fleas on bucks or non-pregnant does are more mobile and will move to pregnant does if able.

Pathology: the rabbit flea, *S. cuniculi,* is the main vector of myxomatosis in rabbits. It may commonly be found on the face and attached to the pinnal margin of cats and dogs which hunt or frequent rabbit habitats.

6.7.3 *Echidnophaga*

This genus contains 20 species distributed through Eurasia, Africa and Australia. The hosts are usually rodents, marsupials, carnivores

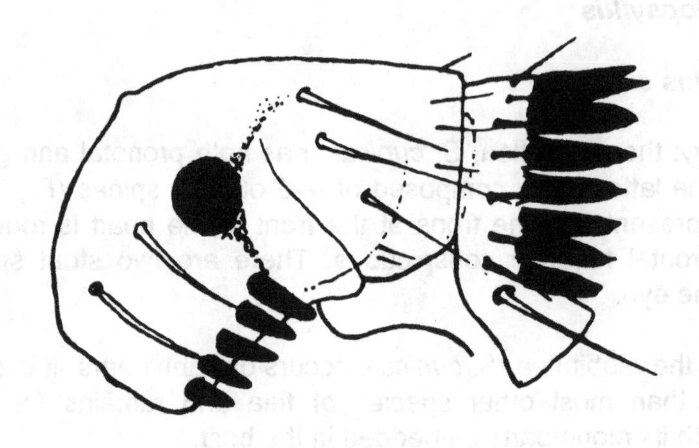

Fig. 6.9 Head and pronotum of the rabbit flea, *Spilopsyllus cuniculi* (reproduced from Urquhart *et al.*, 1987).

and warthogs. The genus contains one cosmopolitan species of veterinary importance, *Echidnophaga gallinacea.*

Echidnophaga gallinacea

The sticktight flea, *E. gallinacea,* is a burrowing flea important mainly in domestic poultry. However, it may also attack cats, dogs, rabbits and humans. It is most common in tropical areas throughout the world, but may also be found in many sub-tropical and temperate habitats.

Morphology: the adult sticktight flea is small, at about 2 mm in length. The head is sharply angled at front (frons). There are no genal or pronotal ctenidia (Fig. 6.10). On the head behind the antenna there are two setae and, in the female, usually a well-developed occipital lobe. The thoracic segments are narrowed dorsally. The mouthparts appear large, extending the length of the forecoxae, and project from the head conspicuously. The maxillary laciniae are broad and coarsely serrated (Fig. 6.10).

Life cycle: after host location, females aggregate on bare areas, often the head, comb or wattles. Newly emerged adults are active and move towards sunlight, which helps them accumulate on the wattles of cocks or hens. After feeding, females burrow into the skin where they attach firmly with their mouthparts. Each female may remain attached for between 2 and 6 weeks. Copulation then takes place. The skin around the point of attachment may become ulcerated. Eggs are laid in the ulceration or dropped to the ground. If laid in the ulceration, larvae hatch, emerge from the skin and drop to the ground to complete their development. The larvae feed on chicken manure. The entire life cycle requires 30–60 days.

Pathology: primarily important as a parasite of birds, the adult sticktight flea is an especially serious pest of chickens. However, it may also be found on rats, cats, dogs and larger insectivores. Infestations on dogs may be persistent if continually exposed to a source of infestation, and fleas are found on the poorly haired areas of the ventrum, scrotum, interdigital and periorbital skin. Infestation of poultry is around the head and may lead to ocular ulceration, caused by self-trauma, resulting in blindness and starvation. This species

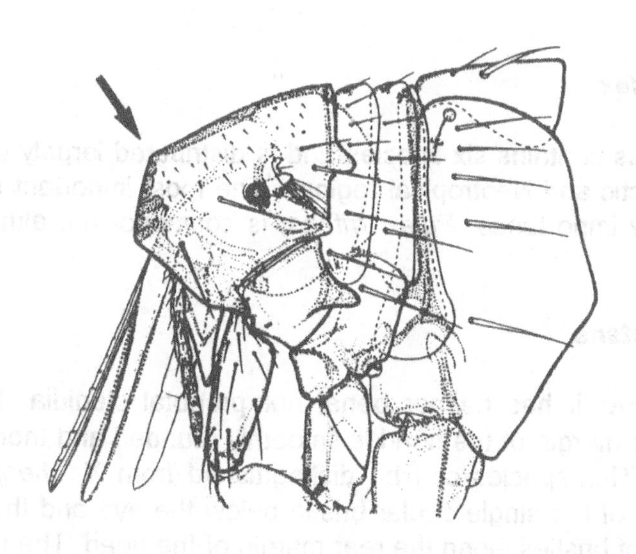

Fig. 6.10 The sticktight flea, *Echidnophaga gallinacea*, female head and thorax (arrow marking angulation of the frons) (reproduced from Lewis, 1993).

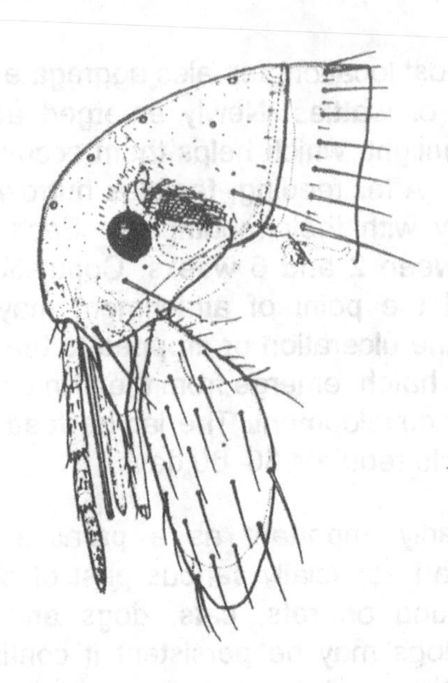

Fig. 6.11 The human flea, *Pulex irritans*, male head and pronotum (reproduced from Lewis, 1993).

may also affect humans.

6.7.4 *Pulex*

This genus contains six species and is distributed largely throughout the Nearctic and Neotropical regions. The most important species of veterinary importance, *Pulex irritans*, is cosmopolitan, although now rare.

Pulex irritans

Morphology: it has neither genal nor pronotal ctenidia (Fig. 6.11). The outer margin of the head is smoothly rounded and there is a pair of eyes. This species can be distinguished from *X. cheopis* by the presence of the single ocular bristle below the eye and the absence of a row of bristles along the rear margin of the head. The metacoxae have a patch of short spines on the inner side. The maxillary laciniae extend about halfway down the forecoxae, which distinguishes this species from the closely related *Pulex simulans* found on Hawaii (where the laciniae extend for at least three-quarters the length of the forecoxae).

Life cycle: the life cycle is typical: egg, three larval stages, pupa and adult. Originally the principal hosts of this species were pigs.

Pathology: *P. irritans* is less important from a veterinary than a medical perspective and less common than the cat or dog fleas. Although described as the human flea, throughout most of its range *P. irritans* is probably more common on pigs than on humans or dogs.

6.7.5 *Xenopsylla*

This genus of 77 species is distributed throughout the tropical and sub-tropical regions of the Old World. Its hosts are rats. There is a single species of veterinary importance in temperate habitats, the Oriental rat flea, *X. cheopis.*

Xenopsylla cheopis

The distribution of the Oriental rat flea, *X. cheopis,* largely follows that of its primary host the black rat, *Rattus rattus.*

Morphology: *X. cheopis* resembles *P. irritans* in that both genal and pronotal ctenidia are absent (Fig. 6.12). The head is smoothly rounded anteriorly. The maxillary laciniae reach nearly to the end of the forecoxae. Eyes are present. The segments of the thorax appear relatively large and the pleural ridge is present in the mesopleuron of the thorax. There is a conspicuous row of bristles along the rear margin of the head and a stout ocular bristle in front of the eye.

Life cycle: the life cycle of *X. cheopis* is typical: egg, three larval stages, pupa and adult. Humidities above 60–70% and temperatures above 12°C and are required for life-cycle development in this species.

Pathology: it is the chief vector of the pathogen *Yersinia pestis,* causing plague and murine typhus in humans. It is also an intermediate host of helminths such as *Hymenolepis diminuta* and

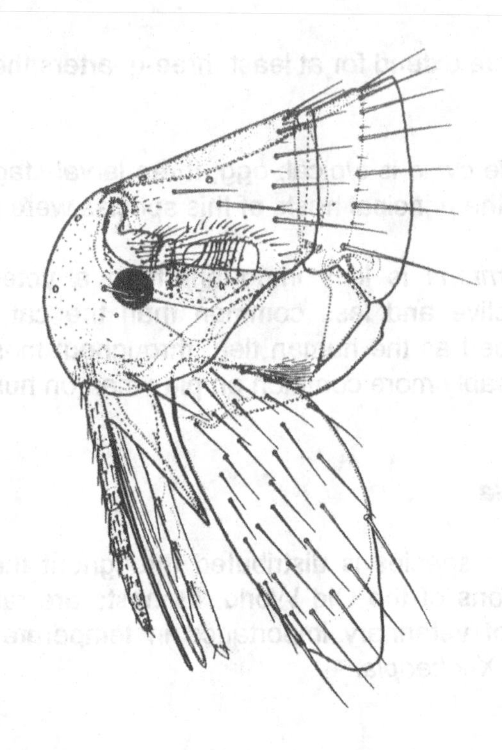

Fig. 6.12 The Oriental rat flea, *Xenopsylla cheopis*, male head (reproduced from Lewis, 1993).

H. nana.

6.8 CERATOPHYLLIDAE

6.8.1 *Ceratophyllus*

This genus is largely Holarctic in distribution and contains 62 species. The hosts are primarily squirrels and other rodents. There are two species of veterinary importance because they feed on birds, particularly poultry.

Ceratophyllus niger

The western chicken flea, *Ceratophyllus niger* is common throughout the western USA, Canada and Alaska.

Morphology: a genal ctenidium is absent and the pronotal ctenidium has more than 24 spines (Fig. 6.13). Eyes are present and the head bears a row of three strong setae below the eye. The pleural ridge is present in the mesopluron of the thorax. The body is elongated and about 4 mm in length, considerably larger than the sticktight flea of poultry, *E. gallinacea*.

Life cycle: the life cycle is typical: egg, three larval stages, pupa and adult. Unlike the sticktight flea, *E. gallinacea*, the adult does not attach permanently to its host. Adults and larvae are found primarily in chicken droppings.

Pathology: the western chicken flea is important primarily as a parasite of poultry, but also may be found on rats, cats, dogs and humans.

Ceratophyllus gallinae

The European chicken flea, *Ceratophyllus gallinae,* is the most common flea of poultry and also infests more than 75 species of wild

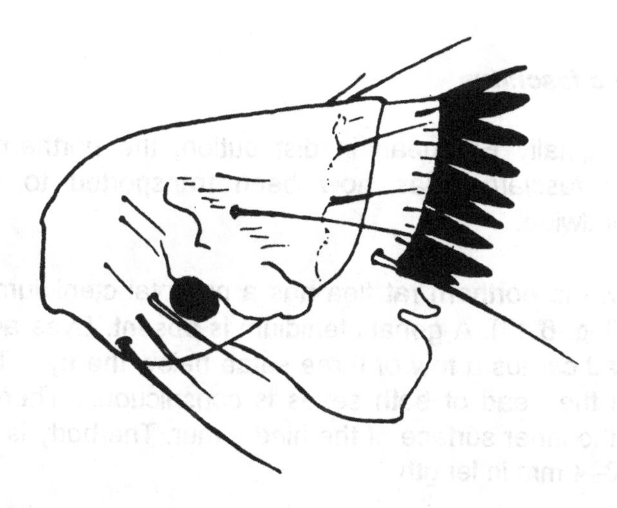

Fig. 6.13 Head and pronotum of chicken fleas of the genus *Ceratophyllus* (reproduced from Urquhart *et al.*, 1987).

birds. It is found predominantly in the Old World but has been introduced into south-east Canada and north-east USA.

Morphology: it is extremely similar in appearance to *C. niger*.

Life cycle: the life cycle is typical: egg, three larval stages, pupa and adult. The flea overwinters in the cocoon and emerges in an old nest in spring as temperatures rise. Adults crawl up trees and bushes in search of a host. They stop periodically and face the brightest source of light, jumping in response to a shadow passing in front of the light. Large numbers may occur in the nests of passerine birds and they may complete their life cycle during the period of nest occupation by these birds.

Pathology: feeding activity may cause irritation, restlessness and, with heavy infestations, anaemia. Adult *C. gallinae* may also feed on humans and domestic pets.

6.8.2 *Nosopsyllus*

This genus is largely Palaearctic in distribution. There are four sub-genera and about 50 species. The one cosmopolitan species of veterinary significance is *Nosopsyllus fasciatus*.

Nosopsyllus fasciatus

Although originally European in distribution, the northern rat flea, *Nosopsyllus fasciatus*, has now been transported to temperate habitats worldwide.

Morphology: the northern rat flea has a pronotal ctenidium with 18–20 spines (Fig. 6.14). A genal ctenidium is absent. Eyes are present and the head carries a row of three setae below the eye. The frontal tubercle on the head of both sexes is conspicuous. There are 3–4 bristles on the inner surface of the hind femur. The body is elongated and about 3–4 mm in length.

Life cycle: the life cycle is typical: egg, three larval stages, pupa and adult. Life-cycle development may be completed at temperatures as

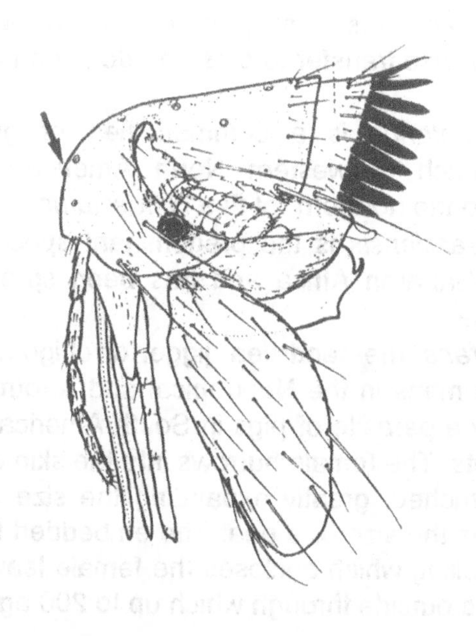

Fig. 6.14 The northern rat flea, *Nosopsyllus fasciatus*, male head (reproduced from Lewis, 1993).

low as 5°C. The larvae of this species may pursue and solicit faecal blood meals from adult fleas. The larvae grasp the adult in the region of the sensilium using their large mandibles. Adults respond by defecating stored semi-liquid blood which is then imbibed by the larvae.

Pathology: its main hosts are rodents, particularly the Norway rat *Rattus norvegicus*. However, it has also been found on house mice, gophers, humans and many other hosts. It is not thought to be an important vector of plague. It is known to be a vector of *Hymenolepis diminuta* in parts of Europe, Australia and South America.

6.9 FLEA SPECIES OF MINOR VETERINARY INTEREST

- *Archeopsylla erinacei* occurs on hedgehogs in Europe and North America and may be transferred to dogs and cats following contact.

- *Leptopsylla segnis,* the European mouse flea, occurs on the house mouse and rats. It may be involved as a weak vector of plague. It may also transfer to cats and dogs and occasionally bite humans.
- *Diamanus montanus* is a common flea of ground squirrels, throughout much of western North America. It is similar in appearance to the northern rat flea *N. fasciatus*.
- *Xenopsylla brasiliensis* is the predominant species of rat flea in parts of sub-Saharan Africa and has been spread to India and South America.
- *Tunga penetrans,* the sand flea, jigger or chigoe, is an important parasite of humans in the Neotropical and Afrotropical regions. It was originally a parasite of pigs in South America and may cause death in piglets. The female burrows into the skin of the feet where it remains attached, greatly expanding the size of the abdomen until it reaches the size of a pea. The embedded female produces a nodular swelling which encloses the female leaving only a small opening to the outside through which up to 200 eggs are passed.

6.10 FURTHER READING

Bibikora, V.A. (1977) Contemporary views on the interrelationships of fleas and the pathogens of human and animal diseases. *Annual Review of Entomology*, **22,** 23–32.

Dryden, M.W. (1989) Biology of the cat flea, *Ctenocephalides felis felis*. *Companion Animal Practice – Parasitology/Pathobiology*, **19**, 23–7.

Dryden, M.W. and Rust, M.K. (1994) The cat flea: biology, ecology and control. *Veterinary Parasitology*, **52**, 1–19.

Gullan, P.J. and Cranston, P.S. (1994) *The Insects. An Outline of Entomology*, Chapman & Hall, London.

Holland, G.P. (1964) Evolution, classification and host relationships of Siphonaptera. *Annual Review of Entomology*, **9**, 123–46.

Humphries, D.A. (1967) The mating behaviour of the hen flea, *Ceratophilus gallinae* (Schrank) (Siphonaptera: Insecta). *Animal Behaviour*, **15**, 82–90.

Humphries, D.A. (1968) The host finding behaviour of the hen flea *Ceratophilus gallinae* (Schrank) (Siphonaptera). *Parasitology*, **58**, 403–14.

Lewis, R.E. (1972) Notes on the geographic distribution and host preferences in the order Siphonaptera. Part 1. Pulicidae. *Journal of Medical Entomology*, **9**, 511–20.

Lewis, R.E. (1993) Fleas (Siphonaptera), in *Medical Insects and Arachnids* (eds R.P. Lane and R.W. Crosskey), Chapman & Hall, London, pp. 529–75.

Rothschild, M. (1965) Fleas. *Scientific American*, **213**, (6), 44–53.

Rothschild, M. (1975) Recent advances in our knowledge of the order Siphonaptera. *Annual Review of Entomology*, **20**, 241–59.

Rothschild, M., Schlein, Y., Parker, K. and Sternberg, S. (1972) The jump of the oriental rat flea *Xenopsylla cheopis* (Roths.). *Nature*, **239**, 45–8.

Soulsby, E.J.L. (1982) *Helminths, Arthropods and Protozoa of Domesticated Animals*, Baillière Tindall, London.

Traub, R. (1972) The relationships between the spines, combs and other skeletal features of fleas (Siphonaptera) and the vestiture, affinities and habits of their hosts. *Journal of Medical Entomology*, **9**, 601.

Traub, R. and Starcke, H. (1980) *Fleas*, A.A. Balkema, Rotterdam.

Urquhart, G.M., Armour, J., Duncan, J.L., Dunn, A.M. and Jennings, F.W. (1987) *Veterinary Parasitology*, Longman Scientific & Technical, London.

7

Lice (Phthiraptera)

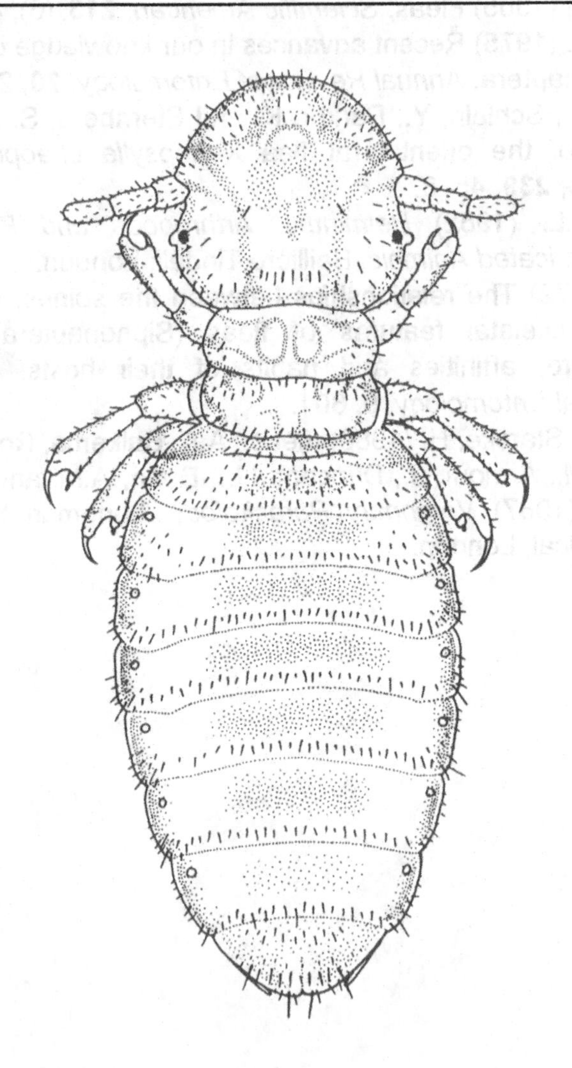

Adult female chewing louse, dorsal view (reproduced from Gullan and Cranston, 1994).

7.1 INTRODUCTION

The lice (order Phthiraptera) are superbly adapted and highly successful insect ectoparasites of birds and mammals. Most species of mammals and birds are infested by at least one species of louse.

In complete contrast to most fleas or ticks, lice spend their entire lives on the host and are highly host specific, many species even preferring specific parts of their host's body. They usually only leave their host to transfer to a new one. To allow them to survive as permanent ectoparasites, lice show a large number of adaptations which enable them to maintain a life of intimate contact with their host. They are small insects, about 0.5–8 mm in length, dorsoventrally flattened, wingless and possess stout legs and claws for clinging tightly to fur, hair and feathers. They feed on epidermal tissue debris, parts of feathers, sebaceous secretions and blood. They usually vary in colour from pale beige to dark grey, but they may darken considerably on feeding.

The Phthiraptera is a small order with about 3500 described species, of which only about 20–30 are of major economic importance. The order is divided into four sub-orders: **Anoplura**, **Amblycera**, **Ischnocera** and **Rhynchophthirina**. However, the Rhynchophthirina is a very small sub-order, including just two African species, one of which is a parasite of elephants and the other a parasite of warthogs. The Amblycera and Ischnocera are usually discussed together and described as the **Mallophaga** which, in older textbooks, is accorded status as a sub-order in its own right. Mallophaga literally means 'wool eating' and the Amblycera and Ischnocera are known as **chewing lice**. The Anoplura are described as **sucking lice**. The description 'biting lice', sometimes used to describe the Anoplura, is a misnomer, because all lice bite.

7.2 MORPHOLOGY

Lice are clearly recognizable as insects since they have a segmented body divided into a head, thorax and abdomen (Fig. 7.1). They have three pairs of jointed legs and a pair of short antennae. All lice are dorsoventrally flattened and wingless. The sensory organs are poorly developed; the eyes are vestigial or absent.

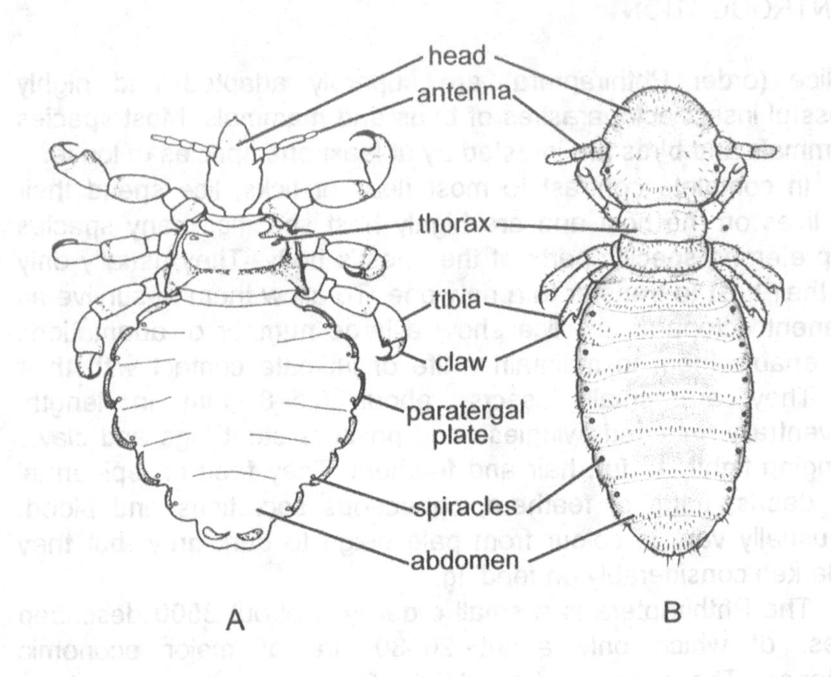

Fig. 7.1 Dorsal view of adult female (A) sucking louse, *Haematopinus* (reproduced from Walker, 1994) and (B) chewing louse, *Damalinia* (reproduced from Gullan and Cranston, 1994).

Adult Amblycera and Ischnocera are usually about 2–3 mm in length. They have large, rounded heads on which the eyes are reduced or absent (Fig. 7.1B). In the Amblycera the four segmented antennae are protected in antennal grooves, so that only the last segment is visible. In the Ischnocera the antennae are three– to five-segmented and are not hidden in grooves. Both sub-orders have distinct, mandibulate mouthparts which are typical of chewing insects, composed of a labrum, a pair of mandibles and a pair of maxillae attached laterally to the labium, which is reduced to a simple broad plate. In the Amblycera the mandibles lie parallel to the ventral surface of the head and cut in a horizontal plane. There is a pair of maxillary palps, which are two- to four-segmented. In the Ischnocera, the mandibles lie at right angles to the head and cut vertically. There are no maxillary palps. Species from both these sub-orders usually feed on fragments of keratin in skin, hair or feathers. However, they will take blood exuding from scratches in the skin and some are able to pierce the skin. At least the first two segments of the thorax are usually visible (Fig. 7.1B). The single pair of thoracic spiracles are on the ventral side of the mesothorax. Typically there are six pairs of abdominal spiracles, but the number may be reduced. The three pairs of legs are weak and slender and end in either one or two claws, depending on the species.

The sucking lice are usually small insects; adults are about 2 mm long on average, but some species of Anoplura may be as small as 0.5 mm in length or as large as 8 mm in length. Characteristically the head is small, but narrow and elongated (Fig. 7.1A). The antennae have five segments. The eyes are reduced or absent. The Anoplura have highly modified mouthparts, quite different to those of other insects, which are highly adapted for piercing the skin of their hosts (Fig. 7.2). They are composed of three **stylets** in a ventral pouch which form a set of fine cutting structures. The true mouth, known as the **prestomum**, opens at the anterior extremity of the ventral pouch. The prestomum is usually lined with fine teeth. During feeding the prestomum is inverted and the teeth help to secure the louse to the host's skin. The stylets are then used to puncture the skin and blood is sucked into the prestomum by a muscular cibarial

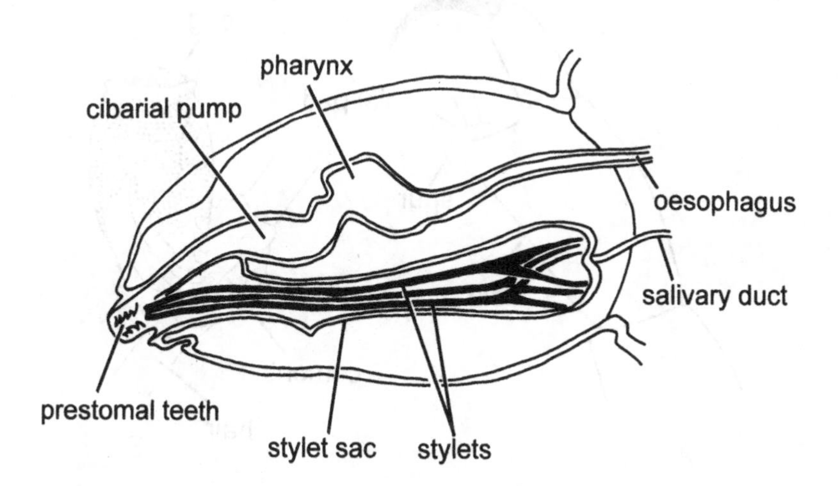

Fig. 7.2 Longitudinal section through the mouthparts and head of an anopluran louse.

pump. The mouthparts are usually retracted into the head when not in use, so that all that can be seen of them is their outline in the head or their tips protruding. There are no palps.

The thoracic segments of Anoplura are usually fused together and are difficult to distinguish. There is one pair of spiracles on the mesothorax and six pairs on segments 3–8 of the abdomen. The abdomen has nine visible segments. The abdomen may have sclerotized **paratergal plates** along the sides. The Anoplura have what appear to be 'crab-like' claws on each leg (Fig. 7.3). These are composed of a single claw projecting from the tarsus. This closes on to a projection from the tibia called a **tibial spur**. This structure enables the lice to cling to hairs.

The female has two pairs of lateral gonopods, giving the abdomen a blunt-ended shape, whereas the sclerotized genitalia of the male give a more pointed posterior tip.

7.3 LIFE HISTORY

The female lays batches of 50–100 eggs on the host, usually

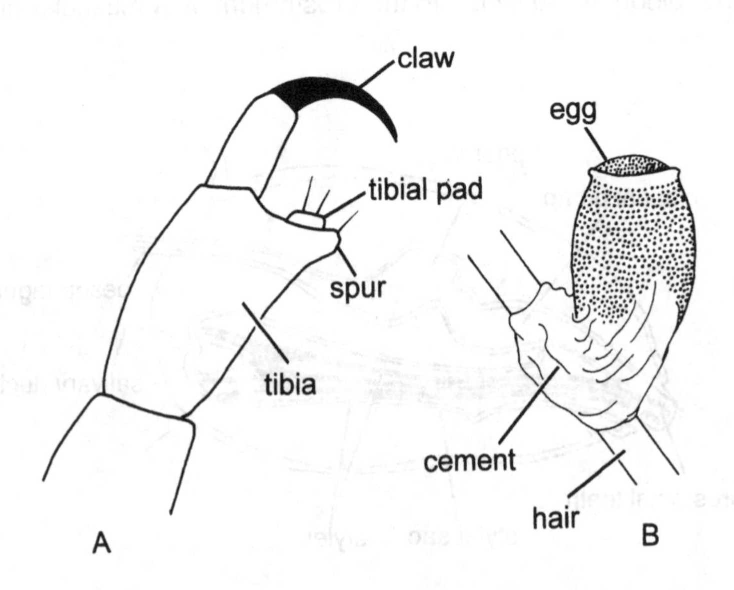

Fig. 7.3 Detail of (A) the tarsus and claw and (B) an egg attached to a hair, of an anopluran louse, *Haematopinus* (reproduced from Walker, 1994).

cementing them firmly to individual hairs or feathers (Fig. 7.3B). The nymph, which closely resembles a smaller version of the adult, hatches from the egg within 1–2 weeks of oviposition.

Over the course of between 1 and 3 weeks the nymphs feed and moult through 3–5 stages, eventually moulting to become a sexually mature adult. The entire egg-to-adult life cycle can be completed in as little as 4–6 weeks. Adults probably live for up to a month, during which they produce small numbers of eggs, probably in the region of one every 1 or 2 days. Some species of lice may be facultatively parthenogenic, which greatly increases their potential rate of population growth.

The Amblycera are thought to be more primitive than the Ischnocera. They exist as body lice of birds and mammals and show adaptations which enable them cling to smooth surfaces. In contrast the Ischnocera are adapted for clinging to hair or feathers. They are more diverse and, generally, more host specific than the Amblycera. The Anoplura are considered to be the most advanced sub-order of lice and are, usually, highly host specific and are found exclusively on eutherian mammals (there are no native anopluran lice of marsupials or monotremes).

Lice usually are unable to survive for more than 1–2 days off their host and tend to remain with a single host animal throughout their lives. Most species of louse are highly host specific and many species specialize in infesting only one part of their host's body. Transfer between hosts is most commonly brought about by close physical contact between individuals.

Lice respond to warmth, humidity and chemical odours. Many receptors are located on the antennae but heat and humidity receptors are located over the entire body. Lice have a tightly defined band of humidity and temperature preferences and respond to humidity or temperature gradients by showing increased rates of turning in favourable microclimates which tends to keep them in favourable areas. In addition, they usually move away from direct light and towards dark objects.

7.4 PATHOLOGY

Heavy louse infestation is known as **pediculosis**. In medical entomology, lice are most well known as vectors of diseases such as typhus and louse-borne relapsing fever. However, lice are of veterinary interest primarily because of the direct damage they can cause to their hosts.

The effect of lice is usually a function of their density. A small number of lice may present no problem and in fact my be a normal part of the skin fauna. However, they have massive potential for increase. For example, the number of *Damalinia ovis* on a sheep have been recorded as increasing from about 4000 in autumn to more than 400,000 in spring. Heavy louse infestations may cause pruritus, alopecia, excoriation and self-wounding. The disturbance caused may result in lethargy and loss of weight gain or reduced egg production. Severe infestation with sucking lice may cause anaemia. Some species of lice may act as intermediate hosts to the tapeworm, *Dipylidium caninum*, and the pig louse, *Haematopinus suis*, may spread swine pox. Heavy infestations are usually associated with young animals or older animals in poor health.

Transfer of lice from animal to animal or from herd to herd is usually by direct physical contact. Because lice do not survive for long off their host, the potential for animals to pick up infestations from dirty housing is limited, although it cannot be ignored. Occasionally, lice also may be transferred between animals by attachment to flies.

Louse infestation is more common in cattle than other domestic animals. In temperate seasonal habitats, heaviest louse infestations occur in late winter and early spring, when the coat is thickest, forming a protective humid environment. This may be exacerbated by winter housing, if the animals are in poor condition and particularly if animals are deprived of the opportunity to groom themselves properly by being housed in stanchions. Most of the population of lice may be lost when winter coat is shed. Louse infestation may also be indicative of some other underlying problem, such as malnutrition or chronic disease. In sheep, lice may be a problem in housed flocks and in heavily fleeced breeds.

Louse infestation in pigs is very common. It occurs most often in the folds of the neck and jowl and around the ears. Light infestation causes only mild irritation. Pediculosis in pigs leads to scratching and skin damage. In horses, light infestations are most commonly found in the mane, base of the tail and submaxillary space. As the population of lice increases, the infestation may spread over the body. As with other animals, the lice spread by contact and their presence leads to irritation, restlessness and rubbing. Long-eared and long-haired breeds dog and cat are especially prone to infestation, although heavy infestations are most usually seen in neglected, underfed animals.

More than 40 species of chewing lice occur on birds. Infestation can cause severe irritation, leading to feather damage, restlessness, cessation of feeding and birds may pluck their feathers. Loss of weight and possibly death may result. Infestation is especially common on young birds and in barn or free-range flocks.

7.5 CLASSIFICATION

The classification of the Phthiraptera is complex and uncertain and remains the subject of considerable debate, with different authorities lumping or splitting the various families and genera to varying degrees. Traditionally the Phthiaptera has usually been divided into two main orders or sub-orders, the **Anoplura** and the **Mallophaga**. However, the Mallophaga is not a monophyletic group and more recent taxonomic studies suggest that the order Phthiraptera should properly be divided into four sub-orders: **Amblycera**, **Ischnocera**, **Anoplura** and **Rhynchophthirina**.

The Amblycera includes six families, of which species of the family **Menoponidae**, containing the genera ***Menacanthus*** and ***Menopon***, are of major veterinary importance on birds. The amblyceran family **Boopidae** contains species that occur on marsupials and a species of the genus ***Heterodoxus*** that may be of importance on dogs. Species of the genera ***Gyropus*** and ***Gliricola*** in the family **Gyropidae** may be important parasites of guinea-pigs.

The Ischnocera includes the three families, two of which, **Philopteridae** and **Trichodectidae** are of major veterinary importance. The Philopteridae contains the genera ***Cuclotogaster***, ***Lipeurus***, ***Goniodes*** and ***Goniocotes***, which are important ectoparasites of domestic birds. The Trichodectidae contains the genera ***Damalinia*** (referred to in older literature as *Bovicola*), ***Felicola*** and ***Trichodectes***, which are ectoparasites of mammals.

There are about 490 species in the sub-order Anoplura. Two families contain species of major veterinary importance, the **Haematopinidae** and **Linognathidae**. The family Haematopinidae contains a single genus of veterinary importance, ***Haematopinus***. The family Linognathidae contains two genera of veterinary importance, ***Linognathus*** and ***Solenoptes***. The anopluran families **Polyplacidae** (genus *Polyplax*) and **Hoplopleuridae** contain species which are parasites of rodents, and the family **Echinophthiridae** contains species which are parasites of marine mammals. The Anoplura also contains the two families of greatest medical interest, the **Pediculidae** and **Pthiridae**, which will not be discussed in this text.

As described previously, the Rhynchophthirina is a very small sub-order including just two species, one of which is a parasite of elephants and the other of warthogs.

7.6 RECOGNITION OF LICE OF VETERINARY IMPORTANCE

The identification of lice is complex and the features used to describe many genera are obscure. However, because lice in general are highly host specific, in many cases information relating to the species of host and the site of infestation will provide a reliable initial guide to identification. The various species of lice are usually found in all geographical regions of the world in which their host occurs.

7.6.1 Guide to the genera of lice of veterinary interest

1 Head broad, equal or almost equal in width to abdomen **2**

Head elongated, much narrower than abdomen **12**

2 Antennae hidden in antennal grooves; antennae four-segmented; maxillary palps present **Amblycera 3**

Antennae not hidden in grooves; antennae three- to five-segmented; maxillary palps absent **Ischnocera 6**

3 On birds **4**

On mammals **5**

4 Small lice, adults about 2 mm in length; abdomen with sparse covering of medium-length setae (Fig. 7.4B); found on thigh or breast feathers; on birds, especially poultry

***Menopon* spp. (Menoponidae)**

Large lice, adults about 3.5 mm in length; abdomen with dense covering of medium-length setae (Fig. 7.4A); found on

the breast, thighs and around vent; on birds, especially poultry
***Menacanthus* spp. (Menoponidae)**

5 On guinea-pigs; oval abdomen, broad in middle; six pairs of abdominal spiracles are located ventrolaterally within poorly defined spiraclar plates (Fig. 7.6A)
***Gyropus* spp. (Gyropidae)**

On guinea-pigs; slender body, with sides of the abdomen parallel; five pairs of abdominal spiracles located ventrally within distinct sclerotized spiraclar plates (Fig. 7.6B)
***Gliricola* spp. (Gyropidae)**

On dogs; relatively large, adults about 3 mm in length; abdomen with a dense covering of thick, medium and long setae (Fig. 7.5) ***Heterodoxus* spp. (Boopidae)**

6 On birds; antennae five-segmented; tarsi with paired claws
Philopteridae 7

On mammals; antennae three-segmented; tarsi with single claws **Trichodectidae 10**

7 Hind legs similar in length to first two pairs **8**

Hind legs at least twice as long as first two pairs; body long and narrow; head with small narrow projections in front of antennae; first segment of antennae considerably longer than following four segments (Fig. 7.7B); on poultry; distribution, worldwide ***Lipeurus* (Philopteridae)**

8 Three long bristles projecting from each side of the dorsal surface of the head; rounded body; adult about 2 mm in length (Fig. 7.7A); on poultry
***Cuclotogaster* (Philopteridae)**

Two long bristles bristles projecting from each side of the dorsal surface of the head **9**

9 Head with prominent angles and a distinct hollow margin posterior to the antennae; adult about 5 mm in length (Fig. 7.8B); on poultry **_Goniodes_ spp. (Philopteridae)**

Head lacking prominent angles; adult about 2 mm in length (Fig. 7.8A); on poultry **_Goniocotes_ (Philopteridae)**

10 **Head rounded anteriorly** **11**

Head sharply angled anteriorly; legs small; abdomen smooth, with only three pairs of spiracles (Fig. 7.9B); on cats

Felicola spp. (Trichodectidae)

11 Setae of abdomen large and thick (Fig. 7.10); on dogs

Trichodectes spp. (Trichodectidae)

Setae of abdomen small or of medium length (Fig. 7.9A); on mammals **_Damalinia_ spp. (Trichodectidae)**

12 Distinct ocular points present behind the antennae; all legs of similar size; adult up to 5 mm in length; distinct paratergal plates visible on abdominal segments; ventral surface of the thorax with a dark-coloured plate (Fig. 7.11)

Haematopinus spp. (Haematopinidae)

No ocular points behind the antennae; forelegs small **13**

13 Two rows of ventral setae on each abdominal segment **14**

One row of ventral setae on each abdominal segment; paratergal plates absent; spiracles on tubercles which protrude from the abdomen; distinct five-sided sternal plate on the ventral surface of the thorax (Fig. 7.12B); on cattle

Solenoptes spp. (Linognathidae)

14 Paratergal plates absent; ventral sternal plate of thorax is narrow or absent (Fig. 7.12A); on cattle, sheep, goats and dogs **_Linognathus_ spp. (Linognathidae)**

Paratergal plates present; ovoid sternal plate on the ventral surface of the thorax (Fig. 7.13); on rodents

Polyplax spp. (Polyplacidae)

7.7 AMBLYCERA

Amblycera are ectoparasites of birds, marsupials or New World mammals. They have large, rounded heads on which the eyes are reduced or absent. They are chewing lice with mouthparts consisting of distinct mandibles on the ventral surface of the head and a pair of 2–4 segmented maxillary palps. Adults are medium-sized or large lice, usually about 2–3 mm in length. The four-segmented antennae are protected in antennal grooves, so that only the last segment is visible.

7.7.1 Menoponidae

The family Menoponidae contains several species, of which two are of major veterinary importance: *Menacanthus stramineus*, the chicken body louse, and *Menopon gallinae,* the shaft louse. These species are ectoparasites of birds, particularly poultry.

Menacanthus stramineus

Morphology: the chicken body louse or yellow body louse, *M. stramineus,* is relatively large, and adults are about 3.5 mm in length (Fig. 7.4A). The palps and four-segmented antennae are distinct. The abdomen has a dense covering of medium-length setae.

Life cycle: the entire life cycle occurs on the host. The eggs are glued to the base of the feathers in dense clusters, particularly around the vent. The eggs hatch in 4–7 days into nymphs, which develop through three stages. Each nymphal stage lasts approximately 3 days. The lice eat the barbs and barbules of the feathers. The egg-to-adult life cycle requires about 2–3 weeks.

Pathology: *M. stramineus* is the most common and destructive louse found on poultry. After introduction into a flock the lice spread from bird to bird by contact. It is an extremely active species and infestation can result in severe irritation, causing the development of inflamed and scab-covered skin. Ultimately infestation may result in decreased hen weight, decreased clutch size and death in young birds and chicks. It is most common on the breast, thighs and around the vent. Populations may reach as many as 35,000 lice per bird. Although it has chewing mouthparts it can cause anaemia by puncturing feather quills and feeding on the blood that oozes out.

Menopon gallinae

Morphology: the shaft louse, *M. gallinae* is pale yellow in colour. It is a small louse; adults are about 2 mm in length. It has small palps and a pair of antennae, folded into grooves in the head (Fig. 7.4B). The antennae have four segments and the abdomen has a sparse covering of small to medium-length setae.

Life cycle: adult females lay their eggs in clusters at the base of a feather. Eggs hatch into nymphs, which pass through three stages

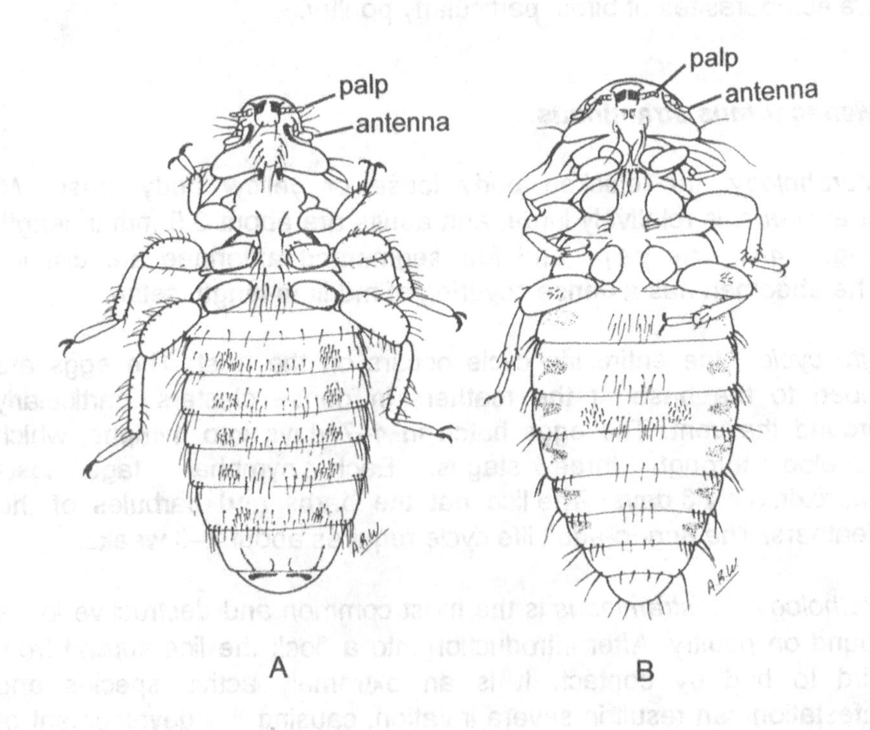

Fig. 7.4 Adult female (A) *Menacanthus* and (B) *Menopon*, in ventral view (reproduced from Walker, 1994).

before moulting to become sexually mature adults. Individuals are highly mobile and move rapidly.

Pathology: *M. gallinae* occurs largely on the thigh and breast feathers of chickens. Although common, it is rarely a severe pathogen. It may also infest turkeys and ducks.

7.7.2 Boopidae

The amblyceran family Boopidae contains species which occur on marsupials and a single species, *Heterodoxus spinigera,* which may be of importance on dogs and other Canidae.

Heterodoxus spinigera

Morphology: *H. spinigera* is a large yellowish-coloured louse; adults are about 5 mm in length, with a dense covering of thick, medium and long setae (Fig. 7.5). It can easily be distinguished from other lice infesting domestic mammals, since the tarsi end in two claws, as opposed to one in the Anoplura and Trichodectidae.

Life cycle: the life cycle is typical, with eggs giving rise to three nymphal stages and the reproductive adult. However, little detail is known.

Pathology: largely confined to warmer parts of Africa, Australasia and the Neotropical and Nearctic regions. The symptoms of infestation are variable. Light infestation may have no obvious effects, with pruritis and dermatitis evident at heavier parasite loads.

7.7.3 Gyropidae

The family Gyropidae contains two species which may be common ectoparasites of guinea-pigs, *Gyropus ovalis* and *Gliricola porcelli.* Species of this family may be distinguished from other families of chewing lice because the tarsi of the mid and hind legs have either one or no claws.

Gyropus ovalis* and *Gliricola porcelli

Morphology: *Gliricola porcelli* is 1–2 mm in length and 0.3–0.4 mm in width (Fig. 7.6B). The maxillary palps have two segments. The five pairs of abdominal spiracles are located ventrally within distinct, sclerotized spiraclar plates. *Gyropus ovalis* is less elongate in shape than *Gliricola porcelli*; it is 1–2 mm in length and 0.5 mm in width (Fig. 7.6A). The maxillary palps have four segments. The six pairs of abdominal spiracles are located ventrolaterally within poorly defined spiracular plates.

Life cycle: the life cycles are typical: egg, three nymphal stages and reproductive adult.

Pathology: these two species are commonly found on guinea-pigs. Originally endemic to the Nearctic, they have now been transported worldwide with the spread of their host. Light infestations may cause little problems. Heavier infestations may result in pruritus and alopecia.

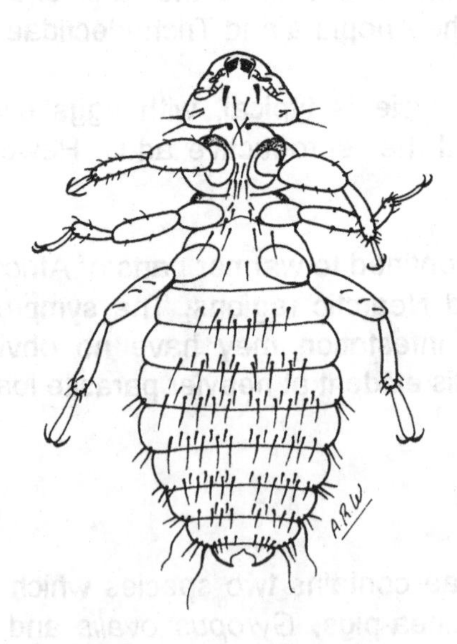

Fig. 7.5 Adult female *Heterodoxus*, ventral view (reproduced from Walker, 1994).

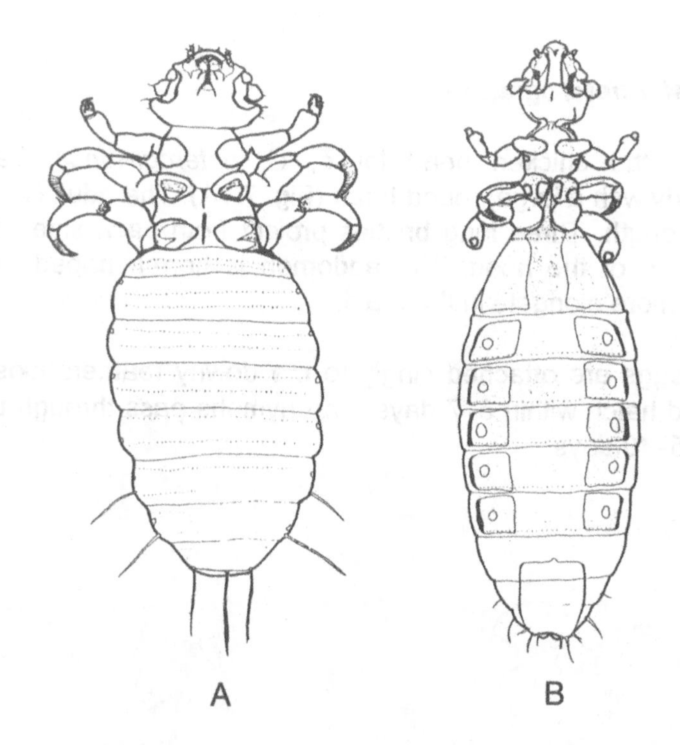

Fig. 7.6 Adult female (A) *Gyropus ovalis* and (B) *Gliricola porcelli,* in ventral view (reproduced from Ronald and Wagner, 1979).

7.8 ISCHNOCERA

The Ischnocera includes the three families, two of which are of major veterinary importance: Philopteridae, on domestic birds, and Trichodectidae, on mammals. The Philopteridae have five-segmented antennae and paired claws on the tarsi. The Trichodectidae have three-segmented antennae and single claws on the tarsi.

7.8.1 Philopteridae

The Philopteridae contains four genera and five species which are important ectoparasites of domestic birds, *Cuclotogaster heterographus, Lipeurus caponis, Goniodes dissimilis, Goniodes gigas* and *Goniocotes gallinae.*

Cuclotogaster heterographus

Morphology: the chicken head louse, *C. heterographus*, has a rounded body with a large, round head (Fig. 7.7A). The adult is about 2.5 mm in length. Three long bristles project from each side of the dorsal surface of the head. The abdomen is barrel-shaped in the female and more elongate in the male.

Life-cycle: eggs are attached singly to the downy feathers close to the skin and hatch within 5–7 days. The nymphs pass through three stages in 25–40 days.

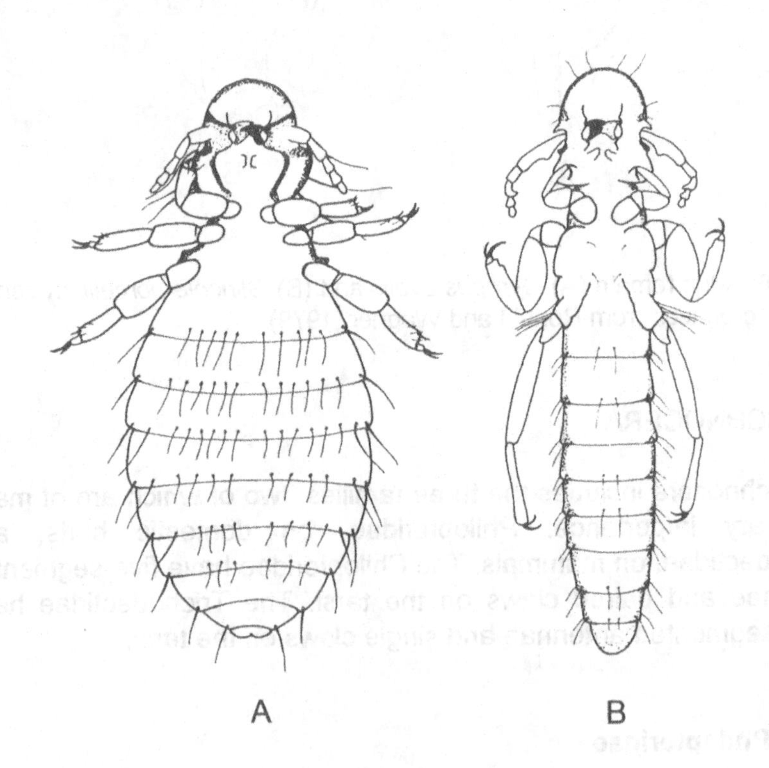

Fig. 7.7 Adult (A) female *Cuclotogaster* and (B) male *Lipeurus,* in ventral view (reproduced from Walker, 1994).

Pathology: *C. heterographus* occurs on the skin and feathers of the head and neck, where the lice feed on tissue debris. Infestations of young birds and chicks may be pathogenic and sometimes fatal.

Lipeurus caponis

Morphology: the wing louse, *L. caponis*, is an elongated, narrow species, about 2.2 mm in length and 0.3 mm in width (Fig 7.7B). The legs are narrow and, characteristically, the hind legs are about twice as long as first two pairs. There are characteristic small angular projections on the head in front of the antennae.

Life cycle: eggs are attached to the feathers and hatch in 4 to 7 days. The nymphs pass through three stages in 20–40 days. Adults are relatively inactive and may live for up to 35 days.

Pathology: *L. caponis* occurs on the underside of the wing and tail feathers. Pathogenic effects usually are slight in healthy animals. Young birds may be susceptible to heavy infestation, especially where underlying disease or malnutrition is debilitating.

Goniodes dissimilis* and *Goniodes gigas

Morphology: *Goniodes* are large lice, about 3 mm in length. They are brown in colour. The head is concave posteriorly, producing marked angular corners at the posterior margins (Fig. 7.8B). The head carries two large bristles projecting from each side of its dorsal surface. The antennae have five segments.

Life cycle: the life cycle is typical. Eggs are attached to the feathers. The nymphs pass through three stages, followed by the reproductive adult.

Pathology: in small number they have little effect on the host. *Goniodes dissimilis* is more abundant in temperate habitats and *G. gigas* in tropical areas.

Goniocotes gallinae

Morphology: the fluff louse, *G. gallinae,* is one of the smallest lice found on poultry, at about 1–1.5 mm in length. The head is rounded and carries two large bristles projecting from each side of its dorsal surface (Fig. 7.8A). The antennae have five segments.

Life cycle: the life cycle is typical. Eggs are attached to the downy feathers close to the skin. The nymphs pass through three stages, followed by the reproductive adult.

Pathology: *G. gallinae* is a small louse of poultry which occurs on the down feathers anywhere on the body. It is generally of little pathogenic significance.

7.8.2 Trichodectidae

The Trichodectidae contains the three genera – *Damalinia*, *Felicola* and *Trichodectes* – which are ectoparasites of mammals.

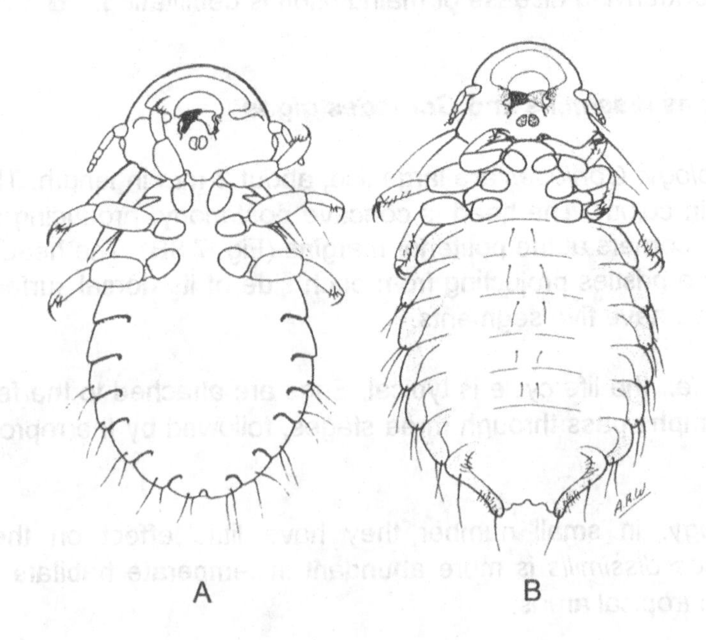

Fig. 7.8 Adult female (A) *Goniocotes* and (B) *Goniodes,* in ventral view (reproduced from Walker, 1994).

Damalinia

There are a number of morphologically similar host-specific species, the most important of which are *Damalinia ovis* on sheep and *Damalinia bovis* on cattle. Also of interest are *Damalinia equi* on horses and *Damalinia caprae* on goats.

Morphology: *D. ovis* is a pale-coloured louse and *D. bovis* is a small, reddish-brown species with dark transverse bands on the abdomen. The head is rounded with a pair of three-segmented antennae. The tarsi carry a single claw (Fig. 7.9A).

Life cycle: the life cycle is typical. Eggs give rise to three nymphal stages, followed by the reproductive adult. The egg-to-adult life cycle requires about 30–35 days. Adult females produce about 30 eggs, each of which is glued singly to the hair or wool next to the skin. *Damalinia bovis* may be facultatively parthenogenic, allowing it to build up populations rapidly.

Pathology: in general, light infestations by *Damalinia* may result in little damage or a chronic dermatitis. However, heavy infestations may result in intense irritation, pruritus, excoriation and alopecia.

In cattle *D. bovis* favours the top of the head, neck, shoulders, back and rump of both dairy and beef cattle. As infestations increase, the lice may spread down the sides and may cover the rest of the body. The lice aggregate in patches and the feeding activity may cause lesions at these sites. *Damalinia bovis* is the only chewing louse found on cattle in the USA.

Infestation of sheep by *D. ovis* can be problematic. This species is mobile and can spread over the entire body, causing considerable irritation, restlessness, interrupted feeding and loss of condition. Exuded serum from bite wounds causes wool matting. Rubbing leads to wool loss. Wounds may attract blowflies.

Damalinia equi is most prevalent on the head, mane, base of the tail and shoulder.

Felicola subrostrata

There is a single species of importance in this genus, *Felicola subrostrata*, which is an ectoparasite of domestic cats.

Morphology: the shape of the head is very characteristic, being triangular and pointed anteriorly (Fig. 7.9B). Ventrally there is a median longitudinal groove on the head which fits around the individual hairs of the host. The antennae have three segments; the legs are small and end in single claws. The abdomen has only three pairs of spiracles and is smooth with few setae.

Life cycle: the life cycle is typical. Eggs give rise to three nymphal stages, followed by the reproductive adult. The egg-to-adult life cycle requires about 30–40 days.

Pathology: this is the only louse that commonly occurs on cats. It is usually of little pathogenic importance, except in elderly or chronically ill animals. It is more problematic in long-haired breeds, and pathogenic populations may develop under thickly matted or neglected fur.

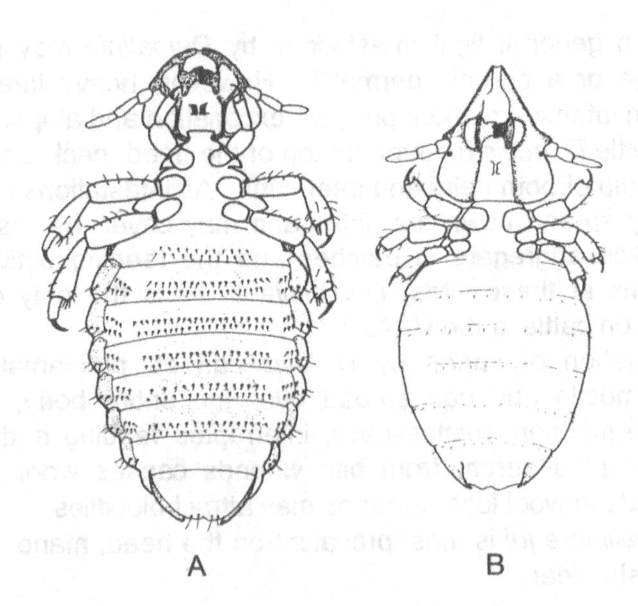

Fig. 7.9 Adult female (A) *Damalinia* and (B) *Felicola,* in ventral view (reproduced from Walker, 1994).

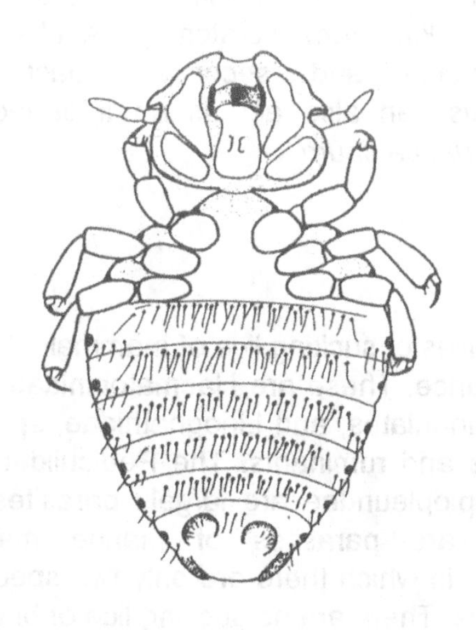

Fig. 7.10 Adult female, *Trichodectes*, in ventral view (from Walker, 1994).

Trichodectes canis

There is a single species of veterinary importance in this genus, *Trichodectes canis*, which is an ectoparasite of domestic dogs. Other species of this genus infest other Canidae.

Morphology: *T. canis* is a small louse, about 1 mm in length. The head is broader than long and the antennae have three segments (Fig. 7.10). The tarsi bear single claws and the abdomen has spiracles on segments 2–6 and many large, thick setae.

Life cycle: the life cycle is typical. Eggs give rise to three nymphal stages, followed by the reproductive adult. The egg-to-adult life cycle requires about 30–40 days.

Pathology: *T. canis* can be a harmful ectoparasite of dogs, particularly puppies. It is most commonly found on the head, neck and tail attached to the base of hairs. It is a highly active species and infestation produces intense irritation around predilection sites. Damage to the skin from scratching results in inflammation, excoriation, alopecia and secondary bacterial involvement. *Trichodectes canis* can also act as an intermediate host of the tapeworm *Dipylidium caninum*.

7.9 ANOPLURA

There are six families of sucking lice of mammals, two of which are of veterinary importance. These are Haematopinidae, species of which are parasites of ungulates, and Linognathidae, species of which are parasites of dogs and ruminants. The Pediculidae are parasites of primates; the Hoplopleuridae are largely parasites of rodents. The Echinophthiridae are parasites of marine mammals and the Neolinagnathidae, in which there are only two species, are parasites of elephant shrews. There are no sucking lice of birds.

7.9.1 Haematopinindae

There is a single genus of interest, *Haematopinus*.

Haematopinus

Twenty-six species have been described in the genus *Haematopinus* of which three are of veterinary importance in temperate habitats: the hog louse, *Haematopinus suis,* the horse sucking louse, *Haematopinus asini*, and the short-nosed cattle louse, *Haematopinus eurysternus*. The tail louse, *Haematopinus quadripertusus*, and the buffalo louse, *Haematopinus tuberculatus*, may infest cattle and buffalo in tropical and sub-tropical habitats, but only the former species is of pathogenic significance.

Morphology: all the species of *Haematopinus* are large lice, about 4–5 mm in length. They possess prominent angular processes, known as ocular points or temporal angles, behind the antennae (Fig. 7.11). Eyes are absent. The thoracic sternal plate is dark and well developed. The legs are of similar sizes each terminating in a single large claw which opposes the tibial spur. Distinct sclerotized paratergal plates are visible on abdominal segments 2 or 3 to 8.

Life cycle: each female produces 1–6 eggs per day. These are glued to the hairs or bristles of the host. Eggs hatch in 1–2 weeks. Nymphs reach maturity in about 2 weeks and the adults live for about 2–3 weeks.

Pathology: *H. suis* is the only species of louse found on pigs and is very common. It is a large (5–6 mm in length) grey louse and usually occurs in the folds of the neck and jowl and around the ears. Light infestation causes only mild irritation. Pediculosis in pigs leads to scratching and skin damage. Transfer is usually by contact but *H. suis* may survive for up to 3 days off its host. Hence, transfer also can occur when animals are put into recently vacated dirty accommodation.

On horses, *H. asini* is most commonly found on the head, neck, back and inner surface of the upper legs. It is a large, yellow–brown-coloured louse which resembles *H. asini*, but has a longer, more robust head. Light infestations may be asymptomatic. In heavy infestations, animals may be restless, lose condition and there may also be anaemia.

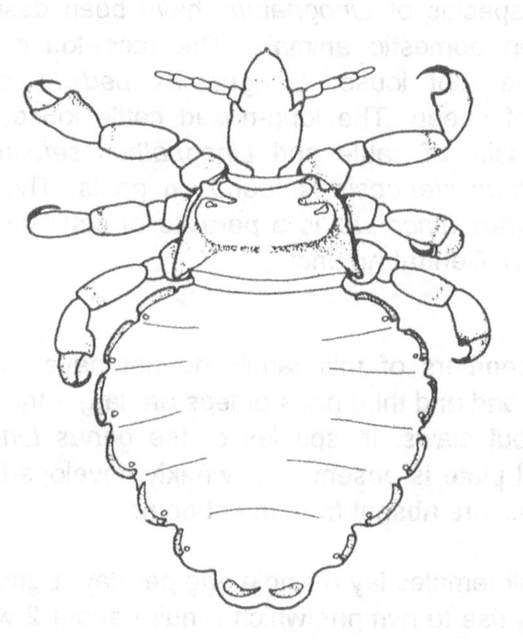

Fig. 7.11 Adult female *Haematopinus*, dorsal view (reproduced from Walker, 1994).

On cattle, infestations of the short-nosed cattle louse, *H. eurysternus,* are most frequently reported on mature animals, reaching their peak in winter, and may cause anaemia and loss of condition. The short-nosed cattle louse occurs on domestic cattle worldwide and is generally considered to be the most important louse infesting cattle. It occurs anywhere on the body of the host. The cattle tail louse, *H. quadripertusus*, is predominantly tropical or sub-tropical in distribution and is found largely in the long hair around the tail. The normal hosts of *H. quadripertusus* are zebu cattle, *Bos indicus*.

7.9.2 Linognathidae

There are two genera of interest in the family Linognathidae, *Linognathus* and *Solenoptes*.

Linognathus

More than 50 species of *Linognathus* have been described, six of which occur on domestic animals. The face louse, *Linognathus ovillus,* and the foot louse, *Linognathus pedalis,* are important ectoparasites of sheep. The long-nosed cattle louse, *Linognathus vituli,* is a parasite of cattle and *Linognathus setosus* parasitizes dogs. *Linognathus stenopsis* is found on goats. The African blue louse, *Linognathus africanus,* is a parasite of cattle in Africa, India, USA and parts of Central America.

Morphology: members of this family do not have eyes or ocular points. The second and third pairs of legs are larger than the first pair and end in stout claws. In species of the genus *Linognathus* the thoracic sternal plate is absent or is weakly developed (Fig. 7.12A). Paratergal plates are absent from the abdomen.

Life cycle: adult females lay a single egg per day. Eggs hatch in 10–15 days, giving rise to nymphs which require about 2 weeks to pass through three nymphal stages. The egg-to-adult life cycle requires about 20–40 days.

Pathology: two species of *Linognathus* are commonly found on sheep, the face louse, *L. ovillus,* and the foot louse, *L. pedalis. Linognathus ovillus* is usually found on the ears and face of sheep. The preferred sites for *L. pedalis* are the feet, legs and scrotum. On the host *L. pedalis* is more sedentary than *L. ovillus* and tends to occur in aggregations. At high densities, however, both species may spread over the entire body. Populations peak in spring and lambs may be particularly susceptible to infestation. Transfer is usually through contact with infested animals. However, *L. pedalis* can survive for several days off the host so infestations may be picked up off contaminated pasture. *Linognathus ovillus* occurs worldwide and *L. pedalis* is common in the USA, South America, South Africa and Australasia.

On cattle, the long-nosed cattle louse, *L. vituli,* is commonly found on the shoulders, neck and rump. It is a small louse (about 2 mm in length) and is more common on dairy cattle and can be problematic particularly in young animals.

Linognathus setosus is a common and widespread parasite of dogs, particularly the long ears of breeds such as the spaniel, basset and Afghan hounds. It may cause anaemia and is usually of greater pathogenic significance in younger animals.

Solenoptes

Only one species in the genus *Solenoptes* is of veterinary importance, the little blue cattle louse, *Solenoptes capillatus,* which is found on cattle.

Morphology: *S. capillatus* is the smallest of the anopluran lice found on cattle (about 1.2–1.5 mm in length). Eyes and ocular points are absent. There are no paratergal plates on the abdomen. The second and third pairs of legs are larger than the first pair and end in stout claws. In contrast to species of *Linognathus*, the thoracic sternal plate is distinct (Fig. 7.12B).

Life cycle: eggs give rise to three nymphal stages, followed by the reproductive adult. The egg-to-adult life cycle requires about about 5 weeks.

Pathology: *S. capillatus* is an important ectoparasite of cattle in the

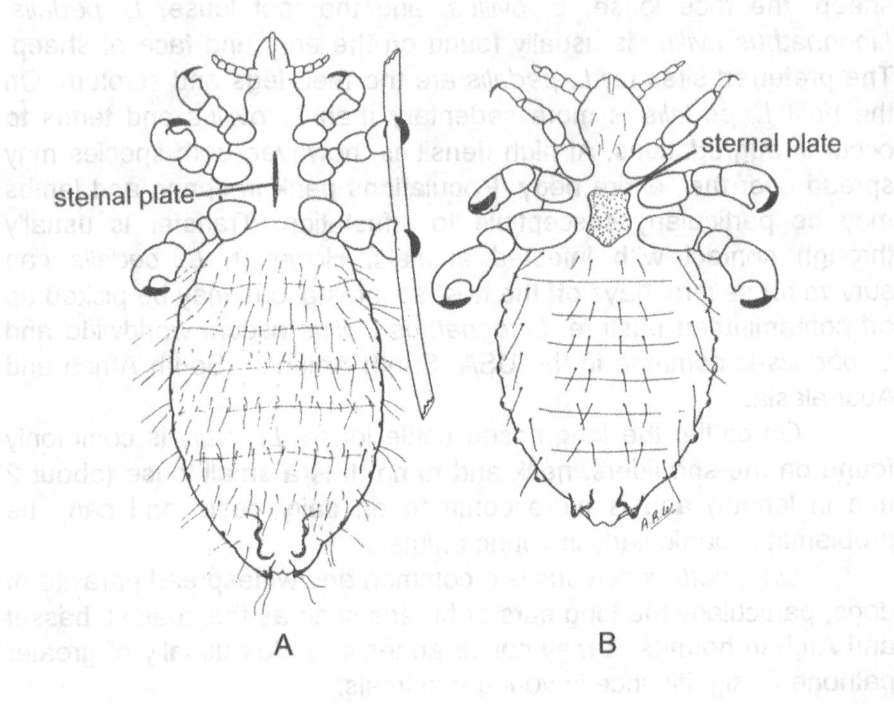

Fig. 7.12 Adult female (A) *Linognathus* and (B) *Solenoptes*, in ventral view (reproduced from Walker 1994).

southern states of the USA and Australia. It usually occurs in aggregations on the muzzle, neck, shoulders, back and tail.

7.9.3 Polyplacidae

Lice of the genus *Polyplax* infest rodents.

Morphology: these lice have prominent five-segmented antennae, no eyes and no ocular points (Fig. 7.13). There is a distinct sternal plate on the ventral surface of the thorax. The forelegs are small and the hind legs are large with large claws and tibial spurs.

Life cycle: eggs give rise to three nymphal stages, followed by the reproductive adult.

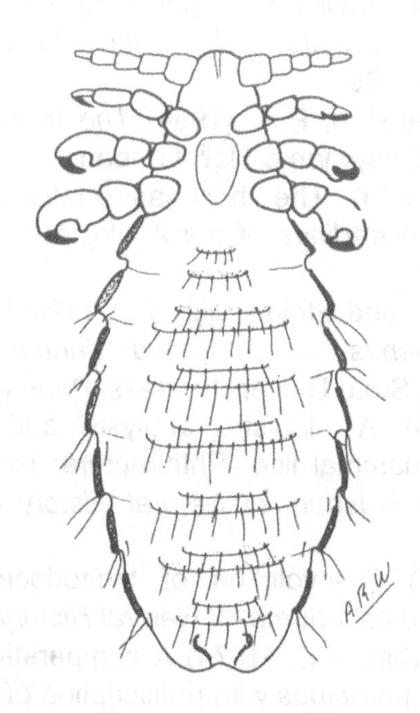

Fig. 7.13 Adult female *Polyplax* in ventral view (reproduced from Walker, 1994).

Pathology: they are ectoparasites of rodents and may cause problems in laboratory colonies. The spined rat louse, *Polyplax serrata,* may transmit various pathogens.

7.10 FURTHER READING

Barker, S.C. (1994) Phylogeny and classification, origins and evolution of host associations of lice. *International Journal for Parasitology*, **24**, 1285–91.

Clay, T. (1970) The Amblycera (Phthiraptera: Insecta). *Bulletin of the British Museum of Natural History (Entomology),* **25**, 73–98.

Cleland, P.C., Dobson, K.J. and Meade, R.J. (1989) Rate of spread of sheep lice (*Damalinia ovis*) and their effects on wool quality. *Australian Veterinary Journal*, **66**, 289–99.

Emerson, K.C. (1956) Mallophaga (chewing lice) occurring on the domestic chicken. *Journal of the Kansas Entomological Society*, **29**, 63–79.

Gullan, P.J. and Cranston, P.S. (1994) *The Insects. An Outline of Entomology*, Chapman & Hall, London.

Hopkins, G.H.E. (1949) The host associations of the lice of mammals. *Proceedings of the Zoological Society of London*, **119**, 387–604.

Kim, K.C., Pratt, H.D. and Stojanovich, C.J. (1986) *The Sucking Lice of North America, an Illustrated Manual for Identification*, Pennsylvania State University Press, Pennsylvania.

Lyal, C.H.C. (1985) A cladistic analysis and classification of trichodectid mammal lice (Phthiraptera: Ischnocera). *Bulletin of the British Museum of Natural History (Entomology)*, **51**, 187–346.

Lyal, C.H.C. (1987) Co-evolution of trichodectid lice and their mammalian hosts. *Journal of Natural History*, **21**, 1–28.

Meleney, W.P. and Kim, K.C. (1974) A comparative study of cattle-infesting *Haematopinus* with rediscription of *H. quadripertusus* Fahrenholz, 1919 (Anoplura: Haematopinidae). *Journal of Parasitology*, **60**, 507–22.

Nelson, B.C. and Murray, M.D. (1971) The distribution of the Mallophaga on the domestic pigeon (*Columba livia*). *International Journal for Parasitology*, **1**, 21–9.

Ronald, N.C. and Wagner, J.E. (1979) The arthropod parasites of the genus *Cavia*, in *The Biology of the Guinea Pig* (eds J.E. Wagner and P.J. Manning), Academic Press, New York, pp. 201–9.

Walker, A. (1994) *The Arthropods of Humans and Domestic Animals*, Chapman & Hall, London.

8

The diagnosis and control of ectoparasite infestation

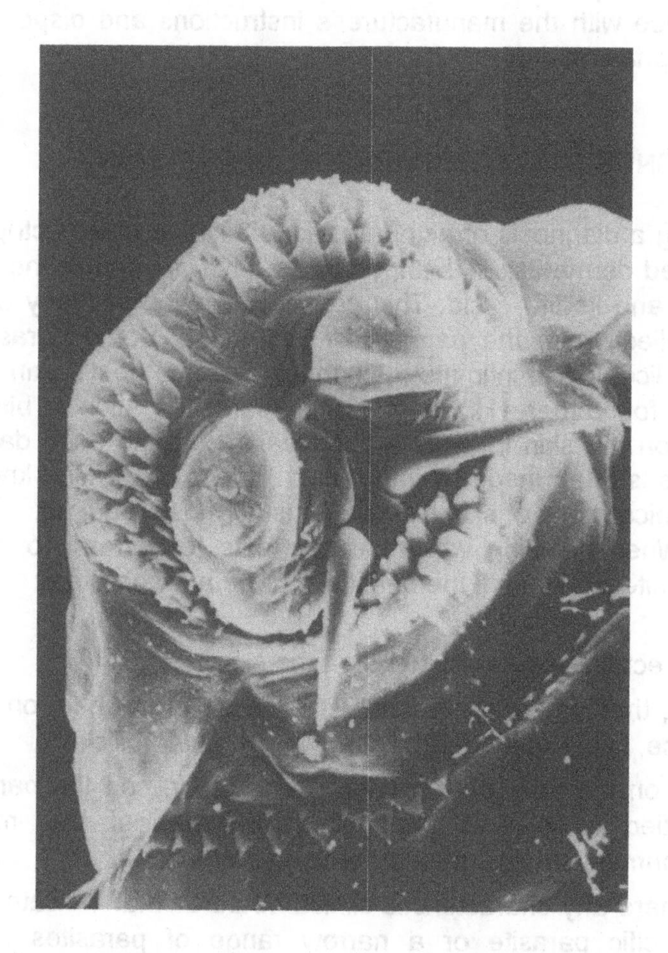

Mouth-hooks and head of a first-stage larva of the sheep nasal bot, *Oestrus ovis*.

8.1 INTRODUCTION

This chapter covers the diagnosis and treatment of ectoparasite infestation and associated dermatoses of domestic animals. It includes discussion of ectoparasite control and the more widely used types of ectoparasiticides, but not a comprehensive list of all ectoparasiticides available worldwide. The principles of ectoparasite control and ectoparasiticide use have been emphasized so that the reader can apply these using locally available ectoparasiticides. It should be emphasized that all insecticides should be used in strict accordance with the manufacturer's instructions and disposed of in an appropriate manner.

8.2 DIAGNOSIS OF ECTOPARASITE INFESTATION

In making a diagnosis of ectoparasitic infestation or an ectoparasite-associated dermatosis it is important to have an idea of the parasite involved and its life cycle. This can be achieved in many cases by direct collection of the parasite or its faeces. Some parasites, for example lice, live in intimate relationship with the host's skin and can easily be found there. However, visiting parasites, such as biting flies, may be on the skin for only a short period of time each day and a diagnosis is often made by implication. Hence, a working knowledge of the clinical signs of skin disease is usually also required.

When deciding which test or tests to perform to diagnose ectoparasite infestation, the questions to be answered are:

- Is the ectoparasite likely to be on the skin?
- If it is, then where on the skin is it living; on the hair, on the skin surface, in the epidermis or dermis, in the hair follicles?
- If not on the skin, but an intermittent visitor, can the parasite be identified while it is feeding or in areas of the immediate environment such as nest or bedding material?
- Are there any characteristic clinical features that indicate whether a specific parasite or a narrow range of parasites could be involved?

The diagnostic tests performed to collect a particular parasite or find evidence of its presence will depend on the answers to these questions.

8.2.1 Hair examination

Many of the larger ectoparasites, such as blowfly larvae or ticks, may be collected directly off a host using appropriately sized forceps. Small specimens may be picked up with the end of a moistened paintbrush. Unattached mites and ticks can be removed by combing or brushing the host animal over a white enamel tray or sheet of paper. Brushing over moistened, white blotting paper or paper towel may help to identify flea infestation, since the dislodged flea faeces will stain red on the paper.

Hairs, collected by coat brushing and plucking, should be mounted in mineral oil, such as liquid paraffin, and examined microscopically for evidence of ectoparasites. Eggs of some parasites, such as lice and *Cheyletiella* spp., may be found attached to the hair shaft. Adult ectoparasites such as lice and various mites may also be found by this method. The hair bulb and lower third of the hair shaft should be examined for evidence of the follicular mite *Demodex*. This is a useful technique for the diagnosis of demodicosis, particularly when lesions occur on the feet. In cases of alopecia it may be useful to examine the upper portion of the hair for evidence of fracture, which occurs with self-induced alopecia due to pruritus.

To ensure that the mouthparts are not left behind, embedded living ticks may be removed most effectively by dabbing the tick and the surrounding skin with alcohol. This relaxes the tick, allowing it to be pulled out intact. Alternatively, the tick can be covered with a layer of petroleum jelly, which prevents respiration and, after about 30 minutes, the tick will drop off.

8.2.2 Acetate strip examination

Short strips of acetate tape (e.g. Sellotape or Scotch 3M tape) can be applied repeatedly to either the hair coat or clipped skin surface. Material and parasites in the coat or on the surface of the skin become attached to the tape which is then mounted on to glass slides and examined microscopically. This is a useful technique for identifying mites.

8.2.3 Superficial skin scraping (epidermal surface examination)

This is a routine technique. It is important to remove coat hair by gentle clipping (the hair can be mounted and examined separately). The surface of the skin can then be scraped using a blunt scalpel blade and the material mounted in 10% potassium hydroxide (KOH) or liquid paraffin. Alternatively a few drops of liquid paraffin can be applied and spread over the skin-scraping site and then scraped with a blunt scalpel blade. The emulsion of superficial epidermis and liquid paraffin is then spread over a microscope slide, covered with a glass cover slip and examined under the microscope. It is important to note that 10% KOH should not be applied to the skin directly. This technique can be used to identify surface mites and multiple scrapings should be taken to increase the likelihood of ectoparasite detection.

The material collected by dry skin scraping can be digested in 10% KOH for 20 minutes and centrifuged to concentrate mites which can be collected from the surface.

8.2.4 Deep skin scraping (deep epidermal examination)

This is an extension of the superficial skin scraping technique and should be performed following or during squeezing of the skin between the thumb and forefinger. After removing the superficial layers the procedure is repeated until capillary ooze occurs. This technique is useful in the diagnosis of burrowing and deep follicular mites such as *Sarcoptes scabiei* and *Demodex* spp. Multiple sites should be scraped to maximize detection of ectoparasites.

8.2.5 Collection of free-living ectoparasites

Mobile free-living mites, ticks and fleas can be extracted from bedding, nests and faecal material by a careful search or by shaking the material through a tier of sieves of decreasing mesh-size. They may also be swept from vegetation using a hand net. Most commonly used for collecting ticks, however, is a blanket drag. This is a woollen blanket or cotton towel, about 1 m square, attached to a bar at one side. The drag is pulled across low-lying vegetation and questing ticks attach to the cloth.

Adult flies can be collected using hand nets, usually consisting of a deep bag of fine mesh netting with a circular, wire-stiffened, opening on a pole. Flies may be picked off as they visit their hosts or baits of rotting carrion or faeces, using either a hand-net or, more simply, by inverting a glass tube over them as they feed or rest. A wide variety of baited traps are also used to attract and catch adult flies.

8.2.6 Biopsy and histopathology

Although these indirect techniques are not as useful as direct identification for the diagnosis of ectoparasite dermatoses, they may be valuable in some circumstances, such as insect and arthropod bite lesions. Histological changes often associated with ectoparasites include:

- An eosinophil-rich dermal infiltrate.
- Collagen degeneration, usually associated with dermal eosinophil infiltration.
- Focal dermal necrosis, which occurs in tick-bite lesions.
- Eosinophilic folliculitis and furunculosis, which may be associated with mosquito-bite lesions in cats and are now thought to occur in the dog associated with arthropod-bite lesions.
- Eosinophilic pustule formation, which may occur in cases of flea-bite dermatitis, *Sarcoptes scabiei* infestation and cheyletiellosis in the dog.
- Lymphocytic mural folliculitis, which has been associated with demodicosis in the dog.
- Eosinophilic granulomas, which commonly occur in nodular skin disease in the horse associated with insect bites.

8.3 THE CHEMICAL CONTROL OF ECTOPARASITES

The drugs and chemicals used against ectoparasites, known generally as ectoparasiticides, contain a very restricted range of elements: carbon, hydrogen, oxygen and nitrogen are almost universal, sulphur occurs in some, while fluorine, chlorine, iodine and phosphorous occur occasionally. Ring structures are very common. All drugs have at least three names. A systemic chemical name (e.g. 2-cyclopropylamino-4, 6-diamino-*s*-triazine), a simpler generic name (e.g. cyromazine) and a trade name (e.g Vetrazin®). In this chapter all chemicals are referred to by their generic name.

8.3.1 Ectoparasiticides: early compounds

Early substances used against ectoparasites were found largely by trial and error or were derived from those developed for horticulture. The latter were often highly toxic, being based on arsenic and mercury, as well as tar, petroleum and nicotine. Carbon disulphide and rotenone (extract of derris) were widely used against ectoparasites. The poisoning of livestock and dairy products was not uncommon and consequently such chemicals have largely been replaced by pesticides that are inherently less toxic.

Sulphur, either as sublimed sulphur (lime sulphur, flowers of sulphur) or precipitated sulphur (monosulfiram), is a traditional topical ectoparasiticide. Sulphur has been incorporated in ointments, powders and shampoos. Sulphur shampoos are still used in veterinary dermatology for their keratolytic and keratoplastic properties in dry scaling dermatoses rather than for an antiparasitic effect.

8.3.2 Ectoparasiticides: neurotoxins

Most ectoparasiticides act as neurotoxins at central nervous system synapses, axons or neuromuscular junctions, leading to spastic or flaccid paralysis.

- **Organochlorines:** chlorinated hydrocarbons have been used in arthropod pest control since the early 1920s; hexachloroethane and hexachlorophene were largely replaced by the salicylanilides.

These were followed, in the 1940s, by DDT and the cyclodienes. However, these chemicals are persistent within the environment and the use of compounds such as DDT, γ-BHC and dieldrin is now tightly controlled or prohibited. Nevertheless, the less persistent methoxychlor, toxaphene and lindane still find appreciable use against biting flies and lice in some parts of the world, where long-term persistence is considered important.

- **Organophosphates:** these compounds replaced the chlorinated hydrocarbons as insecticides and have been a major class of pesticide since the 1950s. Organophosphates are potent cholinesterase inhibitors. The more commonly encountered organophosphates include diazinon, bromocyclen, cythioate, dioxathion, fenthion, malathion, chlorphenvinphos, chlorpyrifos, dichlorvos and phosmet. They can be used against a variety of ectoparasites, including fleas, lice, mites, ticks, keds and flies. Products for use on the animal and in the environment are available and include formulations for use in dips, sprays, spot-ons and collars. Those most often incorporated in environmental preparations are chlorpyrifos, dichlorvos, iodofenphos, diazinon and malathion. These compounds are readily broken down and do not have the residual problems associated with the organochlorines. However, recent concern over neurotoxic side-effects associated with organophosphates has led to greater regulation of dipping procedures which, in addition to the disposal costs, has led to a reduction in their use in farm animals.
- **Carbamates:** these also have anticholinesterase activity but are considered to be less toxic than the organophosphates and include carbaryl, bendiocarb and propoxur.
- **Triazepentadienes** (formamidines): the only member of this group used in veterinary medicine is amitraz. It is used as a dip for the control of ticks on cattle and sheep, lice on cattle and pigs, sarcoptic mange on pigs and demodectic mange on dogs.
- **Phenylpyrrazoles:** this is a relatively new group of ectoparasiticides, currently represented only by fipronil, which acts by inhibiting the neurotransmitter γ-aminobutyric acid (GABA). The effect is highly specific to the invertebrate GABA receptor, making it very safe to the mammalian host and allowing its use on young and pregnant animals. Fipronil, has activity against fleas, ticks, lice, *Trombicula* spp. and *Cheyletiella* spp. The chemical is lipophilic and diffuses into the sebaceous glands of the hair follicle

which then act as a reservoir, giving fipronil a long residual activity. There is evidence that it has an extremely rapid knockdown effect which occurs before fleas have the opportunity to feed. Hence, it may be especially useful in cases of flea allergic dermatitis. It is available as a spray and spot-on preparations.

- **Pyrethrins:** these are synthetic compounds based on the natural compound pyrethrum. This is extracted from the dried and powdered heads of *Chrysanthemum cinereanaefolium*. It is probably the oldest pesticide known and was first discovered and used in China in the fifth century AD. The plant was first introduced into Europe in the late nineteenth century. The active components of pyrethrum are pyrethrins. Pyrethrins act on the central nervous system to give a rapid knockdown. However, they have poor residual activity and are quickly degraded by sunlight. Pyrethrin is used in insecticidal powders and shampoos which often contain other ingredients, such as piperonyl butoxide, with synergistic effects. Pyrethrins are used against lice and fleas, although for control of the latter frequent application and environmental control is also required.
- **Pyrethoids:** these are synthetic chemicals based on pyrethrin which have a similar mode of action but with greater stability. Degradation by UV light occurs but microencapsulation has prolonged their activity to weeks. The microcapsules adhere to the insect ectoskeleton and the pyrethroid is absorbed through the chitin to produce its toxic effect. Photostable pyrethroids used include permethrin, cypermethrin, deltamethrin and fenvalerate. Their activity is enhanced when combined with piperonyl butoxide which acts as a synergist. Pyrethroids are used against fleas, flies, keds, lice and ticks. They are available in shampoos, sprays, powders and flea collars. Permethrin and fenvalerate are both available in environmental products in the USA.
- **Macrocyclic lactones:** the past decade has seen the introduction of a new group of natural products, the macrolide antibiotics, which have a 16-membered macrocyclic lactone ring. These include the avermectins and the milbemycins. The major avermectins are ivermectin, abamectin and doramectin. The major milbemycins are milbemycin, nemadectin and moxidectin. The avermectins, ivermectin and abamectin are fermentation metabolities of the actinomycete *Streptomyces avermytilis*. Doramectin is a genetically engineered avermectin. The

milbemycins are synthetically derived from nemadectin, a natural fermentation product of the actinomycete *Streptomyces cyanogriseus*. These chemicals have a broad spectrum of activity against arthropods and nematodes and low vertebrate toxicity. They are highly effective against parasitic nematodes and are also active against endoparasites and blood-feeding ectoparasites.

8.3.3 Ectoparasiticides: insect growth regulators

Insect growth regulators (IGRs) interfere with various aspects of arthropod growth and development, acting principally on embryonic, larval and nymphal development and disrupting metamorphosis and reproduction. As a result, IGRs do not usually kill the target pests directly and, therefore, require more time to reduce ectoparasite populations than conventional insecticides. The fact that they act on insect-specific phenomena provides IGRs with a high degree of selectivity between insects and vertebrates. IGRs can be divided into three categories: juvenile hormones, chitin synthesis inhibitors and 'others'.

- **Juvenile hormones and juvenile hormone analogues:** species-specific juvenile hormones are produced by most insects at some stage in their lives. They prevent metamorphosis until the larva is fully grown. At certain stages juvenile hormone must be present, at others it must be absent, to permit normal development. Removal of juvenile hormone from young larvae induces early pupariation and the emergence of dwarfed adults, whereas implants in mature ones may postpone or suppress metamorphosis altogether. The presence of juvenile hormone may also prevent egg hatch. A number of juvenile hormones have been identified, their structures established and, from these, derivatives known as the juvenile hormone analogues (JHA) or juvenile hormone mimics have been developed. One of the oldest but still the most widely used of these is methoprene. The newer IGR, fenoxycarb, has been formulated with pyrethroids or organophosphates in sprays and washes for use on-animal and in the environment. Pyriproxifen, which is active against fleas at very small doses, has recently been released in the USA and is available in spray, collar and wash formulations.

- **Chitin synthesis inhibitors:** an alternative target for IGRs is the arthropod cuticle, in particular the amino-sugar polysaccharide chitin, which is an important element in the structure of the arthropod exoskeleton. Chitin is relatively uncommon in other animals and therefore any interaction with the synthesis or deposition of chitin offers a potential means to control the development of arthropods selectively. Several chemical substances are known inhibitors of chitin synthesis, including the benzoylphenylureas (BPUs). The primary effect of BPU treatment is the disruption of chitin synthesis, by prevention of the production of chitin microfibrils. In general, the insect dies during or immediately following the time when chitin synthesis is critical to survival; usually at egg hatch or at moult. This class of chemicals includes compounds such as diflubenzuron and triflumuron. Also, of recent interest is lufenuron, an orally administered benzoylphenylurea derivative used for flea control, which enters the female flea from the blood while feeding and leads to the production of sterile eggs by interference with chitin production.
- **Other insect growth regulators**: the third group of IGRs, the triazine derivatives, currently contains only one representative, cyromazine. Like the BPUs, cyromazine interferes with moulting and pupation, but without acting directly on chitin synthesis. Another difference concerns the spectrum of activity. While BPUs usually act against a wide range of insects, cyromazine shows considerable specificity for the larvae of higher Diptera and is reported to have little effect on the larvae of most other types of insect.

8.3.4 Repellents

Several compounds appear to have insect repellent activities and are incorporated in commercial repellents. These include:

- Pyrethrin
- Diethyltoluamide (DEET)
- Ethanhexadiol
- Dimethyl phtholate
- Butopyronoxyl

These chemicals may be of use when treating dermatoses associated with flying insects such as *Culicoides* spp. in horses.

8.3.5 Desiccants

Recently, desiccants have been used to alter the environmental microclimate and thereby control fleas and free-living mites present in the environment, such as house-dust mites. In particular, the chemical sodium polyborate applied annually to carpets and furnishings in entire houses may give complete flea control. The product claims to be of low toxicity, without odour and also reduces the number of house-dust mites within the environment, which is an important consideration when treating dogs with atopic dermatitis and concurrent flea allergy.

8.4 MODE OF ECTOPARASITICIDE APPLICATION

Ectoparasiticides can be delivered to the parasite by: topical preparations applied to the host's coat, systemic preparations and environmental preparations. The choice of ectoparasite treatment or control technique is influenced by the pathogenesis of the dermatosis produced, the mode of action and efficacy of drugs available and, critically, the life cycle and habits of the parasite in question.

8.4.1 Topical preparations

Topical preparations available include dips, sponge-ons, sprays, powders, mousses, collars and ear tags. They contain many of the chemicals described above. Topical preparations can be classified as parasiticides, repellents and mechanical agents. Topical formulations of neurotoxic insecticides usually give rapid knock-down but, with intermittently available parasites, lengthy residual activity is critical to ensure that the host is protected at periods when the parasite is available.

8.4.2 Systemic preparations

Systemic preparations can be divided into injectable, oral and topically applied products, all of which are delivered to the ectoparasite during its feeding activity on the skin. However, there are a number of drawbacks to the use of systemic treatments, particularly with slow-acting IGRs. First, if the ectoparasiticide takes some time to be effective, treatment needs to be given before parasite populations reach a critical density. Second, free-ranging animals, such as cats, may continually reaquire infestations of parasites such as fleas from feral cats, wild animals and neighbours' untreated cats, maintaining the environmental supply of eggs, larvae and newly emerged fleas. Finally, if the clinical signs are the result of a hypersensitivity reaction to antigens injected by the parasite during feeding, then an ectoparasiticide delivered via the host's blood may not be effective in controlling the dermatitis. In cases of hypersensitivity, an ideal parasiticide prevents feeding and thereby the development of clinical disease.

Systemic ectoparasiticides are of particular value in the control of ectoparasites that spend all, or a considerable portion, of their life cycle on the host, where rates of reinfestaion are relatively low and where the parasite load is an important cause of disease or loss of production. In many cases, systemic ectoparasiticides are best used as one component of an ongoing, integrated control programme.

8.4.3 Environmental preparations

For ectoparasites that are free-living in one or more life-cycle stages, or are present on the host for only short periods, such as the ticks, fleas and flies, parasiticides also may be directed at the free-living stages in the environment. However, parasites may escape chemical treatment of the environment, either by being in difficult to reach sites, such as the base of a carpet, or in a resistant life-cycle stage, such as the pupa or egg. Hence, the residual activity of any insecticide or acaricide is critical so that the ectoparasite will be affected as it changes location or moves from one life-cycle stage to another. Treatment of the environment will be most effective for localized ectoparasite species, so that the insecticide or acaricide can be focused towards selected sites. An ideal environmental ectoparasiticide should target the parasite specifically and have low toxicity for all other species. Examples of more environmentally discriminating products include the IGRs.

8.5 PROBLEMS WITH CHEMICAL CONTROL

The use of ectoparasiticides is associated with risks of side-effects or poisoning resulting from overdose, species sensitivity, breed sensitivity or an interaction between administered medicines. In addition, the treatment of the wider environment with insecticides almost inevitably leads to unwanted effects on non-target organisms.

8.5.1 Poisoning and environmental contamination

Many of the organochlorines and organophosphates are highly toxic to vertebrates and accidental poisonings, resulting from careless use, overdosing or inappropriate treatment, are common. Cats are particularly highly sensitive to lindane. Small or young dogs and cats are commonly overdosed with organophosphate insecticides used for flea treatment, particularly where flea collars are ingested, or used inappropriately or simultaneously with other organophosphate treatments. The pyrethroids are generally believed to have a wide margin of safety with mammals but are toxic to crustaceans and fish. Nevertheless, in cats and small dogs neurotoxic effects have been recorded, particularly with permethrin, fenvalerate, tetramethrin, chrysanthemate and deltamethrin, although in most cases these have been associated with overdosing.

Environmental contamination and effects on non-target animals have been well documented in the case of the organochlorine insecticides; growing concern is associated with the organophosphates. The disposal of pesticides may create problems, particularly when large volumes of liquids, such as organophosphate sheep dips, are involved. As more is learnt of the long-term risk from disposal sites, proper and legal disposal has become more difficult and of greater public concern.

The faeces of cattle provides a rich and unique habitat for over 200 species of invertebrate in temperate climates. Many of these species, particularly the coprophagous beetles, perform a vital role in the nutrient cycle of pastures, assisting with the conversion of the cattle dung into humus. In recent years it has been shown that a number of drugs given to livestock may be eliminated in faeces or urine in a largely undegraded and still toxic form. For example, dichlorvos, coumaphos and cruformate administered in normal doses have deleterious effects on insects in the faeces of cattle and horses for several days after treatment. This phenomenon may be considered advantageous in some cases and in-feed organophosphates have been used to eradicate the pest species of fly that breed in livestock dung. However, insecticidal material in animal dung may also have a range on undesirable effects on non-target organisms.

Insecticidal material in dung may kill non-target insects and the loss of these insects may prevent normal dung decomposition. Such effects have been shown to be particularly pronounced for the avermectins, although not the milbemycins. Avermectins in cattle and horse dung are particularly toxic at low doses to beetle and cyclorrhaphous fly larvae and the mortality of these insects may delay dung-pat decomposition. Nevertheless, the long-term impact of these effects is the subject of considerable dispute. There is no evidence that avermectins affect earthworms, and earthworms are also important agents in dung-pat decomposition. It is widely argued that the presence of earthworms and the abrasive effects of winter weather, in combination with uneven patterns of avermectin use between farms and herds, will result in minimal environmental contamination from the widespread use of avermectins. This debate has yet to be resolved either way.

8.5.2 Resistance

At the recommended doses modern anthelminthics and insecticides are highly effective at removing susceptible individuals, but they can impose strong selection pressure for the development of resistance. Often selection by one type of chemical hastens the development of resistance against other previously effective compounds; cross- and multiple-resistance in ectoparasites is now a growing problem, necessitating drug switching and the use of alternative chemicals. The use of slow-release technology might well play a significant part in increasing the rate of development of resistance. There can be little doubt that resistance to existing chemicals is unlikely to be reversed and, indeed, will become more widespread, and that new compounds developed in the future will also select for resistance. Probably the most optimistic prognosis is that appropriate management will allow the rate development of resistance to be reduced.

8.6 NON-CHEMICAL CONTROL OF ECTOPARASITES

In addition to the use of ectoparasiticidal chemicals, increasing attention is being given to the development and application of simple and inexpensive, non-chemical control technologies. These usually work by modifying some aspect of the parasite's environment, on or off the host, either to increase ectoparasite mortality or reduce its fecundity. These non-chemical techniques, in general, serve to reduce or suppress parasite populations, rather than bringing about their total elimination. As such, they should be seen as valuable components of a general ectoparasite management programme. The range of techniques that are currently the subject of development is vast and will only be covered here briefly.

8.6.1 Physical control

Modification of the parasite's off-host environment may significantly reduce ectoparasite abundance. For example, many of the dipteran pests of cattle and horses have larval stages which develop in animal dung. Management of dung, therefore, is of prime importance in their control and considerable success can be achieved simply by removing dung regularly from pastures or feed lots and dispersing it in such a way that it no longer attracts ovipositing flies. Biting and non-biting flies also can be effectively controlled through simple procedures such as the removal of moist bedding and straw, food wastes, heaps of grass cuttings and vegetable refuse in which they breed.

Similarly, changing the suitability of the on-host environment may help reduce the susceptibility of the host to ectoparasite attack. For example, in sheep, minimizing pasture worm burdens to reduce diarrhoea, tail-docking (amputation of the tails), crutching or dagging (the regular shearing of soiled wool from around the breech), all help to minimize the incidence of sheep blowfly strike by *Lucilia sericata* or

L. cuprina by reducing wool soilage, thereby reducing the availability of oviposition sites and suitability of the fleece for larval survival.

Tick- and mite-infested pasture can be avoided in some circumstances; grazing practices which reduce contact with ticks, such as pasture spelling, which removes all major hosts for over a year, may cause the tick population to collapse. Harvest mites, *Trombicula* spp., are commonly found associated with fruit trees and chalky soils; dogs affected by this mite can be exercised away from such areas. Forage mites may cause a dermatitis in various species including horses, and reduced exposure to infested hay may be an effective control measure; hay stored in lofts above stables could be moved elsewhere or animals can be prevented access to hay barns. Dogs having access to poultry houses may develop dermatitis caused by the poultry mite *Dermanyssus gallinae* and simply excluding the pet may solve the problem. Midges such as *Culicoides* spp. feed on horses, potentially causing sweet itch, at particular times of day and simply stabling during morning and evening may prevent or reduce the problem.

8.6.2 Barriers

There are various types of physical barrier that can be employed to protect the host from ectoparasites. These may be fine mesh screens on windows, plastic strips on milking parlour doors or brow tassles for protection from flying insects. Such techniques may often be used in conjunction with an insecticide; one traditional treatment for the control of sweet itch in horses is to apply a mixture of liquid paraffin and benzoyl benzoate to the mane and tail base.

8.6.3 Biological control

Organisms which are predators, parasites, competitors and pathogens of the ectoparasite can be used as biological controls. For example, the nematode *Steinernema carpocapsae* has been shown experimentally to parasitize and kill the pupal stage of *Ctenocephalides felis* and may be of use as part of a flea control programme where outdoor environmental treatment is important. House flies have been controlled in poultry facilities by the release of pupal parasites. The use of imported dung-burying beetles to remove pastureland dung and to increase the rate of removal of cattle pats has been attempted in Australia, Canada and the USA, to prevent the emergence of dung-breeding flies. Considerable interest is currently being given to the use of the bacteria *Bacillus thuringiensis* as a biological control agent, particularly for lice. Nevertheless, the use of these techniques can be a complex and costly operation and, as yet, cannot be attempted routinely.

8.6.4 Vaccination

Various types of vaccines are being developed for use against ectoparasites. In particular, vaccines comprising parasite gut wall antigens show the greatest promise, particularly for ticks. The vaccinated host develops antibodies directed against the parasite's gut wall. The parasite ingests these antibodies in the blood feed and dies.

Discovery of the host humoral response to antigens of *L. cuprina* has stimulated considerable interest in the production of a vaccine. Vaccination of sheep with partially purified extract of *L. cuprina* larvae can result in a marked reduction in growth when larvae are fed on sheep. Experimental vaccines have been produced based on serine proteases secreted by larvae and larval membrane proteins. However, to date, no effective commercially available vaccine has been developed.

8.6.5 Trapping

Arthropods use a complex interaction of olfactory, visual and tactile cues to locate their hosts. If these cues can be identified and isolated, they can be selectively incorporated into trapping device at levels that produce exaggerated responses from the targets. Walk-through traps have been developed for the control of *Haematobia irritans*, stable flies and face flies. The development of traps for tsetse flies in Africa has been highly successful, identifying and exploiting appropriate visual shapes and colours in combination with host-mimicking chemical odours to attract and catch flies.

A screwworm adult suppression system (SWASS) has been used to attract *C. hominivorax* in North America. This combines an insecticide (2% dichlorvos) with a synthetic odour cocktail known as

'Swormlure' to attract and kill adult flies. Field trials with the SWASS gave a 65–85% reduction in an isolated wild *C. hominivorax* population within 3 months. However, environmental concerns about the release of large quantities of dichlorvos has resulted in the SWASS being largely abandoned as a control technique.

Traps baited with synthetic chemicals or carrion have been developed to attract and kill the sheep blowflies *L. sericata* and *L. cuprina*. At present, however, traps are generally of more value for sampling than control. Nevertheless, such techniques hold considerable promise for future development.

8.6.6 Sterile insect technique

By obtaining a high proportion of matings with fertile wild females, male insects, sterilized with radiation and released into a wild population, can eventually drive the population of wild flies to extinction, provided that:

(1) males are released in sufficiently high numbers so that they outnumber the wild fertile flies;
(2) the released males are fully competitive with wild males; and
(3) the release area is isolated, to protect against immigration.

Probably the most successful example of the use of this technique has been the eradication of the New World screwworm fly, *C. hominivorax,* from the south-western USA, Mexico and North Africa.

Management schemes based on sterile-insect work have also been tried in the control of haematophagous flies like *Haematobia irritans* and *Stomoxys calcitrans*. However, there are severe constraints to the use of this technique. The extensive rearing facilities required to breed and release large number of insects make application expensive, and the additional aggravation caused by the release of billions of blood-feeding, disease-carrying adult insects into the field, makes effective control problematic.

8.6.7 Modelling and forecasting

Models may be of particular value in helping to predict the seasonal patterns of abundance of particular ectoparasites and their economic consequences. The development of such predictive models may allow veterinarians, farmers and entomologists to use ectoparaciticides prophylactically. They may also allow ectoparasiticide treatment to be integrated with non-chemical control techniques, to build effective parasite suppression programmes.

8.7 CATTLE

8.7.1 Mites

Sarcoptic mange

Clinical features: sarcoptic mange is caused by the mite, *Sarcoptes scabiei,* and occurs in cattle worldwide. It causes a pruritic dermatosis with papules, crusts, excoriation, secondary alopecia and lichenification. The female mite burrows in the stratum corneum of the epidermis parallel to the surface and deposits eggs and faeces within the burrows. The pathogenesis of the cutaneous lesions is thought to involve a hypersensitivity reaction most likely to faecal antigens. The lesions tend to occur on the face, neck, shoulders and across the rump.

Diagnosis: diagnosis is confirmed by identification of mites, eggs and faecal pellets in skin scrapings.

Treatment: topical treatment with organophosphates (e.g. coumaphos, phosmet, diazinon or malathion), usually with two applications covering a period equal to two life cycles, is effective. The organochloride γ-BHC is also very effective but environmental and toxicity concerns have limited its use. Traditional treatments include sulphur preparations such as lime sulphur (2%) applied every 10 days on three occasions; these can be used on lactating cattle. The use of other products on lactating and beef cattle vary with regard to milk and carcass withdrawal times and all should be used as directed by the manufacturers. Systemic treatment with ivermectin at a dose of 200 μg/kg for sarcoptic mange has become popular in recent years. A single injection is effective. However, it should be noted that fatal adverse reactions to ivermectin occur in Murray grey cattle.

Psoroptic mange

Clinical features: psoroptic mange is caused by *Psoroptes ovis* and leads to a pruritic dermatosis with papules, crusts, excoriation, secondary alopecia and lichenification. The lesions occur in skin folds, on the whithers, shoulders, neck, rump, base of tail and perineum. In severe cases the lesions may become generalized with haematological and blood biochemical changes (mild anaemia, lymphopenia, neutrophilia, eosinophilia, elevated fibrinogen and globulins). The disease occurs worldwide. The species *Psoroptes natalensis* affects cattle in South Africa and South America and has been introduced into Europe.

Diagnosis: diagnosis is confirmed by observation and identification of mites, eggs and faecal pellets in skin.

Treatment: topical organophosphates (e.g. coumaphos, phosmet or toxaphene in beef cattle) applied twice, to cover two mite life cycles are effective. The highly effective organochloride γ-BHC is now no longer used due to environmental concerns. The triazapentadiene, amitraz, has also been used successfully in the treatment of psoroptic mange in cattle. Systemic treatment with a single injection of ivermectin at a dose of 200 μg/kg is effective.

Chorioptic mange

Clinical features: this is caused by the mite *Chorioptes bovis,* which can live off the host for up to 2 months and may be considered a commensal since it is found on normal cattle. Numbers of mites increase during the winter and dermatological disease may occur as the result of a hypersensitivity. The clinical signs are pruritus, erythema, non-follicular papules, crusts, excoriation and secondary alopecia. The lesions are found on the feet, hindlegs, udder, perineum, scrotum and tail. The neck and head may also be affected in some cases. Chorioptic mange leads to reduced milk yield and loss of condition.

Diagnosis: diagnosis is confirmed by identification of mites, eggs and faecal pellets in skin scrapings.

Treatment: topical organophosphates (e.g. coumaphos, phosmet, crotoxyphos), applied at intervals of approximately 2 weeks, are effective in the treatment of bovine chorioptic mange. Lime sulphur (2%) applied weekly for 4 weeks has also been used effectively.

Demodectic mange

Clinical features: demodectic mange is causes by *Demodex* spp. of mites, *D. bovis* (eyelids and body), *D. ghanesis* (eyelids) and an unnamed species (eyelids and body), all of which are thought to be normal skin commensals. Clinical signs include follicular papules and pustules, usually over the withers, lateral neck, back and flanks. These may become generalized in severe cases. Concomitent pyoderma may occur, leading to furunculosis with ulceration and crust formation. The disease is of economic significance due to hide damage.

Diagnosis: diagnosis is confirmed by identifying mites on skin scrapings or hair pluckings. Examination of the hair bulb and shaft mounted in liquid paraffin reveals numerous mites. Histopathology of skin biopsy will also reveal large numbers of mites within the hair follicles or free within the dermis if a furunculosis is present. Other histological features include a perifolliculitis, mural folliculitis or folliculitis involving mixed inflammatory cells but with numerous mononuclear cells.

Treatment: in many cases demodicosis spontaneously resolves and treatment is unnecessary. The organophosphate trichlorfon, used on three occasions 2 days apart, has been advocated.

Psorergatic mange

Clinical features: psorergatic mange occurs in the USA, Canada, Australia, New Zealand and South Africa, and is caused by the itch mite *Psorergates bos*. It has been suggested that the mite may be a normal commensal since it has been isolated from normal skin. It usually produces a mildly pruritic dermatosis in cattle with patchy alopecia and scaling.

Diagnosis: diagnosis is confirmed by finding mites, eggs and faecal pellets in skin scrapings.

Treatment: if treatment is required, systemic ivermectin can be used although topical lime sulphur (2%) has also been reported to be effective.

Poultry mite infestation

Clinical features: occasionally the poultry mite *Dermanyssus gallinae* may cause dermatitis in cattle. The mite, which is free-living in the environment, causes a pruritus with papule and crust formation especially on the ventrum, limbs and muzzle.

Diagnosis: diagnosis is confirmed by detection of mites in skin scrapings, acetate strips or within the immediate environment.

Treatment: the most effective treatment is avoidance of the infested environment which, alternatively, can be treated with a pesticide; several pesticides have been suggested, including malathion, coumaphos, methoxychlor, diazinon and lime sulphur.

8.7.2 Ticks

Clinical features: various types of tick worldwide feed on cattle and can cause dermatitis. Apart from species of the genera *Argas* and *Otobius*, only the hard ticks are of veterinary importance. These tend to feed in particular sites, but they can also be found anywhere on the body. The ears, face, neck, axillae, groin, distal limbs and tail of cattle are favoured sites. The cutaneous signs associated with tick feeding in cattle include papules, pustules, ulceration and alopecia. The tick mouthparts penetrate the epidermis and become lodged in the dermis where haemorrhage, collagen degeneration and a wedge-shaped area of necrosis occur. Tick feeding can introduce cutaneous bacteria into the skin, causing abscesses, or into the circulation, leading to bacteraemia and septicaemia. The other potential systemic effects of ticks are tick paralysis caused by neurotoxins or the transmission of micro-organisms such as *Rickettsia rickettsii*, the cause of Rocky Mountain fever in USA.

Diagnosis: diagnosis is based on clinical examination and the collection and identification of ticks from the skin.

Treatment: ideally the cattle should be removed from the infested pasture, although this is not always practical and topical therapy is then used to reduce the tick population. The life cycle of the particular ticks involved will influence the application regime, with multiple-host ticks requiring prolonged insecticidal programmes, whereas control of one-host ticks may only require treatment for only a few weeks of the year.

Organophosphates and pyrethroids have been used to control tick populations but, with increasing concern over environmental contamination, there has been a decline in the use of the former. Resistance to various topical insecticides is also an increasing problem. The following compounds have been used in the control of ticks on cattle; amitraz, chlorpyrifos, chlorfenvinphos, coumaphos, crotoxyphos, cypermethrin, cyprothrin, deltamethrin, diazinon, dichlorvos, dioxathion, γ-BHC, flumethrin, malathion, permethrin, phosmet, propetamphos and trichlorfon. These are available in various formulations including sprays, dips and slow-release ear tags. Systemic treatment with 200 μg/kg ivermectin every 2–4 weeks also has been shown to prevent tick engorgement and reproduction. Management measures include separation of cattle from infested pasture and cultivation of the infested land.

8.7.3 Flies

Blood-feeding flies

Clinical features: blood-feeding flies, particularly stable flies, horn flies and tabanids, can cause severe disturbance and annoyance to cattle, leading to reduced weight gain, reduced milk production and hide damage. Fly bites may cause pruritic papules and wheals. Blood-feeding flies may also be important vectors of viral, bacterial and protozoal diseases and filaroid nematodes.

Diagnosis: diagnosis is based on clinical signs and observation of adult flies feeding.

Treatment: minimizing the effects of blood-feeding flies is difficult and involves the use of fly avoidance strategies, repellents, topical insecticides and, in severe cases, treatment involves the use of systemic glucocorticoids.

Control of breeding sites (animal dung, vegetable matter, still-water pools) will help to reduce biting fly numbers. Topical treatment of animals with organophosphates (coumaphos, chlorfenvinphos, diazinon, dioxathion, fenchlorvos, malathion, phosmet, stirofos, trichlorfon) and pyrethrins (permethrin, cyfluthrin, cypermethrin) may give temporary protection. However, maintaining residual activity on the host is difficult. Self-treatment systems, such as insecticidal dust bags and pyrethroid ear tags, may be an effective way of maintaining insecticide levels on an animal. Regular application of repellents will also help to prevent fly bites. Residual insecticide preparations can be used within housing and on fly breeding sites. Housing in accommodation secured from fly entry will help to minimize fly activity around stock.

Nuisance flies

Clinical features: the activity nuisance flies, such as the face fly, house flies and other muscids, lead to disturbance and irritation. These flies may also be mechanical vectors of disease.

Diagnosis: diagnosis is based on observation and identification of feeding flies.

Treatment: topical organophosphates (malathion, coumaphos, crotoxyphos, trichlorfon) and pyrethroids (permethrin, cypermethrin) are effective in reducing fly numbers. Ear tags impregnated with cypermethrin or permethrin are also useful in the control of these flies.

8.7.4 Myiasis

Hypodermiasis

Clinical features: third-stage larvae of warble flies, *Hypoderma* spp., produce painful nodular lesions approximately 3 cm in diameter with a central hole in the skin of the back. Infestation is seen most usually in young animals. If accidentally ruptured, or the larva dies within the skin, anaphylaxis and death may occur. Hide damage is the main economic effect of warbles. If larvae become lodged within the spinal cord, acute posterior paralysis without systemic signs may occur. Flying adults of *Hypoderma bovis* cause annoyance and fright with running (gadding) to avoid the flies, resulting in loss of production.

Diagnosis: diagnosis is based on the appearance of the characteristic cutaneous swellings along the back.

Treatment: warble fly larvae in cattle can be effectively controlled using systemic organophosphate preparations, applied as sprays, pour-ons or spot-ons. However, while systemic organophosphates can kill migrating larvae, they are relatively ineffective once larvae are inside their warbles. Topical therapy using ivermectin in a 0.5% solution applied at a dosage of 500 μg/kg, and systemic therapy with ivermectin at a dosage of 200 μg/kg, have been reported to be effective. Similarly, moxidectin and doramectin may also be effective against warble fly larvae.

Dermatobia

Clinical features: *Dermatobia hominis*, also known as the torsalo, berne or human bot fly, is a serious pest of cattle in Central America. The larvae create boil-like swellings where they enter the skin. The cutaneous swellings can be pruritic and the exit holes may attract other myiasis flies. Infestation may result in damage to the hide and reduction in meat and milk production.

Diagnosis: diagnosis is made on the identification of warbles in the skin and identification of larvae within wounds.

Treatment: historically, control of *D. hominis* has been achieved by the application of various insecticides, at 2–4 week intervals, to livestock to kill larvae in warbles and to prevent reinfestation. Suitable insecticides include toxaphene (camphechlor), DDT/γ-BHC mixtures, crufomate, fenthion and trichlorophon. Intramuscular injection of closantel (10–12.5 mg/kg) has also been found to give effective control. However, ivermectin, abamectin and doramectin, applied topically or by injection, are now commonly used.

Cutaneous myiasis

Clinical features: cutaneous myiasis of cattle is most commonly caused by the obligate screwworms: *Cochliomyia hominivorax* (Nearctic and Neotropical regions), *Chrysomya bezziana* (Oriental and Afrotropical regions) and *Wohlfahrtia magnifica* (eastern Palaearctic). Myiasis occurs largely as a consequence of skin damage due to trauma; castration or dehorning wounds are common oviposition sites, as are the navels of newly born calves. Eggs may also be deposited in body orifices, such as the nostrils, eyes, mouth, ears, anus and vagina. Larvae-filled lesions may be ulcerated, cavernous and painful. Secondary bacterial infection, toxaemia and dehydration lead to death. Other Calliphoridae, including species of *Lucilia*, *Chrysomya*, *Cochliomyia* and *Calliphora* may be secondary invaders of screwworm wounds, but only rarely initiate myiasis of cattle.

Diagnosis: diagnosis is made on the basis of clinical signs and identification of larvae within wounds.

Treatment: the current recommended treatment for wounds infested by *C. hominivorax* is a mixture of the organophosphates coumaphos (5%) and chlorfenvinphos (2%) powder in a vegetable oil base. For *C. bezziana* a range of insecticides has been shown to be effective. Ivermectin, doramectin and orally administered closantel have also proved highly effective in the treatment of *C. bezziana* and *C. hominivorax* infestations.

8.7.5 Fleas

Clinical features: fleas rarely affect cattle; however, infestation with the cat flea, *Ctenocephalides felis felis*, has been reported. Pruritic papular lesions occur at the sites of feeding.

Diagnosis: diagnosis is based on collection by combing and identification of fleas.

Treatment: topical treatment using an organophosphates or pyrethroids along with environmental spraying will control fleas.

8.7.6 Lice

Clinical features: lice cause a pruritic scaling dermatosis with focal secondary alopecia and excoriations. The sucking lice, *Haematopinus eurysternus*, *Linognathus vituli* and *Solenoptes capillatus,* are found especially on the head, neck, whithers, tail, groin, axillae and ventrum. The chewing louse, *Damalinia bovis*, is usually found on the neck, withers and tail. Heavy infestations of sucking lice may cause anaemia.

Diagnosis: diagnosis is based on clinical signs and identification of lice in coat brushings. Characteristic lice eggs may also be found attached to hairs.

Treatment: lice spend their entire life on the host, which makes treatment easy and effective. Topical organophosphates (e.g. chlorfenvinphos, coumaphos, chlorpyrifos, dichlorvos, diaxathion, diazinon, malathion, crotoxyphos, trichlorfon, phosmet and propetamphos), in spray and pour-on formulations, are effective. The pyrethoids, permethrin and cypermethrin, are also effective. A single injection of ivermectin at a dose of 200 µg/kg has been shown to be effective in the treatment of sucking lice.

8.8 SHEEP

8.8.1 Mites

Sarcoptic mange

Clinical features: infestation by *Sarcoptes scabiei* is a rare condition in sheep, but may occur occasionally, leading to pruritus, excoriation, crusts, lichenification and secondary alopecia. Lesions are found on the face, ears and legs.

Diagnosis: diagnosis is confirmed by identification of mites, eggs and faecal pellets on skin scrapings.

Treatment: topical treatment with organophosphates (e.g. coumaphos, phosmet or diazinon) should be repeated at an interval of 10–14 days. Topical preparations are most effective if used after shearing. Systemic treatment with ivermectin, at a dose rate of 200 μg/kg given subcutaneously, should also be effective.

Psoroptic mange (sheep scab)

Clinical features: this is one of the most common dermatological diseases of sheep and is caused by *Psoroptes ovis.* Mite populations differing in virulence are believed to exist and debate continues about their specific or sub-specific status; *Psoroptes cuniculi,* which occurs in the ear, is considered by some to be a strain of *P. ovis* rather than a separate species.

Sheep scab is thought to involve a type I hypersensitivity to faecal antigens deposited on the host's skin, leading to severe pruritus, excoriation, pustules, wool matting and wool loss. Severely affected animals show hyperaesthesia with lip smacking, seizures and eventually death. The disease is usually most severe during the winter months, becoming latent or mild in the summer, during which time mites are most abundant in the body folds of the axilla, periorbital skin and ears.

Diagnosis: diagnosis is based on history and clinical examination and confirmed by identification of mites, eggs and faecal pellets on skin scrapings. ELISA technology for diagnosis of sheep scab is currently being investigated and may lead to the development of a commercial assay.

Treatment: topical treatment with a variety of acaracidal compounds is effective in treating psoroptic mange. In the past γ-BHC was commonly used as a dip formulation but environmental concerns have curtailed its use. Organophosphates have also been used, although resistant strains of *Psoroptes ovis* have been reported. Effective chemicals include coumaphos, phosmet, diazinon, propetamphos, flumethrin and lime sulphur. Dipping is required to soak animals sufficiently in the acaricide. Systemic therapy with two injections of ivermectin at 200 μg/kg 7 days apart is effective. Doramectin (300 μg/kg) requires only a single subcutaneous injection. In the future, alternative forms of treatment, such as vaccines, may become available.

Chorioptic mange

Clinical features: chorioptic mange is caused by *Chorioptes bovis* and occurs in Europe, Australia and New Zealand, but has been eradicated from the USA. Clinical disease occurs in the winter and presents as a pruritic dermatosis with papules, crusts, excoriation, secondary alopecia and lichenification. The lesions usually occur on the lower hind limbs, ventral abdomen and scrotum. Scrotal dermatitis in rams can lead to testicular atrophy and infertility.

Diagnosis: diagnosis is confirmed by identification of mites, eggs and faecal pellets in skin scrapings.

Treatment: topical organophosphates (coumaphos and phosmet) used in dips and repeated after 14 days are effective in the treatment of ovine chorioptic mange. Lime sulphur has also been reported to be effective.

Demodectic mange

Clinical features: ovine demodectic mange is an uncommon dermatosis caused by the follicular mite *Demodex ovis*. The clinical signs are alopecia and scaling, especially on the face, neck and across the shoulders, with occasional involvement of the pinnae, lower limbs and around the coronary bands.

Diagnosis: diagnosis is confirmed by identification of mites, eggs and faecal pellets on skin scrapings or hair pluckings.

Treatment: ovine demodicosis has been successfully treated with trichlorfon as a 2% whole-body dip every 2 days on three occasions.

Trombiculidiasis

Clinical features: mites of the genus *Trombicula* occasionally affect sheep, leading to papular lesions usually on the lower limbs, ventrum, face and muzzle with variable degrees of pruritus and wool loss. In the initial stages small clusters of orange larval mites can be seen on the skin, although the lesions may persist after the larval mites have vacated the skin.

Diagnosis: diagnosis is by observation with the naked eye and microscopic identification of the larval mites collected from the skin by surface scraping or acetate strips.

Treatment: in most cases treatment is not necessary since the dermatosis resolves once the exposure to larval mites ceases. In severe infestations topical organophosphates (e.g. coumaphos, chlorpyriphos, malathion or diazinon) or lime sulphur can be used to control the mites. If severe excoriation occurs due to self-trauma, then glucocorticoids can be used to alleviate the signs.

Psorergatic mange (Australian itch)

Clinical features: this a pruritic dermatosis caused by the mite *Psorergates ovis,* which occurs in Australia, New Zealand, Africa and South America. It has been eradicated from the USA and has never been reported in Europe. The fleece is chewed, matted and discoloured, especially along the flanks and over the thighs.

Diagnosis: diagnosis is confirmed by identification of mites, eggs and faecal pellets in skin scrapings. Multiple skin scrapings may be required since the mites are difficult to find.

Treatment: topical therapy by dip or spray applied twice, 14 days apart during the summer using an organophosphate (coumaphos, diazinon and malathion) is effective in controlling this mange mite. Topical 2% lime sulphur is also effective. Systemic treatment with ivermectin given by subcutaneous injection has also been used in the treatment of ovine psorergatic mange.

Forage mites

Clinical features: the free-living forage mites such as *Caloglyphus berlesei* and *Acarus farinae* can cause dermatitis in areas of contact with contaminated food, especially the face, limbs and ventrum.

Diagnosis: diagnosis is based on the history, clinical examination and demonstration of forage mites in skin scrapings and forage.

Treatment: this dermatosis will resolve without therapy once the contaminated forage has been removed.

8.8.2 Ticks

Clinical features: a range of species of tick feed on sheep and cause dermatitis or systemic disease around the world. Ticks tend to feed at specific sites on the body, particularly around the ears, face, neck, axillae, groin and distal limbs, but they also can be found anywhere on the body.

The cutaneous signs associated with tick feeding in sheep include papules, pustules, ulceration and alopecia. Tick feeding can introduce cutaneous bacteria into the skin, causing abscesses or systemically leading to bacteraemia and septicaemia. The other potential systemic effects of ticks are tick paralysis caused by neurotoxins or the transmission of micro-organisms such as *Rickettsia rickettsii*, the cause of Rocky Mountain fever in the USA.

Diagnosis: diagnosis is based on history, clinical examination and the collection and identification of ticks from the skin.

Treatment: as with other farm animal species, sheep should be removed from the infested pasture. This is not always practical and topical therapy is then used to reduce the tick population. The life cycle of the particular ticks involved will influence the application regime, with multiple-host ticks requiring prolonged insecticidal programmes whereas control of one-host ticks may only require treatment for a few weeks of the year.

The following compounds have been used in the control of ticks on sheep: amitraz, chlorpyrifos, chlorfenvinphos, coumaphos, crotoxyphos, cypermethrin, cyprothrin, deltamethrin, diazinon, dichlorvos, dioxathion, γ-BHC, flumethrin, malathion, permethrin, phosmet, propetamphos, stirofos and trichlorfon. These are available in various formulations, including sprays, dips and slow release ear tags.

Control of the spinose ear tick may be achieved by application of an insecticide dust or oil solution directly into the ear. Environmental measures include separation of sheep from infested pasture and cultivation of the infested land.

8.8.3 Flies

Blood-feeding flies

Clinical features: blood-feeding flies, particularly horse flies, black flies, biting midges, stable flies and mosquitoes, can cause severe disturbance and annoyance to sheep. Fly bites may cause pruritic papules and wheals. Blood-feeding flies also important vectors of viral, bacterial and protozoal diseases and filaroid nematodes.

Diagnosis: diagnosis is based on clinical signs and observation of adult flies feeding.

Treatment: involves fly avoidance, repellents, topical insecticides and the use of systemic glucocorticoids in severe cases. For housed animals, control of fly breeding sites by the removal of animal dung, rotting vegetable matter or spilled feed, will help to reduce fly numbers. Similarly, elimination of water pools, manure-contaminated puddles or sewage outflows may help to reduce the abundance of mosquitoes and biting midges.

Topical treatment of animals with organophosphates (e.g. coumaphos, chlorfenvinphos, diazinon, dioxathion, fenchlorvos, malathion, phosmet, stirofos or trichlorfon) and pyrethrins (e.g. permethrin, cyfluthrin or cypermethrin) may give some temporary protection. Residual preparations can be used within housing and on fly breeding sites. Regular application of repellents will also help to prevent bites.

Keds

Clinical features: keds are wingless, tick-like, blood-feeding flies. They cause pruritus especially around the neck, flanks, rump and abdomen, resulting in rubbing, chewing and scratching. The wool becomes broken and stained with ked faeces. Heavy infestations cause anaemia and keds can transmit blue-tongue virus.

Diagnosis: diagnosis is made on the basis of collection and identification of keds from the fleece.

Treatment: keds are easily controlled because they spend their entire life cycle on the host. Many adults and pupae are removed by shearing. Topical insecticides, such as organophosphates (e.g. coumaphos or malathion), given in three applications, 14 days apart, are effective treatments.

Nuisance flies

Clinical features: the activity of nuisance flies, particularly the head fly, *Hydrotaea irritans,* may cause severe irritation and disturbance to sheep. The feeding activity of *H. irritans* may lead sheep to develop self-inflicted wounds during head shaking and rubbing. These flies are also attracted to wounds in fighting rams, especially between the horn and skin. Secondary bacterial infection and myiasis by *Lucilia sericata* may also occur at wound sites.

Diagnosis: diagnosis is made on the basis of clinical signs along with observation and identification of flies feeding on sheep.

Treatment: topical organophosphates (e.g. crotoxyphos and dichlorvos) and pyrethroids (e.g. permethrin or cypermethrin) may be effective in reducing fly numbers.

8.8.4 Myiasis

Nasopharyngeal myiasis

Clinical features: nasopharyngeal myiasis may be caused by larvae of Oestrinae, the most common of which in sheep is the nasal bot fly, *Oestrus ovis* (see section 5.7.1). Infestation may cause rhinitis, characterized by a sticky, mucoid nasal discharge which at times may be haemorrhagic. Histopathological changes in the nasal tissues of infected sheep include catarrh, infiltration of inflammatory cells and squamous metaplasia, characterized by conversion of secretory epithelium to stratified squamous type. Clinical symptoms of infestation may range from mild discomfort, nasal discharge, sneezing, nose rubbing or head shaking. Dead larvae in the sinuses can cause allergic and inflammatory responses, followed by bacterial infection and sometimes death. Larvae may occasionally penetrate the olfactory mucosa and enter the brain – causing ataxia, circling and head pressing.

Diagnosis: diagnosis is made on the basis of clinical signs and identification of larvae.

Treatment: a single oral treatment of ivermectin at 200 µg/kg is systemically active against all three larval stages of *O. ovis*. Rafoxanide administered orally at a dose rate of 10 mg/kg gives a 100% effective cure

Cutaneous myiasis

Clinical features: myiasis of sheep may be caused by the three species of obligate screwworm flies: *Cochliomyia. hominivorax* (Nearctic and Neotropical regions), *Chrysomya bezziana* (Afrotropical, Oriental regions) and *Wohlfahrtia magnifica* (eastern Palaearctic). Myiasis is also caused by the facultative species belonging to the genera *Lucilia*, *Calliphora*, *Phormia and Protophormia*. Of these, myiases caused by *Lucilia sericata* and *Lucilia cuprina* are of greatest worldwide economic importance.

Screwworm myiasis occurs largely as a consequence of skin damage due to trauma; shearing, tail-docking or castration wounds are common oviposition sites. Eggs may also be deposited in body orifices, such as the nostrils, eyes, mouth, ears, anus and vagina.

The main predisposing host factors for sheep myiasis by species of *Lucilia* are faecal and urine soiling, bacterial dermatitis (especially *Dermatophilosis*) and foot rot. The first two of these are more likely to occur in breeds with marked skin folds and long, fine wool. Skin damage around the poll of fighting rams may precede the development of poll strike and bacterial infection of continually wet fleece predisposes to body strike.

Affected animals become depressed, anorexic and usually rub, chew or scratch the affected areas, which may be wet and the wool discoloured. Screwworm infestation may result in cavernous lesions, with progressive necrosis and haemorrhage; *Lucilia* infestation is generally more superficial. Severe secondary bacterial infection of wounds occurs with subsequent toxaemia, septicaemia and death.

Diagnosis: diagnosis is made on the basis of clinical signs and identification of larvae within wounds.

Treatment: the wounds should be cleaned with an antiseptic wash and an insecticidal, preferably larvicidal compound, should be instilled into the lesions. Insecticides such as coumaphos, chlorfenvinphos, diazinon, dichlorvos, fenthion, fenchlorphos, malathion and stirofos are effective. Severely affected animals may need systemic antibiotic and supportive therapy such as rehydration fluids.

Routine spraying or dipping with organophosphates (chlorpyrifos, chlorfenvinphos, coumaphos and diazinon) and pyrethroids (permethrin, cypermethrin) gives some protection against myiasis. More recently topical use of the chitin synthesis inhibitor, cyromazine, has been shown to protect sheep against blowfly strike for up to 8 weeks. Topical ivermectin preparations have also been reported to be effective in the control of myiasis caused by *Lucilia* spp. when used by a hand-jetting technique. Future developments include the possibility of vaccination against *Lucilia* spp.

Genetic variation in fleece rot and body strike susceptibility has ben identified between Marino sheep strains, bloodlines and between individual sheep within flocks. There is therefore the potential for selection for resistance to reduce the incidence of blowfly myiasis.

The reduction of conformational susceptibility to strike by *Lucilia* spp., through removal of wool and skin folds, may also be brought about by mechanical means. Dagging, the removal of faecally soiled wool, and crutching, the regular shearing of wool from around the breech, may both reduce susceptibility to strike by eliminating suitable oviposition sites. Similarly, susceptibility is reduced in ewes following annual shearing. Surgical removal of skin folds around the breech, the 'Mules' operation, is also used for Marino sheep in Australia. The scar tissue formed following this procedure results in a smooth, denuded area of skin, reducing faecal soiling and the development of potential oviposition sites. Tail docking (amputation) may also reduce the incidence of strike in sheep.

8.8.5 Fleas

Clinical features: fleas are a rare cause of dermatitis in sheep; however, infestation with *Ctenocephalides felis felis* has been reported. A papular rash with pruritus and self-excoriation occurs.

Diagnosis: diagnosis is made by collection and identification of fleas from the coat.

Treatment: topical treatment using an organophosphates or pyrethroids along with environmental spraying will control fleas.

8.8.6 Lice

Clinical features: sheep may become infested with both sucking and chewing lice which cause pruritus, self-excoriation and wool damage. Infestation with the foot louse *Linognathus pedalis* causes foot stamping and biting of the limbs. Apart from fleece damage, lice also cause loss of production. The following species of lice affect sheep; *Linognathus ovillus* (face louse), *L. africanus*, *L. stenopsis* (goat louse), *L. pedalis* (foot louse) and *Damalinia ovis*.

Diagnosis: diagnosis is made on the basis of clinical signs and identification of lice in the fleece.

Treatment: lice spend their entire life cycle on the host and are readily killed by most traditional organophosphates and organochlorine topical insecticides applied as a dip or spray. Topical ivermectin, the pyrethroid cypermethrin and the IGR triflumuron have also been shown to be effective against *D. ovis*. Efficacy is improved if used shortly after shearing.

8.9 HORSES

8.9.1 Mites

Sarcoptic mange

Clinical features: infestation by *Sarcoptes scabiei* is a rare condition in horses, but may occur occasionally, producing a pruritic dermatosis with crust formation, alopecia, excoriation and lichenification, especially on the head and neck.

Diagnosis: diagnosis is confirmed by identification of mites, eggs and faecal pellets on skin scrapings.

Treatment: topical treatments using organophosphates (coumaphos, malathion, methoxychlor), the organochlorine γ-BHC or lime sulphur applied twice, 2 weeks apart are effective. Ivermectin applied topically at 200 μg/kg is also effective against sarcoptic mange in horses.

Psoroptic mange

Clinical features: the mite *Psoroptes ovis* (= *P. equi*) can cause a pruritic dermatosis with papules, thick crusts, excoriation and secondary alopecia. The disease has been reported in North America, Australia and Europe. The lesions usually occur on the head, ears, mane, udder, prepuce, axilla and tail. *Psoroptes cuniculi* can be found in normal ears but has been associated with ear pruritus and head shaking.

Diagnosis: diagnosis is confirmed by identification of mites, eggs and faecal pellets in skin scrapings.

Treatment: in generalized cases of psoroptic mange topical treatment using organophosphates (e.g. coumaphos, malathion or methoxychlor), or γ-BHC applied by spray or dip twice, 14 days apart, is effective. In cases of otitis externa the external ear canal should be cleaned if possible with a ceruminolytic agent and a topical otic acaricidal product used.

Chorioptic mange

Clinical features: infestation with the mite *Chorioptes bovis* produces pruritus, crusts, excoriation and secondary alopecia. The lower limbs, caudal pastern and tail are usually affected, although generalized disease has been reported. The mite population increases with a fall in temperature and chorioptic mange usually occurs in winter.

Diagnosis: diagnosis is confirmed by identification of mites, eggs and faecal pellets in skin scrapings.

Treatment: topical treatment using organophosphates (coumaphos, malathion, methoxychlor) applied twice, 14 days apart, is effective in the treatment of chorioptic mange. Topical lime sulphur (2%) is also effective. The use of oral ivermectin as treatment for chorioptic mange has been reported with varying degrees of success.

Demodectic mange

Clinical features: two species of follicular mite affect horses, *Demodex equi* affecting the body and the longer mite, *D. caballi,* affecting the eyelids and muzzle. Clinical signs of scaling and alopecia with or without papules and pustules occur on the face, shoulders, neck and limbs. Equine demodicosis is rare.

Diagnosis: diagnosis is confirmed by identification of mites and eggs in skin scrapings and hair pluckings.

Treatment: there is little information regarding the most effective treatment for equine demodicosis. Investigation and treatment of underlying systemic disease should be performed.

Forage mites

Clinical features: contact with free-living forage mites especially *Pyemotes tritici* and *Acarus farinae* leads to a pruritic dermatosis. The lesions are papules and crusts which occur in areas of contact especially head, muzzle, limbs and ventrum. If forage is stored overhead, then mites falling on to the horse beneath can produce a dorsal distribution of lesions.

Diagnosis: diagnosis is based on the history, clinical examination and demonstration of forage mites in skin scrapings and forage samples.

Treatment: the dermatitis will resolve once the contaminated forage is removed from the environment.

Poultry mite infestation

Clinical features: occasionally the poultry mite *Dermanyssus gallinae* may cause dermatitis in horses. The mite, which lives in the environment, causes a pruritus with papule and crust formation in areas of contact such as the ventrum, limbs and muzzle.

Diagnosis: diagnosis is confirmed by detection of mites in skin scrapings or within the immediate environment.

Treatment: the dermatitis will resolve once the horse is removed from the infested housing or it is treated with an acaricidal compound. Spraying the housing with organophosphates (malathion, coumaphos, methoxychlor, diazinon) or lime sulphur is effective if thorough.

8.9.2 Ticks

Clinical features: various species of tick feed on horses causing dermatitis or systemic disease. They tend to feed in specific sites, especially the ears, face, neck, axillae, groin, distal limbs and tail, but also may also be found anywhere on the body. The cutaneous signs associated with tick feeding in horses include papules, pustules, ulceration and alopecia. The larval form of *Boophilus microplus* causes a local hypersensitivity reaction with papule and wheal formation at the site of feeding.

Tick feeding can cause abscesses by introduction of surface bacteria into the skin or bacteraemia and septicaemia if the bacteria enter the circulation. The other potential systemic effects of ticks are tick paralysis caused by neurotoxins or the transmission of micro-organisms such as *Rickettsia rickettsii*, the cause of Rocky Mountain fever in the USA.

Diagnosis: diagnosis is based on history, clinical examination and the collection and identification of ticks from the skin.

Treatment: horses should be moved to uninfested pasture if practical, but acaricides may be required. Topical organophosphates (e.g. coumaphos, chlorfenvinphos, dioxathion, malathion or stirofos) and pyrethroids (e.g. decamethrin or flumethrin) can be used to reduce tick numbers and prevent complete engorgement. Cultivation of the infested land will also reduce tick numbers.

8.9.3 Flies

Biting midges

Clinical features: biting midges, *Culicoides* spp., are important ectoparasites of horses, causing a cutaneous hypersensitivity reaction either manifested as a ventral papular dermatitis or a dorsal dermatitis known as 'sweet itch' in the UK and 'Queensland itch' in Australia. Cases of sweet itch have a papular dermatosis, with crusting, ulceration and scaling at the base of the tail and along the mane with severe pruritus. The horses rub the base of tail and mane against any suitable object, resulting in skin thickening (lichenification), excoriation and loss of hair. The disease affects young adults and usually shows a seasonal incidence initially. Large numbers of biting midges will cause annoyance and disturbance to horses. *Culicoides* spp. also act as vectors for the filaroid nematodes *Onchocera* spp. and the arbovirus which causes African horse sickness.

Diagnosis: diagnosis is based on clinical signs and observation of *Culicoides* spp. feeding in the morning and evening. Differential diagnosis of the ventral dermatitis includes other biting flies such as *Stomoxys calcitrans* and *Haematobia irritans*, chorioptic mange and dermatophilosis. Skin biopsy reveals an eosinophilic perivascular dermatitis. Blood count may reveal a circulating eosinophilia suggesting a parasitic or allergic disease and intradermal skin testing, using a *Culicoides* antigen, may support the diagnosis.

Treatment: treatment involves midge avoidance, repellents, topical insecticides and the use of systemic glucocorticoids in severe cases. Stabling during the morning and evening in accommodation secure from midges reduces lesions, but their small size makes this difficult to achieve. Barriers applied to the mane and tail such as liquid paraffin mixed with an insecticide are traditional treatments in the UK. Topical insecticides such as pyrethrins (e.g. cyfluthrin or cypermethrin), especially in pour-on formulations, may be useful. The use of medicated bovine ear tags attached to halters have been recommended in an attempt to control these insects.

Horse flies, deer flies and clegs

Clinical features: these flies produce deep, painful papules and wheals with haemorrhagic crust formation. The lesions tend to occur on the ventral abdomen, legs and neck. Horse flies cause considerable disturbance to horses and large numbers can lead to anaemia. They are mechanical vectors for infectious equine anaemia, anthrax, trypanosomiasis and anaplasmosis.

Diagnosis: diagnosis is made on the basis of history and clinical signs supported by observation and identification of feeding flies.

Treatment: control of these flies is extremely difficult. Housing during the day will reduce biting. Ear tags attached to head collars or necklaces impregnated with cypermethrin or cyfluthrin have been reported to be effective in the control of these biting flies.

Forest flies

Clinical features: *Hippobosca equina* (New Forest fly, horse louse fly) is a common biting fly found worldwide that affects horses, and sometimes cattle. The adults may be particularly abundant in warm, sunny weather and feed on blood, especially around the perineum and between the hind legs. The flies remain on the skin for long periods of time causing annoyance and disturbance.

Diagnosis: diagnosis is made on the basis observation and identification of the feeding flies.

Treatment: topical treatment with topical organophosphates (e.g. malathion, coumaphos, methoxychlor or diazinon) or pyrethroids (e.g. deltamethrin or cypermethrin) may be effective.

Other blood-feeding flies

Clinical features: blood-feeding flies, particularly stable flies, horn flies, black, flies and mosquitoes, can cause severe disturbance and annoyance in horses. Fly bites may cause pruritic papules and wheals. The lesions may become haemorrhagic and necrotic. Hypersensitivity reaction to bites can occur. Blood-feeding flies may also be important vectors of viral, bacterial and protozoal diseases and filaroid nematodes.

Diagnosis: diagnosis is based on clinical signs and observation of adult flies feeding.

Treatment: treatment involves fly avoidance, repellents, topical insecticides and systemic glucocorticoids in severe cases. Control of breeding sites (e.g. animal dung, vegetable matter, still-water pools) will help to reduce biting fly numbers. Topical treatment of animals with organophosphates (e.g. coumaphos, chlorfenvinphos, diazinon, dioxathion, fenchlorvos, malathion, phosmet, stirofos or trichlorfon) and pyrethroids (e.g. permethrin, cyfluthrin or cypermethrin) may give some protection. Residual preparations can be used within housing and on breeding places. Necklaces impregnated with the pyrethoid cypermethrin are effective in reducing *S. calcitrans* numbers. Horses with severe dermatitis can be given glucocorticoids or antihistamines to reduce the cutaneous inflammation and pruritus. Regular application of repellents will also help to prevent bites. Stabling at times of peak fly activity, in accommodation secured from fly entry, will help to prevent disturbance.

Nuisance flies

Clinical features: the activity nuisance flies, such as the face fly and house flies, lead to disturbance and irritation and may be mechanical vectors of disease.

Diagnosis: diagnosis is based on observation and identification of feeding flies.

Treatment: control of nuisance flies may be assisted considerably by appropriate hygiene and removal of dung, in which they breed. Topical organophosphates (e.g. crotoxyphos and dichlorvos) and pyrethroids (e.g. permethrin and cypermethrin) may be effective in reducing fly numbers. However, topical or environmental insecticidal treatment usually brings only temporary relief from these fly species. Ear tags attached to head collars or necklaces impregnated with cypermethrin or permethrin are also useful in the control of flies.

8.9.4 Myiasis

Intestinal myiasis

Clinical features: first-stage larvae of horse bots, *Gasterophilus* spp., burrowing into the skin may cause severe irritation. Light infestations of bots in the stomach or intestine are believed to have little pathogenic effect and are tolerated well. However, bots may cause obstruction to the food passing from the stomach to the intestine, particularly when the larvae are in or near the pylorus. The penetration of the mouth-hooks at the site of attachment may result in erosions, ulcers, nodular mucosal proliferation, stomach perforation, gastric abscesses, peritonitis and, in heavy infections, colic, general debilitation and even rectal prolapse. In particular, infestation with *Gasterophilus pecorum* has been associated with swallowing difficulties associated with oesophageal constriction and hypertrophy of the musculature. The oviposition behaviour of gasterophilids may also cause disturbance, panic and self-wounding.

Diagnosis: diagnosis is made on the identification of larvae in the mouth, oesophagus, stomach or in faeces.

Treatment: both trichlorphon and ivermectin paste have been reported to be effective against *Gasterophilus* larvae. Ivermectin has been used as a 1.87% oral paste or a 1% injectable solution.

Cutaneous myiasis

Clinical features: myiasis may be caused by the obligate screwworms *Cochliomyia hominivorax* (Nearctic and Neotropical regions), *Chrysomya bezziana* (Oriental and Afrotropical regions), *Wohlfahrtia magnifica* (eastern Palaearctic) or, more rarely, flies of the genera *Lucilia, Calliphora, Protophormia* or *Phormia* (worldwide). Similar predisposing factors to those discussed for sheep and cattle exist. However, cutaneous myiasis is less common in horses than sheep or cattle. The shorter coat and regular grooming that many horses receive are likely to play a part in preventing myiasis.

Diagnosis: diagnosis is made on the basis of clinical signs and identification of larvae within wounds.

Treatment: routine spraying with organophosphates (chlorpyrifos, chlorfenvinphos, coumaphos and diazinon) and pyrethroids (permethrin, cypermethrin) gives some protection against myiasis. Routine grooming and prompt treatment of cutaneous wounds will help prevent myiasis.

8.9.5 Fleas

Clinical features: fleas only occasionally infest horses. The species *Echidnophaga gallinacea* and *Tunga penetrans* have been reported affecting horses. Clinical signs include papules, crusts, pruritus, excoriations and secondary alopecia.

Diagnosis: diagnosis is based on history, clinical signs and identification of fleas on the horse.

Treatment: topical organophosphates, pyrethroids and fipronil are effective in the treatment of fleas. The source of the infestation and environment should also be treated.

8.9.6 Lice

Clinical features: horses can be infested by both chewing (*Damalinia equi*) and sucking lice (*Haematopinus asini*). The clinical signs associated with lice infestation are pruritus, scale, crusts, excoriation and secondary alopecia. Severe infestation with *Haematopinus asini* produces anaemia. Lice are more common during the winter months when the coat is long.

Diagnosis: diagnosis is made on the basis of clinical signs and identification of lice in the coat.

Treatment: lice spend their entire life cycle on the host and are readily killed by most traditional organophosphates and organochlorine topical insecticides applied as a dip or spray. A spot-on preparation containing cyhalothrin has been reported effective in the treatment of equine lice. A single topical application of fipronil in spray formulation has been used to treat pediculosis caused by *Damalinia equi* and *Haematopinus asini* in the horse.

8.10 PIGS

8.10.1 Mites

Sarcoptic mange

Clinical features: sarcoptic mange is a common dermatosis of pigs with a worldwide distribution. Clinical signs are marked pruritus, erythema, papules, crusts, excoriation and lichenification, particularly on the ears, flanks, abdomen and rump.

Diagnosis: diagnosis is based on history, clinical examination and identification of mites, eggs and faeces in skin scrapings.

Treatment: systemic ivermectin at a dose of 300 μg/kg is the treatment of choice for porcine sarcoptic mange. Other effective topical treatments are organophosphates (e.g. malathion, coumaphos, diazinon, fenchlorvos, chlorfenvinphos, phosmet or trichlorfon), γ-BHC, amitraz and the newer avermectins and moxidectin. It is important to apply the topical acaricide to the entire body and all in-contact animals should also be treated.

Demodectic mange

Clinical features: dermatitis may occur in pigs due to the mite *Demodex phylloides*. The lesions produced are erythema and papules and, if there is secondary bacterial infection or follicular rupture, pustules and nodules. The lesions usually occur on the snout, eyelids, ventral neck, ventrum and thighs.

Diagnosis: diagnosis is confirmed by identification of mites and eggs in skin scrapings and hair pluckings.

Treatment: the topical organophosphate trichlorfon has been reported as effective in the treatment of porcine demodicosis.

Forage mites

Clinical features: contact with species of forage mite, in particular *Acarus farinae*, from contaminated foodstuffs may cause a papular dermatitis. The lesions usually occur in areas of contact, especially around the face, muzzle and ventrum.

Diagnosis: diagnosis is based on the history, clinical examination and demonstration of forage mites in skin scrapings and forage samples.

Treatment: the dermatitis will resolve once the contaminated forage is removed from the environment.

8.10.2 Ticks

Clinical features: tick-related diseases in pigs within the developed countries have, until recently, been rare. Modern housing methods prevent exposure to ticks but, with the return to traditional extensive pig farming, to cater for the consumer demand, ticks and their related diseases may become more important.

Ticks tend to feed at particular sites, especially the ears, face, neck, axillae, groin, distal limbs and tail of pigs, although they can be found anywhere on the body. The cutaneous signs associated with tick feeding in pigs include papules, pustules, ulceration and alopecia. Tick feeding can introduce surface bacteria either into the skin, causing abscesses, or systemically, leading to bacteraemia and septicaemia. The other potential systemic effects of ticks are tick paralysis caused by neurotoxins or the transmission of micro-organisms such as the virus responsible for African swine fever.

Diagnosis: diagnosis is based on history, clinical examination and the collection and identification of ticks from the skin.

Treatment: pigs should be moved to uninfested pasture if practical. However, acaricides may be required. Topical organophosphates (e.g. coumaphos, chlorfenvinphos, diazinon, dioxathion, fenchlorvos, malathion, phosmet, stirofos or trichlorfon), pyrethroids (e.g. decamethrin, flumethrin or cypermethrin) and amitraz can be used to reduce tick numbers and prevent complete engorgement. Systemic ivermectin may also prevent complete tick engorgement thereby reducing tick numbers. Cultivation of the infested land will also reduce tick numbers.

8.10.3 Flies

Blood-feeding flies

Clinical features: blood-feeding flies cause disturbance, dermatitis and, especially in young pigs, anaemia. Fly bites may produce painful papules and wheals with variable pruritus.

Diagnosis: diagnosis is based on the history and clinical signs.

Treatment: control of fly breeding sites will help to reduce biting fly numbers; manure, wet straw, spilled feed and rotting vegetable matter should be removed regularly. The drainage of areas of standing water may help to reduce mosquito numbers. The use of insecticides should only be considered as a supplement to appropriate hygiene.

Topical treatment of animals with organophosphates (coumaphos, chlorfenvinphos, diazinon, dioxathion, fenchlorvos, malathion, phosmet, stirofos, trichlorfon) and pyrethrins (permethrin, cyfluthrin, cypermethrin) may give some protection. Residual preparations can be used within housing and on fly breeding sites. The use of screens on windows and fly traps may be of value.

8.10.4 Myiasis

Cutaneous myiasis

Clinical features: myiasis may be caused by the obligate screwworms *Cochliomyia hominivorax* (Nearctic and Neotropical regions), *Chrysomya bezziana* (Oriental and Afrotropical regions), *Wohlfahrtia magnifica* (eastern Palaearctic). More rarely, flies of the genera *Lucilia, Calliphora, Protophormia* or *Phormia* (worldwide) may cause myiasis of pigs. Similar predisposing factors to those discussed for sheep and cattle exist. However, the lack of hair or wool makes myiasis much rarer in pigs than sheep.

Diagnosis: diagnosis is made on the basis of clinical signs and identification of larvae within wounds.

Treatment: routine spraying with organophosphates (coumaphos, chlorfenvinphos, diazinon, dioxathion, fenchlorvos, malathion, phosmet, stirofos, trichlorfon) and pyrethroids (permethrin, cyfluthrin, cypermethrin) gives some protection against myiasis. Prompt treatment of cutaneous wounds will help to prevent myiasis.

8.10.5 Fleas

Clinical features: fleas occasionally infest pigs. The species *Ctenocephalides felis*, *Ctenocephalides canis*, *Pulex irritans*, *Echidnophaga gallinacea* and *Tunga penetrans* have been reported infesting pigs. Clinical signs include papules, crusts, pruritus and excoriations. Fleas may act as a vector of swine poxvirus.

Diagnosis: diagnosis is based on history, clinical signs and identification of fleas on the pig.

Treatment: topical organophosphates (coumaphos, chlorfenvinphos, diazinon, dioxathion, fenchlorvos, malathion, phosmet, stirofos, trichlorfon) and pyrethroids (permethrin, cyfluthrin, cypermethrin) are effective in the treatment of fleas. The source of the infestation and environment should also be treated.

8.10.6 Lice

Clinical features: pigs may be infested by the sucking louse, *Haematopinus suis,* which lives especially around the ears, axillae and groin. The clinical signs associated with lice infestation are pruritus, scale, crusts, and excoriation. Severe infestation with *H. suis* produces anaemia. *Haematopinus suis* is also a vector for swine poxvirus.

Diagnosis: diagnosis is made on the basis of clinical signs and identification of lice on the skin.

Treatment: *H. suis* spends its entire life cycle on the host and is readily killed by most traditional organophosphates (coumaphos, chlorfenvinphos, diazinon, dioxathion, fenchlorvos, malathion, phosmet, stirofos, trichlorfon) or γ-BHC applied as a dip or spray. Systemic ivermectin and moxidectin are also effective in the treatment of swine pediculosis and have become the standard form of treatment in many countries.

8.11 GOATS

8.11.1 Mites

Sarcoptic mange

Clinical features: sarcoptic mange is a pruritic dermatosis of goats cased by *Sarcoptes scabiei*, seen worldwide. The clinical signs produced are pruritus, papules, crusts, excoriations, secondary alopecia and lichenification, affecting the face, ears, legs and neck.

Diagnosis: diagnosis is based on history, clinical examination and identification of mites, eggs and faeces in skin scrapings.

Treatment: effective topical treatments are organophosphates (e.g. malathion, coumaphos, crotoxyphos or trichlorfon), or lime sulphur. It is important to apply the topical acaricide to the entire body and in-contact animals should also be treated. Systemic ivermectin at a dose of 200 μg/kg is also effective in the treatment of caprine sarcoptic mange.

Psoroptic mange

Clinical features: *Psoroptes cuniculi* can be found in normal caprine ears and may be a commensal. However, in some circumstances it may produce a pruritic otitis externa, which may progress to otitis media and interna. Crusts, alopecia and excroriation of the pinnae and external ear canal occur, with a head tilt if the middle or inner ear have become affected. In some cases more generalized dermatitis occurs with lesions on the face, poll, legs and interdigital skin.

Diagnosis: diagnosis is based on history, clinical examination and identification of mites and eggs in skin scrapings.

Treatment: psoroptic otitis externa can be treated with aural acaracidal preparations. Cleaning the ear with a ceruminolytic ear cleaner to remove crust and cerumen before instilling the acaricide will increase its efficacy. Topical organophosphates (e.g. malathion, coumaphos, crotoxyphos or trichlorfon) applied twice, 14 days apart, or weekly lime sulphur dips for 1 month are also effective. Systemic ivermectin at a dose of 200 μg/kg is also effective in the treatment of caprine psoroptic mange.

Chorioptic mange

Clinical features: this is caused by the mite *Chorioptes bovis* (= *C. caprae*) which produces a pruritic dermatotsis usually in housed animals during the winter. The clinical signs are papules, crusts, excoriation and secondary alopecia on the feet, hind legs, perineum, udder and scrotum. In some cases lesions also occur on the neck and flanks.

Diagnosis: diagnosis is based on history, clinical examination and identification of mites and eggs in skin scrapings.

Treatment: topical organophosphates (e.g. malathion, coumaphos, crotoxyphos or trichlorfon) applied twice, 14 days apart, or a weekly lime sulphur dip for a month are effective in the treatment of caprine chorioptic mange.

Demodectic mange

Clinical features: caprine demodicosis is a relatively common dermatosis caused by the commensal follicular mite *Demodex caprae* which produces follicular papules and nodules on the head, neck, shoulders and flanks. In some cases secondary infection and follicular rupture occurs, producing ulceration and sinus tract formation (furunculosis)

Diagnosis: diagnosis is confirmed by identification of mites and eggs on skin scrapings and hair pluckings.

Treatment: weekly topical washes with malathion, trichlorfon or amitraz have been reported to be effective in the treatment of caprine demodicosis. Any milk must be discarded during treatment. In goats with a few nodules, lancing and cleaning with lugol's iodine or rotenone in alcohol has been advocated.

Mites of minor importance

The mites *Raillietia caprae* and *Raillietia manfedi* are commensals in some caprine ears. *Raillieta caprae* occurs in the USA, Mexico and Brazil while *R. manfredi* occurs in Australia.

8.11.2 Ticks

Clinical features: various species of tick feed on goats and cause dermatitis or systemic disease around the world. They feed particularly on the skin of the ears, face, neck, axillae, groin, distal limbs and tail, but they can also be found in other areas of the body.

The cutaneous signs associated with tick feeding in goats include papules, pustules, ulceration and alopecia. Tick feeding can introduce surface bacteria into the skin, causing abscesses, or systemically leading to bacteraemia and septicaemia. The other potential systemic effects of ticks are tick paralysis caused by neurotoxins or the transmission of micro-organisms such as *Rickettsia rickettsii*, the cause of Rocky Mountain fever in USA.

Diagnosis: diagnosis is based on history, clinical examination and the collection and identification of ticks from the skin.

Treatment: goats should be moved to uninfested pasture if practical although acaricides may be required. Topical organophosphates (e.g. malathion, coumaphos, crotoxyphos or trichlorfon) and pyrethroids (e.g. decamethrin, flumethrin or cypermethrin) can be used to reduce tick numbers and prevent complete engorgement. Systemic ivermectin may also prevent complete tick engorgement, thereby reducing tick numbers. Cultivation of the infested land will also reduce tick numbers.

8.11.3 Flies

Blood-feeding flies

Clinical features: blood-feeding flies can cause disturbance and annoyance in goats. Fly bites may cause pruritic papules and wheals.

Diagnosis: diagnosis is based on clinical signs and observation of *Cullicoides* spp. feeding in the morning and evening.

Treatment: treatment involves fly avoidance, repellents, topical insecticides and systemic glucocorticoids in severe cases. Control of breeding sites (e.g. animal dung, vegetable matter, water pools) may help to reduce biting fly numbers. Topical treatment of animals with organophosphates (e.g. coumaphos, chlorfenvinphos, diazinon, dioxathion, fenchlorvos, malathion, phosmet, stirofos or trichlorfon) and pyrethroids (e.g. permethrin, cyfluthrin or cypermethrin) may give some protection. Residual preparations can be used within housing and on breeding places. Regular application of repellents will also help to prevent bites. Stabling at times of peak fly activity, in fly-secure accommodation may help to prevent disturbance.

Nuisance flies

Clinical features: the activity nuisance flies, such as the face fly and house flies, lead to disturbance and irritation and may be mechanical vectors of disease.

Diagnosis: diagnosis is based on observation and identification of feeding flies.

Treatment: topical organophosphates (e.g. malathion, coumaphos, crotoxyphos or trichlorfon) and pyrethroids (e.g. permethrin or cypermethrin) are effective in reducing fly numbers. Ear tags impregnated with cypermethrin or permethrin also may be useful in the control of nuisance flies.

8.11.4 Myiasis

Cutaneous myiasis

Clinical features: myiasis in goats may be caused by the obligate screwworms *Cochliomyia hominivorax* (Nearctic and Neotropical regions), *Chrysomya bezziana* (Oriental and Afrotropical regions) and *Wohlfahrtia magnifica* (eastern Palaearctic). Cutaneous myiasis of goats by species of *Lucilia*, or other calliphorid blowflies, is less common than in sheep, probably because of the more open hair of goats.

Diagnosis: diagnosis is made on the basis of clinical signs and identification of larvae within wounds.

Treatment: routine spraying with organophosphates (e.g. coumaphos, chlorfenvinphos, diazinon, dioxathion, fenchlorvos, malathion, phosmet, stirofos or trichlorfon) and pyrethroids (e.g. permethrin, cyfluthrin or cypermethrin) gives some protection against myiasis. Prompt treatment of cutaneous wounds will help to prevent myiasis.

8.11.5 Fleas

Clinical features: fleas occasionally infest goats. *Ctenocephalides* spp. have been reported infesting goats. Clinical signs include papules, crusts, pruritus and excoriations.

Diagnosis: diagnosis is based on history, clinical signs and identification of fleas on the goat.

Treatment: topical organophosphates (e.g. malathion, coumaphos, crotoxyphos or trichlorfon) and pyrethroids (e.g. permethrin, cyfluthrin or cypermethrin) are effective in the treatment of fleas. The source of the infestation and environment should also be treated.

8.11.6 Lice

Clinical features: goats can be infested by both chewing (*Damalinia caprae, D. limbata* and *D. crassiceps*) and sucking lice (*Linognathus stenopsis* and *L. africanus*). The clinical signs associated with lice infestation are pruritus, scale, crusts, excoriation and secondary alopecia. Severe infestation with *Linognathus* spp. produces anaemia. Lice are more common during the winter months when the coat is long.

Diagnosis: diagnosis is made on the basis of clinical signs and identification of lice in the coat.

Treatment: lice spend their entire life cycle on the host and are readily killed by most traditional organophosphates (e.g. malathion, coumaphos, crotoxyphos or trichlorfon), pyrethroids (e.g. permethrin, cyfluthrin or cypermethrin) and γ-BHC. Topical insecticides applied as a dip or spray are usually applied twice, 14 days apart.

8.12 DOGS

8.12.1 Mites

Sarcoptic mange

Clinical features: canine sarcoptic mange is an intensely pruritic dermatosis caused by the mite *Sarcoptes scabiei*. Typical lesions are papules, crusts, excoriation and secondary alopecia. The skin often has a mousy odour and dogs become lethargic with weight loss as the disease progresses. The lesions occur on the ears, elbows, hocks and ventral abdomen but may become generalized. Gentle palpation of the pinnal margins often elicits a marked scratch reflex. The pathogenesis of the skin lesions is thought to involve a type I hypersensitivity to faecal antigens while some dogs are thought to be asymptomatic carriers. In most cases the number of mites present on the skin is small but in others large numbers of mites are present. The latter often occurs in immunosuppressed dogs or those on chronic glucocorticoid therapy and has been termed 'Norwegian scabies' by some veterinary dermatologists.

Diagnosis: diagnosis is based on history, clinical examination and identification of mites, eggs and faeces in skin scrapings. In approximately 50% of cases, mites are not found and the diagnosis is based on clinical features and response to therapy. Although mites are occasionally found on histological examination this is unusual and is not a useful diagnostic test. Faecal examination may also reveal mites and a blood count may reveal an eosinophilia. Recently serological measurement of serum antibodies (IgE) directed against faecal antigens has proved a promising aid in the diagnosis of sarcoptic mange.

Treatment: the coat should be clipped, especially around the lesions and keratolytic shampoo used to remove the crusts and scale. A topical acaracidal shampoo or wash should be used every 7–14 days for 4–6 weeks. Suitable scabicidal chemicals are organophosphates (e.g. phosmet or malathion), γ-BHC or amitraz. Systemic ivermectin at 200–400 μg/kg given twice, 14 days apart is effective, but should not be used in collies or collie crossbreds. It is important to treat all in-contact dogs and identify the source if possible. Re-exposure is sometimes a problem. Grooming equipment should also be cleaned thoroughly.

Otodectic mange

Clinical features: the ear mite *Otodectes cynotis* causes otitis externa with intense aural irritation, pruritus and head shaking. Examination of the external ear canal reveals a thick reddish brown exudate with some crust formation. The infestation is very contagious and is a particularly prevalent in juveniles. Although the pathogenesis of the lesions has not been elucidated, a hypersensitivity may be involved and asymptomatic carriers exist. Occasionally ectopic infestation occurs where the mite causes localized dermatitis on the skin especially on the neck, rump and tail.

Diagnosis: examination of the external ear canal using an auroscope reveals small, fast-moving mites. Microscopic examination of aural exudate or superficial skin scrapings in ectopic cases reveals mites and eggs.

Treatment: topical ceruminolytic ear preparations should be used to remove crust and cerumen from the external ear canal. An otic acaracidal product (e.g. thiabendazole or permethrin) should then be used for 2 weeks beyond a clinical cure. Amitraz (1 ml diluted in 33 ml mineral oil) can be used as an otic acaricide. Systemic ivermectin at 200–400 μg/kg given twice, 14 days apart is effective but should not be used in collies or collie crossbreds.

Demodectic mange

Clinical features: canine demodicosis is caused by the long follicular mite *D. canis*, although some workers claim to have identified a second, shorter species in some cases. Demodicosis is thought to occur due to an innate cutaneous immune deficiency (juvenile form) or immunosuppression secondary to other diseases (adult form). The suggested predisposing factors are genetic, age, short hair, poor nutrition, hormonal, stress, endoparasites and debilitating diseases such as neoplasia.

There appears to be breed predispositions to canine demodicosis which vary across the world. These include the West Highland white terrier, English bull terrier, Staffordshire bull terrier, boxer, pugs, Shar Pei, doberman and German shepherd dog. The disease occurs in a juvenile or adult onset form, either localized or generalized and, if concurrent staphylococcal infection occurs, a pustular form. The juvenile form is usually localized without concurrent secondary infection with areas of alopecia, erythema and scaling, especially on head and legs. This form usually resolves without treatment.

The disease tends to become chronic if concurrent staphylococcal infections occurs; this is the so called 'pustular demodicosis'. The signs associated with pustular demodicosis include alopecia, erythema, pustules, crusts, variable pruritus and, in some cases, lymphadenopathy.

Adult-onset demodicosis usually occurs in dogs older than 5 years and may reflect underlying diseases causing immunosuppression, such as hyperadrenocorticism. Lesions include alopecia, scale and crust formation anywhere on the body, but most often on the head and legs. Elderly dogs with spontaneous demodicosis may have underlying systemic neoplasia. A clinical syndrome seen particularly in young adult dogs is pododemodicosis, which is usually a pustular form affecting all four feet. This disease is often chronic with concurrent pyoderma and is often refractory to treatment. Long-term control of both the bacteria and ectoparasite is required. Some cases of pododemodicosis can progress to the generalized form.

Diagnosis: diagnosis of demodicosis is made on the basis of clinical features and the identification of numerous demodectic mange mites on superficial or deep skin scrapings, hair pluckings or biopsy. A significant number of dogs with generalized demodicosis have anaemia. The histopathological findings include perifolliculitis, folliculitis and furunculosis, but if a mural folliculitis is present in sections without intraluminal mites further sections should be cut and examined.

Treatment: juvenile demodicosis usually resolves spontaneously if left untreated. Cases given glucocorticoids may develop pyoderma and become generalized. If a pyoderma is present then systemic antibiotics (cephalosporins are the most effective) and topical antibacterial washes should be used until 7–10 days after resolution. The coat should be clipped and topical acaracidal agents should be used to control the demodicosis. Amitraz is currently the topical chemical of choice, usually used at 0.05% every 7–14 days until two consecutive negative skin scrapings are obtained. Systemic ivermectin given orally at between 400 and 800 μg/kg daily has been reported to be effective. The avermectin milbemycin is available in many countries and is effective in the treatment of canine demodicosis. In a small number of cases the demodicosis recurs several weeks or months after apparently successful treatment. In these cases an underlying disease should be sought and continual long-term therapy may be required.

Trombiculidiasis

Clinical features: larvae of *Trombicula* spp. occasionally affect dogs in the summer and autumn, leading to papular lesions usually on the lower limbs, ventrum, face and muzzle, with variable degrees of pruritus. In the initial stages small clusters of orange larval mites can be seen on the skin although the lesions may persist after the larval mites have vacated the skin. The mites are most often seen in fruit-growing areas on chalky soil.

Diagnosis: diagnosis is made by observation with the naked eye and microscopic identification of the larval mites collected from the skin.

Treatment: the dermatitis should resolve shortly after the larvae have left the skin but acaricidal treatment may be necessary. In severe infestations topical organophosphates (e.g. coumaphos, chlorpyriphos, malathion or diazinon) or lime sulphur can be used to control the mites. If severe excoriation occurs due to self-trauma, then glucocorticoids can be used to alleviate the signs.

Cheyletiellosis

Clinical features: cheyletiellosis is a contagious, variably pruritic infestation caused by *Cheyletiella yasguri* and seen in dogs of any age, but particularly the young. Scaling, dry or greasy, occurs on the back and head along with a papular rash in some cases.

Asymptomatic carriers occur and any in-contact animal should be examined for mites and treated if necessary.

Diagnosis: diagnosis is based on history, clinical examination and identification of mites or eggs. Microscopic examination of hairs collected by coat brushing or plucking may reveal eggs attached to the hair or mites. Superficial skin scrapings and acetate-tape impressions may also reveal mites or eggs. Examination of a faecal sample by the faecal floatation technique may also be of use in diagnosis.

Treatment: cheyletiellosis is easily treated with topical acaricidal compounds such as the carbamates (e.g. carbaryl), organophosphates (e.g. phosmet, chlorpyriphos, malathion or diazinon) and fipronil. All in-contact animals should also be treated and an environmental acaricidal spray used around the home.

Poultry mite infestation

Clinical features: the poultry mite *Dermanyssus gallinae* is an unusual cause of dermatitis in the dog. The mite may come into houses from bird nests in the roof or eaves, but the disease is most often associated with dogs that have access to chicken houses. The clinical signs include variable pruritus, papules, crusting and scaling usually on the back and limbs.

Diagnosis: the diagnosis is made on the basis of the clinical features and finding mites on skin scrapings, acetate-tape strips and coat brushings.

Treatment: treatment is by topical insecticidal products and removal of the source of infestation.

8.12.2 Ticks

Clinical features: various species of ticks can be found on dogs in different parts of the world. They cause local inflammation within the dermis, can act as vectors for various micro-organisms, produce paralysis and, in severe infestations, anaemia.

The spinose ear tick, *Otobius megnini*, found in southern USA causes otitis externa with head shaking and scratching and, in severe cases, convulsions. In Europe, the dog tick, *Ixodes canisuga*, castor bean tick, *Ixodes ricinus,* and hedgehog tick, *Ixodes hexagonus,* are particularly important causes of localized dermatitis and potential vectors of disease in dogs.

Diagnosis: diagnosis is based on history, clinical examination and the collection and identification of ticks from the skin.

Treatment: dogs should be kept away from tick-infested pasture if possible. Topical insecticidal compounds such as carbamates (e.g. carbaryl), organophosphates (e.g. chlorpyriphos, malathion or diazinon) and fipronil can be used to kill ticks on dogs. Individual ticks can be removed manually; however, care should be taken to remove all the mouthparts.

8.12.3 Flies

Many flies can cause nuisance or annoyance to dogs but they are not as great a problem as they are in large, domestic species. There is growing interest in the aetiology and pathogenesis of ulcerative lesions on the bridge of the nose previously thought to be a form of pyoderma, so called 'nasal pyoderma'. The histological changes seen in these cases is of an eosinophilic furunculosis similar to that seen in mosquito-bite dermatitis of the cat. This raises the possibility these nasal lesions in the dog are due to a flying insect or other arthropod, such as spiders.

Sandflies

Clinical features: flies of the genus *Phlebotomus* in the Old World and *Lutomyia* in the New World are of primary importance as vectors of the protozoan *Leishmania* spp. Although these flies feed on dogs, their bite is not usually associated with significant skin disease. However, during blood-feeding, the introduction of the protozoan *Lieshmania* can lead to the development of a chronic exfoliative non-pruritic dermatosis with scaling, especially on the face, pinnae and feet. The organism can also produce a severe systemic disease.

Diagnosis: leishmaniasis is diagnosed on the basis of history, clinical signs and serology. The diagnosis is confirmed by histological identification of the organism in the skin or internal organs.

Treatment: topical treatment with organophosphates or pyrethroids may help reduce sandfly activity. Regular application of repellents such as DEET may also help prevent bites.

Other blood-feeding flies

Clinical features: blood-feeding flies may cause painful papules and wheals, especially on the hairless areas of the head, ears, neck, ventrum and legs.

Diagnosis: diagnosis is made on the basis of history, clinical signs and, ideally, observation and identification of flies while feeding.

Treatment: topical treatment with organophosphates or pyrethroids may help effect temporary relief from biting flies. Regular application of repellents such as DEET may also help to reduce fly feeding activity. In severe cases of fly-bite dermatitis systemic or topical glucocorticoids can be used to treat the dermatitis.

8.12.4 Myiasis

Cutaneous myiasis

Clinical features: myiasis of dogs may be caused by the obligate screwworms *Cochliomyia hominivorax* (Nearctic and Neotropical regions), *Chrysomya bezziana* (Oriental and Afrotropical regions) and *Wohlfahrtia magnifica* (eastern Palaearctic). Cutaneous myiasis may also be caused by species of *Lucilia*, *Calliphora, Protophormia* or *Phormia*, in various parts of the world. Myiasis may occur in old or debilitated dogs, particularly when wounds are left untreated. Stray dogs injured in road traffic accidents may develop myiasis if left unnoticed, paralysed and with severe cutaneous wounds. Long-coated dogs with untreated chronic diarrhoea and faecal coat contamination may develop myiasis, especially in warm, humid weather.

Diagnosis: diagnosis is made on the basis of clinical signs and identification of larvae within wounds.

Treatment: affected wounds are surgically debrided and sprayed or washed with organophosphates (e.g. malathion) or pyrethroids (e.g. permethrin). Prompt treatment of cutaneous wounds will prevent myiasis.

Furuncular myiasis

Clinical features: species of the genus *Cuterebra* are dermal parasites largely affecting rodents and rabbits, although dogs may also be affected. They are found exclusively in the New World. The larvae produce large, subdermal nodules and parasitism by more than one larva is common.

Diagnosis: diagnosis is made on the basis of clinical signs and identification of larvae within or exiting warbles.

Treatment: little information is available relating to the treatment of *Cuterebra*, but ivermectin, moxidectin or doramectin are likely to be effective.

8.12.5 Fleas

Clinical features: the most common flea seen on dogs is the cat flea *Ctenocephalides felis felis*. Several other species have been reported to cause dermatitis, including *Ctenocephalides canis*, *Echidnophaga gallinacea*, *Leptopsylla segnis*, *Pulex irritans* and *Ceratophyllus* spp. Flea-bite dermatitis is one of the most common canine dermatoses. Clinical signs include papules, crusts, pruritus, excoriations and secondary alopecia. The lesions mostly occur on the dorsum, around the tail base, the ventral abdomen and the neck.

Diagnosis: diagnosis is based on history, clinical signs and identification of fleas or flea faeces on the dog. Intradermal skin testing using a flea extract may also be of some value in the diagnosis of flea-bite dermatitis.

Treatment: the treatment of dogs for flea infestation or flea-bite dermatitis involves:

- Application of an adulticide on to the affected animal's coat. Effective products may contain organophosphates, pyrethroids, carbamates or fipronil. Flea products for on-animal use come in spray, shampoo, wash and powder formulations.
- Application of an adulticide to all in-contact animals (dogs and cats).
- Treatment of the environment using products able to kill the egg, larval and pupal stages of the life cycle. Products available include those containing organophosphates, pyrethroids and IGRs. These are available in spray formulations. The IGR, lufenuron, may be given orally to the dog and transferred to the flea in blood at feeding, leading to the production of infertile eggs; this is a useful product if there is no significant reservoir host nearby (e.g. a neighbour's untreated dog). A relatively recent environmental treatment is the desiccant sodium polyborate which is now available in USA and Europe for use in the domestic environment.
- Regular vacuuming to remove flea eggs and larva will also help in a flea control programme.
- Control of small mammals in and around the house that may act as a reservoir host for fleas.
- The affected animal can be given glucocorticoids to alleviate the dermatitis and pruritus.

8.12.6 Lice

Clinical features: dogs can be infested by both the chewing louse, *Trichodectes canis,* and the sucking louse, *Linognathus setosus*. The clinical signs associated with louse infestation are pruritus, scaling, crusts, coat matting, excoriation and secondary alopecia. Severe infestation with *L. setosus* produces anaemia, especially in puppies. Lice are more common in kennels where asymptomatic carriers may act as reservoirs.

Diagnosis: diagnosis is made on the basis of clinical signs and identification of lice in the coat. These can be collected by acetate-tape strips or coat brushings for examination under the microscope.

Treatment: lice spend their entire life cycle on the host and are readily killed by most organophosphates (e.g. chlorpyriphos, malathion or diazinon), pyrethroids (e.g. permethrin), carbamates (e.g. cabaryl), fipronil and γ-BHC. Topical insecticides applied as a dip or spray are usually applied twice, 14 days apart. Severely anaemic animals may require supportive therapy.

8.13 CATS

8.13.1 Mites

Sarcoptic mange

Clinical features: this is a very rare pruritic dermatosis of cats. A small number of cases have been reported in association with feline immunodeficiency virus but the pathogenesis in these cases is unclear.

Diagnosis: diagnosis is based on clinical features and identification of mites in skin scrapings.

Treatment: cats are highly susceptible to organophosphate toxicity. However, malathion dips are effective and relatively safe. A lime sulphur dip every 10 days until remission may also be effective. Systemic ivermectin at 200–300 μg/kg repeated after 14 days may also be used.

Notoedric mange

Clinical features: this is a contagious pruritic dermatosis of the cat caused by the mite *Notoedres cati* which has also been reported to affect foxes, dogs and rabbits. Notoedric mange is rare in the UK but endemic in some parts of the world. Characteristic lesions include papules, yellow crusts, alopecia and lichenification occurring on the pinnae, face, eyelids, neck, elbows, perineum and feet. There is intense pruritus and in many cases peripheral lymphadenopathy.

Diagnosis: diagnosis is based on clinical features and the identification of mites and eggs in skin scrapings.

Treatment: treatment is similar to that described for sarcoptic mange. Malathion or lime sulphur dips may be used until remission occurs. Injections of ivermectin, at 300–400 μg/kg, are reported to be effective in feline notoedric mange.

Otodectic mange

Clinical features: as in the dog, the ear mite *Otodectes cynotis* causes otitis externa with intense aural irritation, pruritus and head shaking. Examination of the external ear canal reveals a thick, reddish-brown exudate with some crust formation. The infestation is very contagious and is a particularly prevalent in feral and young cats. Although the pathogenesis of the lesions has not been elucidated, a hypersensitivity may be involved and asymptomatic carriers exist. Occasionally ectopic infestation occurs where the mite causes localized dermatitis on the skin, especially on the neck, rump and tail.

Diagnosis: examination of the external ear canal using an auroscope reveals small, fast-moving mites. Microscopic examination of aural exudate mounted in liquid paraffin or superficial skin scrapings in ectopic cases reveals mites and eggs.

Treatment: topical ceruminolytic ear preparations should be used to remove crust and cerumen from the external ear canal. An otic acaricidal product (thiabendazole, permethrin) should then be used for 2 weeks beyond a clinical cure. Systemic ivermectin, at 200–400 μg/kg, given twice, 14 days apart, is effective and the adult cat appears to be relatively free of side-effects.

Demodectic mange

Clinical features: demodicosis is a rare feline dermatosis caused by the long follicular mite *Demodex cati* or the short form of *Demodex* (a putative new species which is, as yet, unnamed). Demodicosis due to *D. cati* is usually localized, affecting the periocular skin, eyelids, head and neck with erythema, scaling, crusts and alopecia. Ceruminous otitis externa has also been reported. Generalized demodicosis is very rare but has been reported with variable pruritus, alopecia, scaling, crusting and hyperpigmentation on the head, neck, legs and trunk. Feline demodicosis is usually associated with underlying debilitating diseases such as diabetes mellitus, feline leukaemia virus infection (FeLV) and systemic lupus erythematosus (SLE). The generalized disease has been seen in association with inflammatory bowel disease and concurrent long-term corticosteroid administration.

Demodicosis due to the short form of *Demodex* occurs in the absence of underlying disease, with severe pruritus, alopecia, scaling, excoriation and crusts. The lesions are usually on the head, neck and elbows. Symmetrical alopecia with or without scaling has also been reported with this type of demodicosis.

Diagnosis: demodicosis is diagnosed on the basis of clinical features and identification of mites in skin scrapings and hair pluckings. The short form is found by superficial skin scraping, acetate strips and within the stratum corneum on biopsy sections.

Treatment: some cases spontaneously recover but local lesions can be treated with lime sulphur dip, carbaryl, malathion or amitraz. Sedation, anorexia and depression can be a problem especially with amitraz, particularly in kittens. Fenchlorphos and phosmet dips have also been reported to be effective.

Trombiculidiasis

Clinical features: larvae of *Trombicula* spp. occasionally affect cats in the summer and autumn, leading to papular lesions usually on the lower limbs, ventrum, face and Henry's pocket at the base of the pinnae, producing variable degrees of pruritus. In the initial stages small clusters of orange larval mites can be seen on the skin, although the lesions may persist after the larval mites have vacated the skin. The mites are most often seen in fruit-growing areas on chalky soil.

Diagnosis: diagnosis is by observation with the naked eye and microscopic identification of the larval mites collected from the skin.

Treatment: the dermatitis should resolve shortly after the larvae have left the skin; however, acaracidal treatment may be necessary. In severe infestations topical organophosphates (e.g. malathion) or fipronil can be used to control the mites. If severe excoriation occurs due to self-trauma then glucocorticoids can be used to alleviate the pruritus and inflammation.

Cheyletiellosis

Clinical features: cheyletiellosis is a contagious, variably pruritic infestation caused by *Cheyletiella blakei* and is seen in cats of any age, but particularly the young. The disease is often mild with dry or greasy scaling on the back. In some cases there is a papulocrustous dermatitis affecting the dorsum, often described as miliary dermatitis. Asymptomatic carriers occur and any in-contact animal should be examined for mites and treated if necessary.

Diagnosis: diagnosis is based on history, clinical examination and identification of mites or eggs. Microscopic examination of hairs collected by coat brushing or plucking may reveal eggs attached to the hair or mites. Superficial skin scrapings and acetate-tape impressions may also reveal mites or eggs. Examination of a faecal sample by the faecal floatation technique may also be of use in diagnosis.

Treatment: cheyletiellosis is easily treated with topical acaricidal compounds such as the carbamates (e.g. carbaryl), the organophosphate malathion (note - other organophosphates may be toxic) and fipronil. Systemic ivermectin at 200–400 μg/kg is effective and the adult cat appears to be relatively free of side-effects. All in-contact animals should also be treated and an environmental acaricidal spray used around the home.

Poultry mite infestation

Clinical features: the poultry mite *Dermanyssus gallinae* can cause dermatitis in the cat. The mite may come into houses from bird nests in the roof or eaves but the disease is most often associated with cats that have access to chicken houses or those that regularly raid birds' nests. The clinical signs are variable and include pruritus, papules, crusts and scales, usually on the back and limbs.

Diagnosis: the diagnosis is made on the basis of the clinical features and finding mites on skin scrapings and coat brushings.

Treatment: topical treatment with insecticidal products licensed for flea treatment (dichlorvos, permethrin) is effective. Removal of the source of infestation is required to prevent recurrence.

8.13.2 Ticks

Clinical features: ticks cause local inflammation within the dermis, are potential vectors of various micro-organisms, produce paralysis and, in severe infestations, anaemia. The spinose ear tick, found in southern USA, causes otitis externa with head shaking and scratching and, in severe cases, convulsions. In Europe the dog tick, *Ixodes canisuga*, castor bean tick, *Ixodes ricinus,* and hedgehog tick, *Ixodes hexagonus,* are important causes of localized dermatitis in cats.

Diagnosis: diagnosis is based on history, clinical examination and the collection and identification of ticks from the skin.

Treatment: unlike other domestic species, cats are usually free-ranging and control of their exposure to infested pasture is impractical. Topical insecticidal compounds such as pyrethroids and fipronil can be used to kill ticks on cats.

8.13.3 Flies

Cats are normally free-ranging and are able to avoid flies more easily than other domestic species. Although they may, in theory, be bitten by flying insects, flies are not usually considered clinically important except for the notable exception of feline mosquito-bite hypersensitivity.

Feline mosquito-bite hypersensitivity

Clinical features: mosquitoes have been shown to feed during the night on the faces of cats, leading to a papular dermatitis with cutaneous oedema, epidermal ulceration crusting and alopecia.

Diagnosis: diagnosis is usually based on circumstantial evidence in history and clinical features. Observation of the mosquito feeding confirms the diagnosis. Histological examination of the lesional skin reveals an eosinophilic folliculitis and furunculosis.

Treatment: topical and systemic glucocorticoids are used to treat the lesions but prevention requires confining the affected cat indoors at night and covering any windows with mosquito netting.

8.13.4 Myiasis

Cutaneous myiasis

Clinical features: myiasis may be caused by the obligate screwworms *Cochliomyia hominivorax* (Nearctic and Neotropical regions), *Chrysomya bezziana* (Oriental and Afrotropical regions) and *Wohlfahrtia magnifica* (eastern Palaearctic). Cutaneous myiasis may also be caused by species of *Lucilia*, *Calliphora, Protophormia* or *Phormia*, in various parts of the world. Myiasis may occur in old or debilitated cats, particularly when wounds are left untreated. During warm humid weather cats paralysed and/or after involvement in road traffic accidents, often develop myiasis. Long-coated cats with untreated chronic diarrhoea and faecal coat contamination may develop myiasis, especially in warm, humid weather.

Diagnosis: diagnosis is made on the basis of clinical signs and identification of larvae within wounds.

Treatment: affected wounds are surgically debrided and sprayed or washed with organophosphates (e.g. malathion) and pyrethroids (e.g. permethrin). Prompt treatment of cutaneous wounds will prevent myiasis. Clipping the hair around the perineum, especially in long-coated cats with a tendency toward diarrhoea will also help to prevent the development of myiasis.

Furuncular myiasis

Clinical features: species of the genus *Cuterebra* are dermal parasites largely affecting rodents and rabbits, although cats may also be affected. They are found exclusively in the New World. The larvae produce large, subdermal nodules and parasitism by more than one larva is common.

Diagnosis: diagnosis is made on the basis of clinical signs and identification of larvae within or exiting warbles.

Treatment: little information is available relating to the treatment of *Cuterebra*, but ivermectin, moxidectin or doramectin are likely to be effective.

8.13.5 Fleas

Clinical features: the most common flea seen on cats is *Ctenocephalides felis felis*. The rabbit flea, *Spilopsyllus cuniculi,* may be found on the pinnal margin in cats that hunt rabbits or enter burrows, causing a mild papular dermatitis. Several other species have been reported to cause dermatitis in cats, including *Ctenocephalides canis*, *Echidnophaga gallinacea*, *Spilopsyllus cuniculi*, *Leptopsylla segnis*, *Pulex irritans* and *Ceratophyllus* spp. Flea-bite dermatitis is one of the most common feline dermatoses. The clinical signs are of two types:

- Symmetrical alopecia; this is due to excessive grooming stimulated by pruritus. There are no lesions on the skin itself but the cat has broken hairs usually on the ventrum and hind legs but occasionally also on the back.

- Miliary dermatitis; signs include papules, crusts, pruritus, excoriations and secondary alopecia. The lesions mostly occur on the dorsum, around the tail base and the neck.

Diagnosis: the diagnosis is based on history, clinical signs and identification of fleas or flea faeces on the cat. Intradermal skin testing using a flea extract may also be of some value in the diagnosis of flea-bite dermatitis. In cases of symmetrical alopecia due to excessive grooming, microscopic examination of the hair tips reveals fractured ends.

Treatment: treatment for flea-bite dermatitis is as described in the dog and involves:

- Application of an adulticide on to the affected animal's coat. Examples are products containing organophosphates, pyrethroids, carbamates or fipronil. Flea products for on-animal use come in spray, shampoo, wash, mousse and powder formulations.
- Application of an adulticide to all in-contact animals (dogs and cats).
- Treatment of the environment using products able to kill the egg, larval and pupal stages of the life cycle. Products available include those containing organophosphates, pyrethroids and IGRs. The IGR, lufenuron, may be given orally to the cat and transferred to the flea in blood at feeding, leading to the production of infertile eggs; this is a useful product if there is no significant reservoir host nearby (e.g. a neighbour's untreated cat). A relatively recent environmental treatment is the desiccant sodium polyborate which is now available in the USA and Europe for use in the house.
- Regular vacuuming to remove flea eggs and larva will also help in a flea control programme.
- Control small mammals in and around the house that may act as a reservoir host for fleas.
- The affected animal can be given glucocorticoids to alleviate the dermatitis and pruritus.

8.13.6 Lice

Clinical features: cats are affected by only one species of louse, *Felicola subrostrata*. The clinical signs associated with lice infestation are pruritus, scaling, crusting, coat matting, excoriation and secondary alopecia. Lice are more common in catteries where asymptomatic carriers may act as reservoirs. Kittens are most often affected by lice.

Diagnosis: the diagnosis is made on the basis of clinical signs and identification of lice in the coat.

Treatment: lice spend their entire life cycle on the host and are readily killed by most organophosphates (e.g. malathion or diazinon), pyrethroids (e.g. permethrin), carbamates (e.g. cabaryl) and fipronil. Topical insecticides applied as a dip or spray are usually applied twice, 14 days apart.

8.14 SMALL MAMMALS

This section covers the more important ectoparasites seen in pet rabbits, guinea-pigs, mice, rats, gerbils and hamsters. It is not intended to be an exhaustive list, but should be of use when dealing with these species when kept as pets in the home environment.

8.14.1 Mites

Sarcoptic mange

Clinical features: although rare, rabbits may develop a severe pruritic, crusting dermatitis with excoriation and secondary alopecia due to *Sarcoptes scabiei* infestation.

Diagnosis: diagnosis is on the basis of history, clinical signs and identification of mites in skin scrapings.

Treatment: systemic treatment, using ivermectin orally or by subcutaneous injection at 200–400 μg/kg on three occasions, 7 days apart, is effective. All in-contact animals should also be treated. The cage or housing should be cleaned.

Notoedric mange

Clinical features: the rat ear mite, *Notoedres muris,* causes a pruritic papular and crusting dermatatitis usually around the pinnae, head and neck. Although rare, *Notoedres cati* may also affect rabbits, also causing a pruritic, crusting dermatitis.

Diagnosis: diagnosis is made on the basis of history, clinical signs and identification of mites in skin scrapings.

Treatment: systemic treatment using ivermectin orally or by subcutaneous injection at 200–400 μg/kg, on three occasions, 7 days apart, is effective. All in-contact animals should also be treated. The cage or housing should be cleaned.

Trixacarus caviae

Clinical features: this sarcoptid mite causes a severely pruritic crusting dermatitis with excoriation and secondary alopecia in guinea-pigs. The most severely affected animals may have convulsions when handled or while scratching. Untreated animals die from anorexia and exhaustion. Asymptomatic carriers may act as a source of infestation or develop clinical disease when under stress.

Diagnosis: diagnosis is made on the basis of history, clinical signs and identification of mites in skin scrapings.

Treatment: systemic treatment using ivermectin orally or by subcutaneous injection at 200–400 μg/kg, on three occasions, 7 days apart, is effective. All in-contact animals should also be treated. The cage or housing should be cleaned.

Psoroptic mange

Clinical features: *Psoroptes cuniculi* is a common cause of otitis externa and, occasionally, dermatitis in rabbits. The ear canal becomes full of a crusting, ceruminous exudate and the pinnae become excoriated by scratching. If the mite spreads to the body a pruritic, crusting dermatitis occurs with lesions especially on the ventral abdomen and urogenital region.

Diagnosis: diagnosis is made on the basis of history, clinical signs and identification of mites in the ceruminous exudate of the external ear canal or skin scrapings. Otoscopic examination of the external ear canal may also reveal mites.

Treatment: cleaning the external ear canal with a ceruminolytic agent to remove the exudate should be performed before giving acaracidal treatment. For otitis externa alone, topical ivermectin applied into the ear canal can be used. Systemic treatment using ivermectin orally or by subcutaneous injection at 200–400 μg/kg, on three occasions, 7 days apart, is effective. All in-contact animals should also be treated. The cage or housing should be cleaned.

Myocoptes musculinus

Clinical features: this is a common fur mite of mice but has also been reported to cause a suppurative dermatitis in guinea-pigs. Infestation is often asymptomatic but the mite may cause a scaling, crusting, pruritic dermatitis with secondary alopecia in mice.

Diagnosis: diagnosis is on the basis of history, clinical signs and identification of mites in coat brushings and skin scrapings.

Treatment: systemic treatment using ivermectin orally or subcutaneous injection at 200–400 μg/kg, on three occasions, 7 days apart, is effective. All in-contact animals should also be treated. The cage or housing should also be cleaned.

Chirodiscoides caviae

Clinical features: this is a common fur mite of guinea-pigs which may be asymptomatic or, if present in large numbers, causes a scaling, pruritic dermatitis with secondary alopecia. Concurrent infestation with *Trixicarus caviae* can occur.

Diagnosis: diagnosis is on the basis of history, clinical signs and identification of mites in coat brushings and skin scrapings.

Treatment: any underlying or concurrent disease should be treated. Systemic treatment using ivermectin orally at 200–400 μg/kg, on three occasions, 7 days apart, is effective. All in-contact animals should also be treated. The cage or housing should also be cleaned.

Listrophorus gibbus

Clinical features: this is a fur mite of rabbits which is thought to be a normal coat commensal. If present in large numbers, it produces a pruritic, scaling dermatitis. Underlying debilitating disease such as dental disease may be involved in the development of the fur mite dermatitis.

Diagnosis: diagnosis is made on the basis of history, clinical signs and identification of mites in coat brushings, acetate strips and skin scrapings.

Treatment: any underlying or concurrent disease should be treated. Systemic treatment using ivermectin orally or subcutaneously at 200–400 μg/kg, on three occasions, 7 days apart is effective. All in-contact animals should also be treated. The cage or housing should also be cleaned.

Demodectic mange

Clinical features: demodicosis is a relatively common disease of hamsters (*Demodex criceti* and *Demodex aurati*) and gerbils (*Demodex meriones*). These follicular mites produce folliculitis and subsequent hair loss. The clinical signs are a scaling dermatosis and alopecia. Demodicosis usually occurs as the result of an underlying debilitating disease. Demodicosis is rare in rabbits and guinea-pigs.

Diagnosis: diagnosis is made on the basis of history, clinical signs and identification of mites in coat brushings and skin scrapings.

Treatment: topical treatment with a mixture of ronnel and propylene glycol applied to one-third of the body daily. Topical amitraz treatment has also been reported. Treatment is continued until skin scrapings are negative. Underlying debilitating diseases should be investigated and treated.

Cheyletiellosis

Clinical features: *Cheyletiella parasitivorax* is a common fur mite of rabbits, which is likely to be a normal commensal, and asymptomatic carriers occur. If present in sufficient numbers or if the affected animal has a hypersensitivity reaction to the mite, a variably pruritic, scaling dermatitis occurs. Secondary alopecia occurs in pruritic cases. It is common for in-contact humans to develop a papular rash, especially on the forearms.

Diagnosis: diagnosis is made on the basis of history, clinical signs and identification of mites in coat brushings and skin scrapings.

Treatment: topical acaricides, such as pyrethrin powders and dichlorvos-containing spays are effective against *Cheyletiella* spp. Fipronil spray, which is effective against *Cheyletiella* spp. in dogs and cats, may also be effective in rabbits. Systemic treatment using ivermectin, orally or subcutaneously at 200–400 μg/kg, on three occasions, 7 days apart, is effective.

Psorergates simplex

Clinical features: this is a follicular mite that causes small, white intradermal nodules in mice.

Diagnosis: diagnosis can be made by excision of a mass and microscopic examination of the dermal pouch containing mites.

Treatment: topical malathion powder or methoxychlor dip has been reported. Systemic treatment using ivermectin, orally at 200–400 µg/kg, may be effective.

Myobia musculi / Radfordia affinis / Radfordia ensifera

Clinical features: *Myobia musculi* and *Radfordia affinis* are common fur mites of mice. *Radfordia ensifera* affects rats. They are normally asymptomatic but if present in large numbers they cause a scaling, pruritic dermatitis with excoriation and secondary alopecia. Underlying debilitating disease, overcrowding and poor housing contribute to the pathogenesis of the disease.

Diagnosis: diagnosis is made on the basis of history, clinical signs and identification of mites in coat brushings, acetate-tape strips and skin scrapings.

Treatment: any underlying or concurrent disease should be treated and management problems corrected. Systemic treatment using ivermectin, orally at 200–400 µg/kg, on three occasions, 7 days apart, is effective. All in-contact animals should also be treated. The cage or housing should also be cleaned.

Ornithonyssus bacoti

Clinical features: this is the tropical rat mite which causes dermatitis and pruritus in rats, mice and hamsters worldwide. In severe infestations hosts may become anaemic and die.

Diagnosis: diagnosis is made on the basis of history, clinical signs and identification of mites in coat brushings and skin scrapings.

Treatment: topical acaricides or systemic ivermectin, orally at 200–400 µg/kg, may be effective in the treatment of this parasite.

8.14.2 Flies

Adult flies are not usually a problem in domestic small mammals, although rabbits and guinea-pigs kept outdoors may be attacked by biting midges, mosquitoes, black flies and other blood-feeding flies. Diagnosis may be difficult but fly-associated dermatitis should be considered in cases with papular and urticaria lesions without evidence of resident ectoparasites. In such cases topical repellents, insecticides and mechanical barriers should be employed. Mosquito or fly nets may be used and dichlorvos-impregnated fly strips hung near to the cages in an attempt to control flying insects. Mosquitoes and black flies are vectors of myxomatosis and other viral diseases in rabbits, especially in Australia.

8.14.3 Myiasis

Furuncular myiasis

Clinical features: species of the genus *Cuterebra* are dermal parasites of rodents and rabbits. The larvae produce large subdermal nodules and parasitism by more than one larva is common. They are found exclusively in the New World.

Diagnosis: diagnosis is made on the basis of clinical signs and identification of larvae within or exiting warbles.

Treatment: little information relating to the treatment of infestation by *Cuterebra* spp. is available, but treatment with ivermectin, moxidectin or doramectin is likely to be effective.

Cutaneous myiasis

Clinical features: myiasis may be caused by the obligate screwworms *Cochliomyia hominivorax* (Nearctic and Neotropical regions), *Chrysomya bezziana* (Oriental and Afrotropical regions) and *Wohlfahrtia magnifica* (eastern Palaearctic). Cutaneous myiasis may also be caused by species of *Lucilia*, *Calliphora*, *Protophormia* or *Phormia*, in various parts of the world. Myiasis is a particular problem with guinea-pigs and rabbits kept outdoors. Predisposing factors are long hair and urinary or faecal contamination of the coat. Animals with diarrhoea and perineal contamination during warm, sunny weather are particularly at risk of developing myiasis.

Diagnosis: diagnosis is made on the basis of clinical signs and identification of larvae within wounds.

Treatment: affected wounds are surgically debrided and sprayed or washed with organophosphates (malathion) or pyrethroids (permethrin). The perineum should be clipped in animals with diarrhoea and washed once or twice daily to prevent myiasis. Prompt treatment of cutaneous wounds and good hygiene will prevent myiasis.

8.14.4 Fleas

Clinical features: *Ctenocephalides felis felis* may affect rabbits and guinea-pigs, especially in the same household as an infested cat, causing a pruritic papular dermatitis. The European rabbit flea, *Spilopsyllus cuniculi,* which lives on wild rabbits and within their burrows, is rarely seen on domestic rabbits. The viral infection myxomatosis is spread by *Spilopsyllus cuniculi* in Europe. A wide variety of other species of flea can affect rabbits, including *Pulex irritans*, *Echidnophaga gallinacea*, and *Nosopsyllus fasciatus*.

Diagnosis: diagnosis is based on history, clinical signs and identification of fleas or flea faeces on the animal.

Treatment: topical insecticides, including pyrethroids, organophosphates and fipronil can be used on the animal, and efforts to control environmental sources of fleas as described under flea treatment in cats and dogs should also be instigated.

8.14.5 Lice

Clinical features: lice can affect:

- rabbits (*Haemodipsus ventricosus* and *H. setoni*),

- guinea-pigs (*Gliricola porcelli* and *Gyropus ovalis*),
- mice (*Polyplax serrata*), and
- rats (*Polyplax spinulosa*).

Guinea-pig lice are common in colonies and usually asymptomatic, although heavy infestations produce a pruritic, scaling dermatosis with secondary alopecia. *Polyplax* spp. and *Haemodipsus* spp. produce a pruritic, scaling dermatosis with anaemia and death if untreated.

Diagnosis: diagnosis is made on the basis of clinical signs and identification of lice and their eggs in the coat and on the hairs.

Treatment: lice spend their entire life cycle on the host and are readily killed by most organophosphates (e.g. diazinon, malathion, methoxychlor), pyrethroids (e.g. permethrin) and fipronil. Topical insecticides, applied as a dip or spray, are usually applied twice, 14 days apart. The entire colony must be treated. Systemic ivermectin, at 200–400 μg/kg, can be used effectively to control lice on rabbits, guinea-pigs, mice and rats.

8.15 BIRDS

This section covers the diagnosis and treatment of the most important cutaneous ectoparasites affecting birds.

8.15.1 Mites

Knemidocoptic mange

Clinical features: in domestic poultry *Knemidocoptes mutans* causes scaling and crusting, which becomes honeycombed in appearance, around the legs, leading in some cases to foot deformation. The condition is known as 'scaly leg'.

In parrots, budgerigars and parakeets, *Knemidocoptes pilae*, infests the skin around the beak and sometimes legs, causing proliferative scaling which may be honeycombed in appearance and leads to deformation of the beak, anorexia and death if untreated. The condition is known as 'scaly face' or 'scaly beak'.

The mite *Knemidocoptes laevis gallinae* burrows in the skin around the feathers, causing pruritus and feather loss on the back, neck, wings and around the vent. Poultry, pheasants and geese are affected and the condition is known as 'depluming itch'.

Diagnosis: diagnosis is made on the basis of history, clinical signs and identification of mites in skin scrapings.

Treatment: topical treatment with ivermectin, γ-BHC, organophosphates (e.g. coumaphos, malathion) or carbaryl are effective against *Knemodicoptes* spp. Because of its potential toxicity, γ-BHC should be used with care and applied to the lesions on the head and legs, taking care to avoid oral ingestion. Systemic treatment of birds with ivermectin, at 200 μg/kg intramuscularly is effective. Topical ivermectin applied as a spot-on is also effective.

Trombiculosis

Clinical features: the larval forms of *Trombicula* spp. can infest birds during the summer and autumn, causing pruritus and cutaneous vesicles at the site of biting.

Diagnosis: diagnosis is made on the basis of history, clinical examination and identification of mites feeding on the skin.

Treatment: the dermatitis will resolve once the mites have left the skin, although topical preparations containing acaricides such as coumaphos, pyrethrin and malathion can be used to kill the mites.

***Ornithonyssus* spp.**

Clinical features: the northern fowl mite and tropical fowl mite, *Ornithonyssus* spp., spend their entire life cycle on the host, causing pruritus, plumage damage, weakness, anaemia and death. They affect poultry and wild birds.

Diagnosis: mites can be collected and identified by microscopic examination. The presence of northern fowl mites is most easily determined by examination of the base of the feathers around the vent area on hens. The mites produce a rough and scaly condition on the skin and a darkening of the feathers due to the dried blood and excreta.

Treatment: to minimize the risk of mite infestation all new birds brought into a poultry house should be confirmed as mite free and, since the mites can survive for several days off-host, care must be taken to minimize the introduction of mites from other houses on clothing or equipment. Infested birds should be treated topically with an acaracidal compound such as pyrethroids, carbaryl, malathion or rotenone. Insecticidal sprays used to treat birds should be able to penetrate to the base of feathers. The environment may also be treated with an insecticidal preparation.

Dermanyssus gallinae

Clinical features: the red mite *Dermanyssus gallinae* can cause severe dermatitis in poultry. The clinical signs are pruritus, plumage damage, weakness, reduced egg production, anaemia and death. As with *Ornithonyssus* spp., this mite can also affect wild birds.

Diagnosis: the diagnosis is made on the basis of the clinical features and finding mites on the skin at night. Examination of the bird during the day will not reveal the mites, which feed only at night and spend the day in crevices in the walls.

Treatment: hygiene procedures, as described for *Ornithonyssus* spp., should be adopted to prevent infestation. Topical treatment of the affected birds with insecticidal products such as permethrin, carbaryl, malathion or rotenone may be effective. However, because this mite lives in cracks and crevices in the environment, thorough cleaning and insecticidal treatment of infested poultry houses is essential.

8.15.2 Ticks

Clinical features: there are numerous ticks which may affect birds, including larvae of *Ixodes ricinus, Dermacentor reticulatus, Haemaphysalis* spp. and *Rhipicephalus* spp., and adult *Argas* spp. Avian ticks are usually only a problem in outdoor aviary birds, free-range chickens or recently caught birds. *Argas persicus* affects poultry and wild birds leading to anaemia, debilitation, loss of production and transmits micro-organisms such as *Borrelia anserina, Aegyptianella pullorum* and spirochaetes.

Diagnosis: diagnosis is based on the history, clinical signs and observation of feeding ticks. It should be remembered that *Argas* spp. are nocturnal feeders and examination of birds at night is necessary for diagnosis.

Treatment: topical treatment of the affected birds with insecticidal products such as permethrin, carbaryl, malathion or rotenone is effective. Insecticidal treatment of the environment is essential. Reinfestation from wild bird sources may need to be controlled by separation of captive birds.

8.15.3 Flies

Hippoboscids

Clinical features: these flat, tick-like flies feed on blood, causing anaemia and death in severe infestations. They also transmit *Haemoproteus* spp.

Diagnosis: diagnosis is made by capture and identification of the flies in the plumage.

Treatment: topical insecticides such as permethrin, carbaryl, malathion or rotenone are effective. It is important to keep domestic birds separated from wild birds, which are a source of reinfestation.

Blood-feeding and nuisance flies

Clinical features: nuisance flies, such as houseflies, lesser house flies and *Ophyra* spp., and blood-feeding flies, such as biting midges, mosquitoes and black flies, can cause disturbance, dermatitis and anaemia, particularly in poultry units. They can also act as vectors for protozoal, bacterial and viral infections.

Diagnosis: the diagnosis is based on history, clinical signs and identification of feeding flies, especially at night. Traps can be used to collect and identify flying insects.

Treatment: the most direct and practical means of controlling most nuisance flies in poultry houses is the appropriate management of the accumulated manure, since this forms the primary breeding site for many species, particularly when wet. This may be accomplished by a building design which maximizes the ventilation of the manure, by provision for drainage of water away from the house and by careful maintenance of the animal watering system to minimize leaks.

Manure within a poultry house may be treated with insecticides; pyrethroids appear to be the most widely acceptable chemicals available for this purpose. However, insecticidal treatment is often not sufficiently persistent and cannot penetrate the manure adequately to kill fly larvae. In addition, naturally occurring predator, parasitoid and parasite populations are often substantial and can contribute significantly to fly control. Insecticidal treatment of manure may eliminate these beneficial arthropods. Hence, where possible, fly control should be promoted by measures which maintain populations of beneficial arthropods; during cleaning, the accumulated manure should not be removed simultaneously and, where possible, portions of the manure should be removed in a staggered schedule. Predator, parasitoid and parasite populations may also be augmented by periodic releases in the animal production system.

Fly resting sites in the structure of a poultry house may be treated with a residual insecticide. Fly baits (containing an attractant and an insecticide or IGR) may be used to attract and kill adults.

8.15.4 Myiasis

Cutaneous myiasis

Clinical features: myiasis may be caused by the obligate screwworms *Cochliomyia hominivorax* (Nearctic and Neotropical regions), *Chrysomya bezziana* (Oriental and Afrotropical regions) and *Wohlfahrtia magnifica* (eastern Palaearctic). Cutaneous myiasis may also be caused by species of *Lucilia*, *Calliphora, Protophormia* or *Phormia*, in various parts of the world. Myiasis usually occurs because of plumage contamination, especially around the vent, or infestation of untreated wounds. Populations of facultative blowflies may be maintained in and around poultry houses by the carcasses of dead birds and broken eggs, which should be removed and buried or incinerated.

Diagnosis: diagnosis is made on the basis of clinical signs and identification of larvae within wounds.

Treatment: affected wounds are surgically debrided and sprayed or washed with organophosphates (malathion) and pyrethroids (permethrin). Prompt treatment of cutaneous wounds will prevent myiasis.

8.15.5 Fleas

Clinical features: although not often seen, fleas probably affect all birds, especially while in the nest. Fleas that affect poultry are *Ceratophyllus niger*, *C. gallinae* and *Echidnophaga gallinacea*. *Ceratophyllus* spp. cause irritation, disturbance and anaemia in severe infestations. The sticktight flea, *Echidnophaga gallinacea* burrows into the skin and remains on the host, causing irritation and, in heavy infestations, anaemia.

Diagnosis: diagnosis is made on the basis of history, clinical signs and identification of fleas on the bird.

Treatment: topical insecticidal compounds such as permethrin, carbaryl, malathion or rotenone can be used to kill adult fleas on the bird. Larvae within the environment must be removed by physically cleaning the aviary or chicken house and applying a residual environmental insecticide.

8.15.6 Lice

Clinical features: lice are important ectoparasites of domestic birds, causing irritation, pruritus, scratching and secondary feather damage. Laying hens affected will have reduced egg production and viability. Although there are many important species of avian lice, *Cuclotogaster heterographus, Menacanthus stramineus, Lipeurus caponis, Goniodes dissimilis, Goniodes gigas* and *Goniocotes gallinae* are the most important in domestic birds.

Diagnosis: diagnosis is made on the basis of clinical signs and identification of lice in the plumage and their eggs attached to feathers.

Treatment: lice spend their entire life cycle on the host and are readily killed by most insecticides. Topical insecticidal compounds such as permethrin, carbaryl, malathion or rotenone can be used to kill lice on the bird.

8.16 FURTHER READING

Axtell, R.C. and Arends, J.J. (1990) Ecology and management of arthropod pests of poultry. *Annual Review of Entomology*, **35**, 101–26.

Baron, R.W. and Colwell, D.D. (1991) Mammalian immune responses to myiasis. *Parasitology Today*, **7**, 353–5.

Beck, T., Moir, B. and Meppen, T. (1985) The cost of parasites to the Australian sheep industry. *Quarterly Review of Rural Economics*, **7**, 336–43.

Brander, G.C., Pugh, D.M., Bywater, R.J. and Jenkins, W.L. (1991) *Veterinary Applied Pharmacology and Therapeutics*, Baillière Tindall, London, pp. 497–512.

Cleland, P.C., Dobson, K.J. and Meade, R.J. (1989) Rate of spread of sheep lice (*Damalinia ovis*) and their effects on wool quality. *Australian Veterinary Journal*, **66**, 289–99.

Coles, G. (1994) Parasite control in sheep. *In Practice*, **16**, 309–18.

Drummond, R.O., George, J.E. and Kunz, S.E. (1988) *Control of Arthropod Pests of Livestock: A Review of Technology*, CRC Press, Boca Raton, Florida.

Dryden, M.W. and Rust, M.K. (1994) The cat flea: biology, ecology and control. *Veterinary Parasitology*, **52**, 1–19.

Emerson, K.C. (1956) Mallophaga (chewing lice) occurring on the domestic chicken. *Journal of the Kansas Entomological Society*, **29**, 63–79.

Fadok, V.A. (1984) Parasitic skin diseases of large animals. *Veterinary Clinics of North America, Large Animal Pratice*, **6**, 3–26.

Fallis, A.M. (1980) Arthropds as pests and vectors of disease. *Veterinary Parasitology*, **6**, 47-73.

French, N.P., Wall, R., Cripps, P.J. and Morgan, K.L. (1994) Blowfly strike in England and Wales: the relationship between prevalence and farm management factors. *Medical and Veterinary Entomology*, **8**, 51–6.

Hall, M.J.R. and Wall, R. (1994) Myiasis of humans and domestic animals, in *Advances in Parasitology* (eds J.R. Baker, R. Muller and D. Rollinson), Academic Press, London, Vol. 35, pp. 258–334.

Holmes, D.H. (1984) *Clinical Laboratory Animal Medicine*, The Iowa State University Press.

Keck, G. (1995) Poisening and side-effects from ectoparasiticide treatment of small animals. *Veterinary International*, **7**, 20–31.

Kwochka, K.W. (1987) Fleas and related disease. *Veterinary Clinics of North America, Small Animal Practice*, **17**, 1235–62.

Kwochka, K.W. (1987) Mites and related disease. *Veterinary Clinics of North America, Small Animal Practice*, **17**, 1262–84.

Lofstedt, J. (1983) Dermatologic diseases of sheep. *Veterinary Clinics of North America, Large Animal Practice*, **5**, 427–48.

Loomis, E.C. (1986) Ectoparasites of cattle. *Veterinary Clinics of North America, Food Animal Practice*, **2**, 299–321.

Loomis, E.C. (1986) Insecticides and acaricides for cattle. *Veterinary Clinics of North America, Food Animal Practice*, **2**, 323–8.

Loomis, E.C. (1986) Epidemiology and control of ectoparasites of small ruminants. *Veterinary Clinics of North America, Food Animal Practice*, **2**, 397–426.

Medleau, L. and Rakich, P.M. (1992) Dermatologic diseases, in *Small Animal Medical Therapuetics* (eds M.D. Lorenz, L.M.

Cornelius and D.C. Ferguson), J. B. Lippincott, Philadelphia, pp. 31–84.

Messinger, L.M. (1995) Therapy for feline dermatoses. *Veterinary Clinics of North America, Small Animal Practice*, **25**, 981–1005.

Moriello, K. and Mason, I.S. (1995) *Handbook of Small Animal Dermatology*, Permagon Press, London.

Muller, G.H., Kirk, R.W. and Scott, D.W. (1989) *Small Animal Dermatology*, W.B. Saunders, Philadelphia.

Mundell, A.C. (1990) New therapeutic advances in veterinary dermatology. *Veterinary Clinics of North America, Small Animal Practice*, **20**, 1541–56.

Murnaghan, M.F. and O'Rourke, F.J. (1978) Tick paralysis, in *Arthropod Venoms* (ed. S. Bettini), Springer-Verlag, New York, pp. 419–64.

Paradis, M. and Villeneve, A. (1988) Efficacy of ivermectin against *Cheyletiella yasguri* infestation in dogs. *Canadian Veterinary Journal*, **29**, 633–5.

Radostits, O.M., Blood, D.C. and Gay, C.C. (1994) *Veterinary Medicine – a Textbook of the Diseases of Cattle, Sheep, Pigs and Horses*, Ballière Tindall, London.

Roberson, E.L. (1988) Chemotherapy of parasitic diseases, in *Veterinary Pharmacology and Therapeutics* (eds N.H. Booth and L.E. MacDonald), Iowa State University Press, Ames, Iowa, pp. 877–81.

Sargison, N. (1995) Differential diagnosis and treatment of sheep scab. *In Practice*, **17**, 3–10.

Scott, D.W. (1988) *Large Animal Dermatology*, W.B. Saunders, Philadelphia.

Smith, M.C. (1983) Dermatologic diseases of goats. *Veterinary Clinics of North America, Large Animal Practice*, **5**, 449–55.

Steelman, C.D. (1976) Effects of external and internal arthropod parasites on domestic livestock production. *Annual Review of Entomology*, **21**, 155–78.

Tarry, D. (1985) The control of sheep headfly disease. *Proceedings of the Sheep Veterinary Society*, **2**, 51–6.

Tarry, D.W. (1986) Progress in warbly fly eradication. *Parasitology Today*, **2**, 111–16.

Tarry, D.W., Sinclair, I.J. and Wassall, D.A. (1992) Progress in the British hypodermosis eradication programme: the role of serological surveillance. *Veterinary Record*, **131**, 310–12.

Tenquist, J.D. and Wright, D.F. (1976) The distribution, prevalence and economic importance of blowfly strike in sheep. *New Zealand Journal of Experimental Agriculture*, **4**, 291–5.

Wall, R. (1995) Fatal attraction: the disruption of mating and fertilization for insect control, in *Insect Reproduction* (eds S.R. Leather and J. Hardie), CRC Press, Cambridge, pp. 109–28.

Wharton, R.H. and Norris, K.R. (1980) Control of parasitic arthropods. *Veterinary Parasitology*, **6**, 135–64.

Williams, R.E. (1986) Epidemiology and control of ectoparasites of swine, in *Veterinary Clinics of North America, Large Animal Practice*, **2**, 469–80.

GLOSSARY

A

Abdomen The third major division of an insect body.
Abscess Localized accummulation of pus or purulent matter within a cavity often surrounded by a fibrous connective tissue capsule.
Acalypterae Flies with small or no squamae or calypters and no groove in the second antennal segment.
Acari A sub-class of the class Arachnida, containing the mites and ticks.
Accessory gland A gland opening into the genital chamber.
Adenotrophic viviparity The production of live offspring, where eggs are retained within the female oviduct until the larvae are mature at which stage they are laid and immediately pupate.
Aedeagus The male copulatory organ (penis).
Afrotropical region The biogeographic region which includes all of sub-Saharan Africa.
Alate Possessing wings.
Allergic Suffering from an allergy.
Allergy An altered immunological reactivity on the second or subsequent contact with an antigen; now used when referring to hypersensitivity reactions.
Alopecia Loss of hair.
Ambulacrum Terminal structures of the mite leg, usually composed of paired claws and an empodium.
Ametabolous Lacking metamorphosis; immature stages lacking only genitalia.
Anaemia Literally means 'no blood'. Now used to indicate an reduction or deficiency of red blood cell count.
Anal fold A distinctive fold in the anal area of the wing.
Anal groove A shallow, semicircular groove, posterior or anterior to the anus in ticks.
Anautogenous Requiring an initial protein meal to initiate vitellogenesis and mature eggs.
Annelida A phylum containing the segmented worms (bristleworms, earthworms, leeches).
Anopheline A mosquito belonging to the sub-family Anophelinae.
Anoplura A sub-order of Phthiraptera, known as sucking lice.

Anorexia Complete loss of appetite and not eating.

Antenna A paired segmented sensory organ which protrudes antero-dorsally from the head near the eyes.

Anterior Located towards the front (head end) of an animal.

Anthropophilic Associated with humans.

Antibody A modified serum globulin.

Antigen A molecule which induces the formation of antibody.

Anus The posterior opening of the digestive tract, at the opposite end to the mouth, from which waste products are expelled from the body.

Apical At or towards the apex.

Apodeme An ingrowh of the exoskeleton to which muscles are attached.

Apolysis The separation of the old from the new cuticle during moulting.

Apterous Without wings, used to describe primitive, wingless insects.

Argasid A tick of the family Argasidae, known as the soft ticks.

Arista Part of the antenna of cyclorrhaphous flies which protrudes from the third antennal segment.

Arthropod An animal in the phylum Arthropoda, characterized by the presence of a segmented body, an exoskeleton, jointed limbs, tagmatization, a dorsal blood vessel, a haemocoel and a ventral nerve cord.

Artilodactyla Even-toed ungulates. An order of mammals containing the ruminants (deer, giraffes, sheep, goats, antelopes and oxen) and pigs, hippopotamuses and camels.

Astigmata A sub-order of mites, lacking stigmata.

Autogenous Able to mature eggs as an adult without an initial protein meal.

Axilla The armpit area of birds and mammals underneath the forelegs, arms or wings.

B

Bacteria Single-celled micro-organisms with a simple nucleus intermediate in size between protozoa and rickettsia.

Bacteriaema The presence of bacteria in the bloodstream.

Basimere Part of the external reproductive apparatus of male fleas.

Basis capituli The anterior part of the body of mites and tick formed from the expanded coxae of the first pair of legs, from which the mouthparts project.

Blowfly Name commonly given to flies of the family Calliphoridae, particularly those responsible for cutaneous myiasis.

Boil An inflamed swelling in the skin.

Bovidae Family of mammals containing the sheep, goats, cattle, buffalo and antelope

Bristle A large seta.

C

Calliphoridae A family of dipterous flies, including screwworm flies and blowflies.

Calypter A membranous flap at the base of the wings of Diptera (also commonly known as squamae).

Calypterae Flies with large squamae or calypters and a groove in the second antennal segment.

Canidae A family of mammals, including the dogs, jackals, wolves and foxes.

Capitulum Anterior body of mites and ticks including the mouthparts (also commonly known as the gnathosoma).

Carbamate A synthetic insecticide.

Caudal At or towards the anal end of an animal.

Cell An area of the wing membrane of an insect, partially or completely surrounded by veins.

Ceratopogonidae A family of nematocerous Diptera, including the biting midges *Culicoides*.

Cerci A pair of appendages originating from abdominal segment 11.

Chelicerae The paired piercing component of the mouthparts of mites and ticks.

Chitin The major component of arthropod cuticle, a polysaccharide composed of acetyl-glucosamine and glucosamine sub-units.

Chitin synthesis inhibitor An insecticide that prevents chitin formation.

Class The taxonomic ranking between phylum and order, e.g. Insecta.

Claw A hooked structure at the distal end of the pretarsus, usually paired.

Clypeus Part of the insect head below the frons to which the labrum is attached anteriorly.

Cockle Ridging of the skin of domestic animals caused by the feeding of hippoboscid flies and lice.

Colostrum The milk secreted for 3–4 days after parturition, which is rich in immunoglobulins.

Comb A row of spines, usually describing the row of projections on the head or thorax of fleas; more properly known as the ctenidium.

Compound eye An aggregation of ommatidia, each acting as a single facet of the eye.

Coprophagous Feeding on dung or excrement.

Copulatory protuberance In some species of mite the final nymphal female stage has a pair of these organs at the posterior end of the idiosoma; used for attachment to an adult male.

Copulatory suckers In some species of mite the adult male has a pair of these organs at the posterior end of the idiosoma; used for attachment to the copulatory protuberances of nymphal females.

Cornea Cuticle covering the eye or ocellus.

Cosmopolitan Distributed worldwide.

Costa The most anterior longitudinal wing vein.

Coxa The basal (first) leg segment.

Crepuscular Active at low light intensities at dawn or dusk.

Crop The food storage area of the digestive system, posterior to the oesophagus.

Cross-veins Transverse wing veins that link the longitudinal veins.

Ctenidium A row of spines on the head or thorax of fleas (pleural ctenidia).

Culicine A mosquito of the sub-family Culicinae which includes the majority of mosquito species, including the important genera *Aedes* and *Culex*.

Cuticle The external skeletal structure secreted by the epidermis, composed of chitin and protein.

Cyclodines A class of organochlorine insecticides.

Cyclorrhapha A sub-order of dipterous flies, characterized by the larva forming a puparium from the last larval skin inside which pupation occurs. They also have an antenna composed of three segments, the third of which bears a protruding arista.

Cytokines Soluble molecules which act as signals between cells.

D

Demodecosis Infestation of the skin with *Demodex* mites.

Dermatitis Any inflammatory condition of the skin.

Detritivorous Feeding on organic detritus of plant or animal origin.

Diapause Delayed development controlled by environmental conditions.

Dichoptic The condition in which there is a wide gap between the eyes; typical of female Diptera.

Diptera An order of insects possessing only a single pair of functional wings and a pair of halteres.

Distal At or near the furthest end from the attachment of an appendage.

Diurnal Active during daylight.

Dorsal Upper surface. The side of the body opposite from where the legs project.

Dorsal vessel A longitudinal tube which act as the main pump for haemolymph.

E

Ecdysis The process of casting off the cuticle in the final stage of moulting.

Eclosion Hatch from the egg.

Ectoparasite A parasite that lives externally on its host.

Empodium A pad or seta on the pretarsus.

Encephalitis Inflammation of the brain.

Endemic Native or restricted to a particular geographic area.

Endocuticle The flexible, unsclerotized inner layer of the procuticle.

Endoparasite A parasite that lives internally within its host (see parasite).

Endopterygota Winged insects which have complete metamorphosis, the wings developing only within the pupa.

Engorge To feed fully, usually with blood.

Enzootic A disease present in a natural host within its natural range.

Eosinophil A leucocyte with a multilobular nucleus and eosinophilic intracytoplasmic granules. A member of the granulocyte series of blood leucocytes.

Epicuticle The inextensible outermost layer of cuticle.

Epidemic The spread of disease from its endemic area and/or from its normal host.

Epidermis In vertebrate animals this is the outer layer of skin. In arthropods the epidermis is the inner living layer of the integument which produces the cuticle.

Epiphora Tear overflow.

Epizootic An unusually high number of cases of disease in an epidemic.

Equidae A family of mammals, including the horses and donkeys.

Erythema Redness of the skin due to inflammation.

Excoriation Erosions and ulcerations caused by mechanical damage, especially scratching or biting.

Exocuticle The rigid, sclerotized outer layer of the procuticle.

Exopterygota Winged insects which have an incomplete metamorphosis, the wings developing externally in the immature stages and becoming fully functional in the adults.

Exoskeleton The outer body layer, also known as the integument or cuticle.

F

Facet Outer layer of the ommatidia that compose the compound eyes of arthropods.

Facultative Not obligatory.

Family The taxonomic ranking between order and genus.

Femur In Diptera: the segment of the leg between the trochanter and tibia or, if the trochanter is fused with the femur, between the coxa and tibia. In Acari: the segment between the trochanter and genu.

Festoon Rounded pattern or crenellations in the posterior body of some genera of ticks.

Fore Anterior, towards the head.

Foregut The part of the gut lying between the mouth and the midgut.

Frons The front (medio-anterior) part of the insect head.

Furuncular myiasis Myiasis where individual fly larvae form boil-like infestations below the skin.

Furunculosis Rupture of hair follicle due to severe inflammation.

G

Gadding Panic in livestock, usually cattle, caused by the oviposition behaviour of adult flies, most usually warble flies, *Hypoderma*.

Gena An area at the side of the head of insects.

Genal ctenidia A row of sclerotized spines located on the gena of some genera of flea.

Genitalia Ectodermally derived structures of both sexes associated with reproduction.

Genital pore Opening in the body wall of the genital duct of male or female arthropods (also known as the gonopore).

Genus Taxonomic ranking between family and species.

Glossinidae A family of Diptera, containing a single genus *Glossina* the tsetse flies.

Gnathosoma The anterior section of the body of mites and ticks, incorporating the mouthparts (also known at the hypostome).

H

Haematoma A blood-filled swelling.

Haemocoel The main body cavity of arthropods.

Haemoglobin An iron-containg protein present within red blood cells which carries oxygen around the body.

Haemolymph The fluid filling the haemocoel.

Haemolytic anaemia Anaemia due to lysis of red blood cells.

Haller's organ A pit in the tarsi of the forelegs of ticks which contains sensory chemoreceptors.

Haltere Club-shaped, reduced hindwings of Diptera, used as a sensory aid for balance during flight.

Hard tick A tick of the family Ixodidae. Described as hard because of the hard scutum on the dorsal surface.

Haustellum Sucking mouthparts.

Head The anterior of the three major divisions of an insect body.

Hermaphrodite Possessing both testes and ovaries.

Hind At or towards the posterior.

Hindgut The posterior section of the gut, extending from the midgut to the anus.

Hippoboscidae A family of dorsoventrally flattened Diptera, including the keds and forest flies, which live in close association with their hosts.

Holarctic Both Nearctic and Palaearctic regions.

Holometabolous Development in which there is complete metamorphosis, the body form changing abruptly at the pupal moult.

Holoptic The condition in which there is a narrow gap between the eyes; typical of male Diptera.

Hormone A chemical messenger that regulates some activity at a distance from the organ that produced it.

Host The organism on which a parasite feeds.

Hyperplasia An excessive proliferation of tissue.

Hypersensitivity An exaggerated or inappropriate immune response to antigens, such as saliva from arthropods; a type of allergy causing tissue damage.

Hypopygium A part of the external genital apparatus of male flies, known as the penis.

Hypostome A part of the mouthparts of ticks and mites, situated between the palps.

I

Idiosoma The main body region of ticks and mites.

IgE A particular type of immunoglobulin involved in immediate type I hypersensitivity reactions when bound to mast cells.

IL-1 A type of cytokine which is produced by various cells, including macrophages, B lymphocytes and keratinocytes.

Imago Adult insect.

Inflammation The vascular and cellular tissue response to injury, characterized by pain, heat, redness, swelling and loss of function.

Insect Arthropod of the class Insecta.

Instar A stage in the life cycle of an arthropod, such as egg, larva, pupa, adult.

Integument The epidermis plus cuticle.

Interleukins A group of soluble molecules (cytokines) which act as signals between cells. These were originally identified when produced by leucocytes as signals for other leucocytes; hence interleukins or between leucocytes.

Ischnocera A sub-order of chewing lice, including the genera *Damalinia* on mammals and *Goniodes* on birds.

Ixodidae A family of ticks, known as hard ticks because of the hard scutum on the dorsal surface.

J

Juvenile hormone A hormone released by the corpora allata into the haemolymph, involved in many aspects of insect physiology, including moulting.

Juvenile hormone mimics Synthetic chemicals that mimic the effects of juvenile hormone on development.

K

Kairomone A chemical used in communication, to the benefit of the receiver and which may be to the disadvantage of the producer.

Keratinocytes The epithelial cells which make up most of the epidermis.

Keratitis Inflammation of the cornea.

L

Labella A paired organ (singular: labellum), forming lobes at the apex of the proboscis, derived from the labial palps.

Labial palp A segmented appendage of the labium.

Labium Forming the floor of the mouthparts (a 'lower lip'), often with a pair of palps and two pairs of median lobes.

Labrum Forming the roof of the preoral cavity (an 'upper lip').

Lacina The mesal lobe of the maxillary stipes.

Larva An immature insect life-cycle stage which follows eclosion from the egg. Usually applied to insects with complete metamorphosis.

Larviparous Reproduction in which the egg hatches within the female and the larva is deposited.

Lateral At, or close to, the sides.

Lateral suture A line or ridge in the integument of *Argas* ticks which divides the dorsal and ventral surfaces.
Leg The walking limb of arthropods.
Lesion Any pathological or traumatic deviation from normal tissue.
Leucopenia A reduced number of circulating white blood cells.
Lichenification A thickening and hardening of the skin with exaggeration of the surface wrinkles.
Lymphadonopathy Disease of lymph nodes usually characterized by enlargement.

M

Macrotrichiae A trichoid sensillum; seta or hair; on the wings of insects, particularly ceratopogonid midges.
Maggot A legless, larval insect; usually applied to immature stages of clorrhaphous Diptera.
Malpighian tubules Thin, blind-ending tubule, originating near the junction of the mid- and hindgut, involved in nitrogenous waste excretion and water regulation.
Mamillae Small bumps in the integument of soft ticks such as *Argas* and *Ornithodoros*.
Mandible The jaws in biting and chewing insects. May be a needle-like piercing organ, as in mosquitoes, or tooth-like as in chewing lice.
Mange A skin disease cause by infestation with mites.
Mastitis Inflammation of the udder, usually due to bacterial infection.
Maxilla The second pair of jaws in chewing insects.
Maxillary palp A segmented sensory appendage borne on the stipes of the maxilla.
Mechanical transfer The movement of pathogens by passive transfer, with no biological vector.
Medial Towards the middle (also median).
Mesostigmata A sub-order of mites, with stigmata located above the coxae of the second, third or fourth pair of legs; also known as gamesid mites.
Mesothorax The second segment of the thorax.
Metamorphosis The relatively abrupt change in body form between the immature and sexually mature, adult stage.

Metastigmata A sub-order of the Acari, known as ticks, with stigmata located above the coxae of the second, third or fourth pair of legs and a toothed hyperstome; also known as Ixodida

Metathorax The third segment of the thorax.

Microtrichiae Small extensions of the cuticle on the wings of some insects.

Midge Common name for blood-feeding ceratopogonid flies.

Midgut The middle section of the gut.

Miliary Like millet seed.

Moulting The formation of new cuticle followed by ecdysis.

Mouth-hooks The skeletal mouthparts of higher flies.

Muscidae A family of cyclorrhaphous Diptera, including the house fly and similar species.

Myiasis Infestation of the tissues of a living host by fly larvae.

N

Nearctic region A biogeographic region which includes North America, from northern Mexico to Alaska, and Greenland.

Necrophagous Eating dead and/or decaying animal matter.

Necrosis Tissue or cell death in the living organism.

Neotropical region A biogeographic region including southern Mexico, Central and South America.

Neurotoxin A toxic material which has its effect on nervous tissue.

New World North and South America.

Nocturnal Active at night.

Nomenclature The naming of plants and animals.

Nulliparous A female that has not yet oviposited.

Nymph In mites and ticks the second, and subsequent, immature stages. In insects, usually those with incomplete metamorphosis, all immature stages.

O

Obligate Compulsory; a parasite which cannot survive without its living host.

Ocelli The simple eyes of some adult and nymphal arthropods (singular: ocellus).

Oesophagus The foregut that lies anterior to the pharynx and anterior to the crop.
Oestridae A family of Diptera, containing the obligate myiasis species the bots and warbles.
Old World Europe, Asia and Africa.
Ommatidium A single element of the compound eye.
Ophthalmomyiasis Infestation of the eye with dipterous larvae.
Order The taxonomic ranking between class and family, e.g. Diptera.
Organochlorine A group of organic chemicals containing chlorine, including several insecticides.
Organophosphate A group of organic chemicals containing phosphorus, including several insecticides.
Oriental The biogeographic region which includes Pakistan, India, South-East Asia, southern China, Malaysia, Philippines and western Indonesia.
Ostium A slit-like opening in the dorsal vessel allowing the one-way movement of haemolymph from the pericardial sinus into the dorsal vessel.
Otitis An inflammatory condition of the ear.
Ovariole Ovarian tubes that form the ovary and in which the egg follicles develop prior to ovulation.
Ovary One of the paired gonads of female arthropods, usually composed of a number of ovarioles.
Ovigerous An egg-producing female mite.
Oviparous Reproduction in which eggs are laid.
Ovipositor The organ used for laying eggs.
Ovoviviparity Retention of the developing fertilized egg within the female arthropod; similar to viviparity but with no nutrition of the hatched young.

P

Palaearctic The biogeographic region which includes Europe, Iceland, North Africa, Russia, central Asia, the Middle East, central and northern China, Japan and Korea.
Palps Paired segmented organs associated with the maxilla (maxillary palps) and labium (labial palps); singular: palp.

Papule A small, solid, red, elevated circumscribed cutaneous lesion up to 1 cm in diameter.
Parasite An organism that lives at the expense of another (host) which it does not kill.
Parasitism The relationship between a parasite and its host.
Parasitoid A parasite that kills its host.
Paratergal plates Hardened, sclerotized plates on the lateral surface of the abdomen of some insects, particularly lice.
Parous A female that has laid at least one egg.
Parthenogenesis Development from an unfertilized egg.
Pathogen A parasite which causes disease.
Pedicel Stalked pretarsi of mites, particularly astigmata.
Pediculosis Infestation with lice.
Perianal The area around the anus of mammals.
Pericardial sinus The body compartment that contains the dorsal vessel.
Perineum The area between the anus and the genital opening of mammals.
Peritreme A paired, sclerotized process associated with the stigmata, seen especially in the mesostigmatid mites.
Pharynx The anterior part of the foregut, anterior to the oesophagus.
Pheromone A chemical used in communication between individuals of the same species.
Phlebotominea A sub-family of dipterous flies in the family Psychodidae, known as sandflies.
Phoresy The movement of one animal by attachment to another animal.
Pinna The ear flap (plural: pinnae).
Pleural rod A vertical thickening of the integument of the mesopleuron of fleas; known also as the pleural ridge or meral rod.
Pleuron The lateral region of the body, bearing the limb bases (plural: pleura).
Polychaeta An order of marine, annelid worms with numerous chaetae borne on projections of the body (parapodia).
Polyradiculoneuritis A diffuse lower motor neurone paresis.
Posterior The body of an animal furthest from the head.
Postscutellum A projecting posterior area of the thorax of Diptera underneath the scutellum.

Prepuce The fold of skin around the penis.
Prestomal teeth Structures at the end of the labella of some dipterous flies.
Pretarsus The last segment of the leg of mites.
Proboscis A general term for elongate mouthparts.
Pronotum The upper (dorsal) plate of the prothorax.
Prothorax The first segment of the thorax.
Protozoa Single-celled animals with at least one well-defined nucleus, some of which are pathogenic.
Proventriculus The grinding organ of the foregut.
Proximal At or near the end of attachment of an appendage.
Pruritus Itching, skin irritation.
Pseudotrachea A ridged groove on the ventral surface of the labellum of some higher Diptera.
Ptilinum A sac everted from a suture between the antennae of schizophoran Diptera.
Pulvillus The expanded terminal structure of the pretarsus of some genera of mites, which may be membranous bell- or sucker-like discs (plural: pulvilli).
Pupa The inactive stage between larva and adult in holometabolous insects.
Pupariation The process of puparium formation.
Puparium The hardened skin of the final stage larva of higher, cyclorrhaphous Diptera in which the pupa forms.
Pupation The process of becoming a pupa.
Puritus Intense and persistent itching.
Pustular Forming spots or pustules in the skin.
Pustule A small, circumscribed elevation of the skin filled with pus.
Pyaemia The presence of pus in circulating blood.
Pygidium An area of sensory setae at the posterior of the abdomen of fleas; also known as the sensilium.
Pyoderma A skin disease characterized by the presence of pus or purulence, usually in the form of pustules, and caused by bacteria.
Pyotraumatic Purulent skin disease associated with pruritus.
Pyrethrin One of the insecticidal chemicals present in the plant pyrethrum.
Pyrethroids Synthetic chemicals with similar structure to pyrethrins.
Pyrexia Elevated body temperature or fever.

R

Reservoir An animal infected with a disease pathogen, but not necessarily suffering clinical disease and acting as a source of disease which arthropod vectors transmit to non-diseased animals.

Resilin A rubber-like protein in some insect cuticles, particularly important in the jumping mechanism of fleas.

Retinula cell A sensory cell of the light receptors, ommatidia or ocelli, comprising a rhabdom of rhabdomeres.

Rhabdom The central zone of the retinula, consisting of microvilli filled with visual pigment.

Rhabdomere One of typically eight units, comprising a rhabdom.

Rhinitis Inflammation nasal mucosa.

Rickettsia A group of parasitic micro-organisms intermediate in size between bacteria and viruses, without cellular structure, many of which are pathogenic and transmitted by arthropods.

Rotenone An insecticidal chemical derived from legumes.

S

Saliva A fluid containing a complex mixture of agents often containing digestive enzymes; produced by the salivary glands. In blood-feeding arthropods the saliva may contain anticoagulants.

Saprophagous Feeding on decaying organisms.

Scab Skin disease caused by psororoptic mites, usually applied to sheep infested with *Psoroptes ovis*.

Scape The first segment of the antenna of insects.

Sclerite A plate on the body wall surrounded by membrane or sutures.

Sclerotized Cuticle hardened by cross-linkage of protein chains.

Screwworm Common name given to the larvae of the obligate agents of cutaneous myiasis, *Cochliomyia hominivorax*, *Chrysomya bezziana* and, less commonly, *Wohlfahrtia magnifica*.

Scutellum An area of the thorax of dipterous flies at the posterior dorsal margin between the wings.

Scutum The sclerotized plate on the dorsal surface of ixodid hard ticks, also known as the dorsal shield.

Sebaceous gland A gland in the skin of mammals associated with hair follicles.

Sebum An oily secretion produced by the sebaceous glands of mammals with spreads over the skin and hair.
Seminal vessicle Male sperm storage organs.
Semiochemical Any chemical used in intra- and interspecific communication.
Sensilium An area of sensory setae on the posterior abdomen of fleas; also known as the pygidium.
Sensillum A sense organ.
Seta A long, thin, cuticular extension, produced by an epidermal cell; flexible at the base; may be called a hair; large setae are called bristles.
Simuliidae A family of nematocerous Diptera, including the genus *Simulium*, known as blackflies.
Soft tick Ticks of the family Argasidae, which do not have a hard scutum on the dorsal idiosoma.
Species A group of organisms that can interbreed in natural populations producing fully fertile offspring.
Spermatophore An encapsulated package of spermatozoa.
Spine An unjointed cuticular extension.
Spiracle An external opening of the tracheal system.
Spirochaetosis Infection with spirochaete bacteria, such as *Borrelia*.
Spur A long, sharp, articular, multicellular projection of the integument.
Squama Membranous flaps at the base of the wings of Diptera; typically the alula, the alar squama and the thoracic squama (also known as a calypters).
Squamous A condition in which the skin of the host forms thick scales, as in some forms of demodecosis.
Stadium The period between moults.
Stemma The simple eye of many larval insects.
Sternum The ventral surface of a segment.
Stigmata External openings of the tracheal system in the integument of ticks and mites.
Stipes The distal part of the maxilla.
Stomoxyinae A sub-family of dipterous flies, including the stable fly *Stomoxys calcitrans*.
Striations Fine grooves in the integument of some mites and ticks, forming complex patterns.

Strike Name commonly given to cutaneous myiasis by fly larvae, usually applied to myiasis of sheep by *Lucilia sericata* or *Lucilia cuprina*.

Stylostome A feeding tube produced around the mouthparts of trombiculid mites in the skin of the host.

Subcutaneous Beneath the skin of vertebrates.

Sucker Name commonly given to the pulvillus at the end of the legs of some types of mite.

Suture A groove on the arthropod that may show the fusion of two exoskeletal plates.

Synanthropic Associated with humans.

Systematics The practice of biological classification.

Systemic insecticide An insecticide taken into the body of a host that kills insects feeding on the host.

T

Tabanidae A family of Diptera, including the horse flies, deer flies and keds.

Tagma The group of segments that form a major body unit (head, thorax, abdomen).

Tarsomere A subdivision of the tarsus.

Tarsus The leg segment distal to the tibia, bearing the pretarsus; in insects composed of up to five tarsomeres (plural: tarsi).

Taxonomy The theory and practice of naming and classifying living organisms.

Teneral The condition of a newly emerged, not yet fully mature, adult insect.

Tergum The dorsal surface of a segment.

Thorax The middle of the three major body divisions of insects; composed of the pro-, meso- and metathorax.

Thrombocytopenia Reduced numbers of platelets circulating in the blood.

Tibia In insects – the fourth leg segment following the femur; in Acari – the fifth leg segment following the genu.

Tormogen cell The socket-forming epidermal cell associated with a seta.

Toxaemia Illness caused by poisoning.

Tracheae Tubular elements of the gas exchange system in insects and some Acari.

Tracheole Fine tubules of the gas exchange system in insects and some Acari.

Transovarial transmission The transmission of pathogens between generations via the eggs.

Transtadial transmission The transmission of pathogens between stadia.

Transverse At right angles to the longitudinal axis.

Trichogen cell A hair-forming epidermal cell associated with a seta.

Trochanter In insects and Acari, the second leg segment following the coxa.

Trombiculidae A family of prostigmatid mite, parasitic only in the larval stage, including the harvest mites and chiggers.

Trypanosomiasis A disease caused by *Trypanosoma* protozoans.

Tubercle A large, rounded projection from the surface of an arthropod.

U

Udder The milk-producing organ of domestic animals such as cattle, sheep, pigs and goats.

Urticaria Inflammation and irritation of the skin associated with allergic or similar reactions.

V

Vas deferens The ducts that carry sperm from the testis.

Vector An arthropod that transmits a pathogenic organism.

Vein Tubes of cuticle in a network which support the wings of insects.

Ventral Towards or at the lower surface.

Vertebrate An animal with a skull which surrounds the brain and a skeleton of bone or cartilage, including the spine of vertebral bones surrounding a spinal cord of nerves; includes mammals, birds, fish, reptiles and amphibians.

Virus Nucleic acid within a protein or protein and lipid coat which is unable to multiply outside the host tissues. Many are pathogenic.

Vitellogenesis The process by which oocytes grow by yolk deposition.

Viviparity Producing live offspring.

W

Warble Swelling in skin caused by infection with larvae of flies causing furuncular myiasis.

Wax layer The lipid or waxy layer outside the epicuticle of some arthropods.

Z

Zoogeographic region Areas of the world containing characteristic animal and plant species which have been isolated from each other.

Zoonosis A disease on animals that may be communicated to humans.

INDEX

Page numbers appearing in **bold** refer to figures and those in *italic* refer to tables.